29th International Symposium on Space Terahertz Technology (ISSTT 2018)

Pasadena, California, USA
26 - 28 March 2018

ISBN: 978-1-5108-7474-9

Printed from e-media with permission by:

Curran Associates, Inc.
57 Morehouse Lane
Red Hook, NY 12571

Some format issues inherent in the e-media version may also appear in this print version.

Printed by Curran Associates, Inc. (2019)

International Symposium on Space Terahertz Technology
c/o NRAO Headquarters
520 Edgemont Road
Charlottesville, VA 22903-2475
USA

Phone: (434) 296-0254
Fax: (434) 296-0278

mbishop@nrao.edu

Additional copies of this publication are available from:

Curran Associates, Inc.
57 Morehouse Lane
Red Hook, NY 12571 USA
Phone: 845-758-0400
Fax: 845-758-2633
Email: curran@proceedings.com
Web: www.proceedings.com

ISSTT 2018 in Pasadena, California

(group photo)

Contents

Sponsors

We greatly appreciate the support from our industrial sponsors:

Virginia Diodes, Inc.	www.vadiodes.com
Cernex, Inc	www.Cernex.com
CernexWave	www.Cernexwave.com
LongWave Photonics LLC	www.longwavephotonics.com
Insight Product Co.	www.insight-product.com

Virginia Diodes, Inc.

CERNEX, Inc.
CernexWave

LONGWAVE
PHOTONICS

nsight
Product

We also acknowledge the support of our nonprofit co-sponsors:

The IEEE Microwave Theory and Techniques Society (IEEE MTT-S)
www.mtt.org/

The International Society of Infrared, Millimeter, and Terahertz Waves (IRMMW-THz)
www.irmmw-thz.org/

The Jet Propulsion Laboratory, California Institute of Technology (JPL)
www.jpl.nasa.gov/

Welcome

Dear colleagues,

It was our pleasure to host the 2018 29th IEEE International Symposium on Space Terahertz Technology (ISSTT2018) held on March 26-28, 2018 on the grounds of California Institute of Technology in Pasadena, California, USA. We thanks all the participants for their contributions.

Committees

The **Local Organizing Committee (LOC)** was responsible for the planning and organizing activities. It consisted of the following JPL scientists and engineers:

Daniel Cunnane

Cecile Jung-Kubiak

Boris Karasik (Chair)

Corinne Karpinski

Jonathan Kawamura

Imran Mehdi

Peter Siegel

Jose Siles

The ISSTT2018 website is at www.isstt2018.com and will remain operaitonal until October 2019.

The **Scientific Organizing Committee (SOC)** was responsible for reviewing the abstracts and making recommendations on acceptance and format of presentation for each abstract. This Committee also decided on the place and dates for the next ISSTT2019.

The SOC members were:

Andrey Baryshev	University of Groningen, the Netherlands
Victor Belitsky	Chalmers University of Technology, Sweden
Brian Ellison	Rutherford Appleton Laboratory, United Kingdom
Jian-Rong Gao	SRON, the Netherlands
Gregory Goltsman	MSPU, Russia
Christopher Groppi	Arizona State University, USA
Jeffrey Hessler	Virginia Diodes Inc., USA
Heinz-Wilhelm Hübers	DLR, Germany
Boris Karasik	Jet Propulsion Laboratory, USA
Valery Koshelets	IRE RAS, Russia
Alain Maestrini	Paris Observatory, France
Hiroshi Matsuo	NAOJ, Japan
Imran Mehdi	Jet Propulsion Laboratory, USA
Patricio Mena	University of Chile, Chile
Omid Noroozian	NRAO, USA
Patrick Pütz	University of Cologne, Germany
Christophe Risacher	MPIfR Bonn, Germany
Karl Schuster	IRAM, France
Sheng-Cai Shi	Purple Mountain Observatory, China
Edaward Tong	Smithsonian Astrophysics Observatory, USA
Christopher Walker	University of Arizona, USA
Ghassan Yassin	University of Oxford, United Kingdom

2018 29th IEEE International Symposium on Space Terahertz Technology (ISSTT 2018)

March 26-28, 2018, Pasadena, California, USA

Symposium Schedule

For ISSTT 2018, **109** abstracts have been accepted, **6** of which were Invited presentations, **44** contributed Oral personations, and **59** contributed Poster presentations. The review process was carried out by SOC members with each abstract been reviewed by at least two reviewers and by an LOC member. The conference schedule (see below) consisted of **5** Invited Sessions, **11** Oral Sessions and **1** Poster Session.

SYMPOSIUM SCHEDULE

Time	Sunday, March 25	Time	Monday, March 26		Time	Tuesday, March 27		Time	Wednesday, March 28		Thursday, March 29
		8:00 AM	Registration	60							
		8:30 AM			8:30 AM	Registration	30	8:30 AM	Registration	30	
		9:00 AM	Welcome	20	9:00 AM	Session T1 (1 talk)	30	9:00 AM	Session W1 (1 talk)	30	Tour to JPL 9:30 am - noon
		9:20 AM			9:30 AM			9:30 AM			
			Session M1 (4 talks)	80		Session T2 (4 talks)	80		Session W2 (4 talks)	80	
		10:40 AM	Coffe Break	30	10:50 AM	Coffe Break	30	10:50 AM	Coffe Break	30	
		11:10 AM			11:20 AM			11:20 AM			
			Session M2 (4 talks)	80		Session T3 (4 talks)	80		Session W3 (4 talks)	80	
		12:30 PM	Lunch Break SOC meeting	90	12:40 PM	Lunch Break	80	12:40 PM	Lunch Break	80	
		2:00 PM	Session M3 (1 talk)	30	2:00 PM	Session T4 (2 talks)	60	2:00 PM	Session W4 (1 talk)	30	
		2:30 PM			3:00 PM			2:30 PM			
			Session M4 (5 talks)	100		Session T5 (2 talks)	40		Session W5 (5 talks)	100	
					3:40 PM						
		4:10 PM	Coffe Break	30		Poster session Coffe Break Poster session	110	4:10 PM	Coffe Break	30	
		4:40 PM						4:40 PM			
			Session M5 (4 talks)	80					Session W6 (4 talks)	80	
		6:00 PM			5:30 PM	Break	30				
5:30 PM	Welcome Reception and Registration 5:30 pm - 8:30 pm				6:00 PM	Reception & Banquet 6:00 pm - 9:00 pm		6:00 PM	Wrap-up and farewell	10	
								6:10 PM			
8:30 PM											
					9:00 PM						
Duration (min.)	150			600			570			580	

March 26, 2018 (Monday)

9:00 am Welcome Note

9:10 am ISSTT2018 Intro & Info

Session M1: **Sources and Local Oscillator Systems I**
Chair: *Jian-Rong Gao*

9:20 am **M1.1.** Ali Khalatpour [1], Qing Hu [1], and John L. Reno [2], *[1]Massachusetts Institute of Technology (MIT), Cambridge, Massachusetts 02139, USA; [2]Center for Integrated Nanotechnology, Sandia National Laboratories, Albuquerque, New Mexico 87123, USA.*
"4.7 THz local oscillator for GUSTO."

9:40 am **M1.2.** Heinz-Wilhelm Hübers [1,2], Till Hagelschuer [1], Heiko Richter [1], Martin Wienold [1], Lutz Schrottke [3], Xiang Lü [3], Benjamin Röben [3], Klaus Biermann [3], Holger T. Grahn [3], *[1]German Aerospace Center (DLR), Institute of Optical Sensor Systems, Rutherfordstr. 2, 12489 Berlin, Germany; [2] Humboldt-Universität zu Berlin, Department of Physics, Newtonstr. 15, 12489 Berlin, Germany; [3] Paul-Drude-Institut für Festkörperelektronik, Leibniz-Institut im Forschungsverbund Berlin e. V., Hausvogteiplatz 5−7, 10117 Berlin, Germany.*
"Compact and efficient 4.7-THz local oscillator with a GaAs/AlAs quantum-cascade laser."

10:00 am **M1.3.** Mark B. Taylor, Lorene A. Samoska, Jose E. Velazco, Robert H. Lin, Andy Fung, Alejandro Peralta, *Jet Propulsion Laboratory, California Institute of Technology, 4800 Oak Grove Dr., Pasadena, CA 91109, USA.*
"Spatial Power Combining Amplifier (SPCA) for W-Band Radar in Earth and Planetary Science."

10:20 am **M1.4.** Behnam. Mirzaei [1], J. R. G. Silva [2], D. Hayton [3], W. Laauwen [2], Y. Gan [2,4], Q. Hu [5], A. Khalatpour [5], C. Groppi [6], and J. R. Gao [1,2], *[1] Kavli Institute of NanoScience, Delft University of Technology, Delft, the Netherlands; [2] SRON Netherlands Institute for Space Research, Groningen/Utrecht, the Netherlands; [3] Jet Propulsion Laboratory, California Institute of Technology, Pasadena, CA 91109, USA; [4] Kapteyn Astronomical Institute, University of Groningen, 9747 AD, Groningen, The Netherlands; [5] Massachusetts Institute of Technology (MIT), Cambridge, Massachusetts 02139, USA; [6] School of Earth and Space Exploration, Arizona State University, AZ, USA*
"Prototype 4.7 THz array local oscillator for GUSTO."

10:40 am Coffee Break

Session M2: **SMM and THz Receivers**
Chair: *Heinz-Wilhelm Hübers*

11:10 am **M2.1.** Alain Maestrini [1], L. Gatilova [1,2], J. Treuttel [1], Y. Jin [2], A. Cavanna [2], D. Moro-Melgar [1], T. Vacelet [1], A. Féret [1], S. Caroopen [1], G. Gay [1], F. Dauplay [1], J-M. Krieg [1], P. De Maagt [3], C. Goldstein [4], *[1] LERMA, Observatoire de Paris, PSL Research University, CNRS, UMR*

8112, Sorbonne Universités, UPMC Univ. Paris 06, F-75014 Paris, France; [2] C2N-Marcoussis, Route de Nozay, F-91460 Marcoussis, France; [3] ESA-ESTEC, Keplerlaan 1, PO Box 299, NL-2200 AG Noordwijk, The Netherlands; [4] Centre National d'Etudes Spatiales, 18 avenue Edouard Belin, F-31401 Toulouse cedex 9, France.
"1080-1280GHz Schottky Receiver for JUICE-SWI with 1600-2600K DSB Receiver Noise Temperature."

11:30 am **M2.2.** Sascha Krause, D. Meledin, V. Desmaris and V. Belitsky, *Group for Advanced Receiver Development, Onsala Space Observatory Division, Department of Space, Earth and Environment, Chalmers University of Technology, Gothenburg, 41296, Sweden.*
"Noise performance of a balanced waveguide NbN HEB mixer utilizing a GaN buffer-layer at 1.3 THz."

11:50 am **M2.3.** Wei Miao [1,2], H. Gao [1,2], Z. Lou [1,2], J. Hu [1,2], W. Zhang [1,2], Y. Ren [1,2], K.M. Zhou [1,2], S.C. Shi [1,2], H.Li [3], J.C. Cao [3], and Y. Delorme [4], *[1] Purple Mountain Observatory, CAS, Nanjing, 210008, China; [2] Key Lab of Radio Astronomy, CAS, Nanjing, 210008, China; [3] Shanghai Institute of Microsystem and Information, CAS, Shanghai, 200050, China; [4] Observatoire de Paris, Paris, 75014, France.*
"A fully integrated heterodyne receiver based on a hot electron bolometer mixer and a quantum cascade laser."

12:10 pm **M2.4.** Theodore Reck, Cecile Jung-Kubiak, Maria Alonso-delPino, Goutam Chattopadhyay, *Jet Propulsion Laboratory, California Institute of Technology, Pasadena, CA 91109, USA.*
"A MEMS-based, Dicke-switched radiometer at 560GHz."

12:30 pm Lunch Break & SOC/LOC Meeting

Session M3: **Invited I**
Chair: *Ghassan Yassin*

2:00 pm **M3.1.** Satoshi Ochiai, *National Institute of Information and Communications Technology, Koganei, Tokyo 184-8795, Japan.*
"Results of SMILES and the plan of follow-on THz mission, SMILES-2."

Session M4: **Direct Detectors and Instruments**
Chair: *Peter Day*

2:30 pm **M4.1.** Pierre Echternach, B.J. Pepper, T. Reck, and C.M. Bradford, *Jet Propulsion Laboratory, California Institute of Technology, Pasadena, CA 91109, USA.*
"Single Photon Detection of 1.5THz Radiation with the Quantum Capacitance Detector."

2:50 pm **M4.2.** Omid Noroozian [1,2,3], E. M. Barrentine [2], Jochem J.A. Baselmans [4], A. D. Brown [2], M. Cyberey [3], L. A. Hess [2], A. W. Lichtenberger [3], T. R. Stevenson [2], E. J. Wollack [2], S. H. Moseley [2], *[1] National Radio Astronomy Observatory, Charlottesville, VA 22903, USA; [2] NASA Goddard Space Flight Center, Greenbelt, MD 20771, USA; [3] University of Virginia, Charlottesville, VA 22903, USA; 4 Netherlands Institute for Space Research (SRON), Utrecht, Netherlands.*
"Enabling Technologies for Photon-Counting Spectroscopy with the Origins Space Telescope (OST) in the Mid/Far-Infrared Region."

3:10 pm **M4.3.** Matt Bradford on behalf of the SuperSpec team and the STARFIRE development team, *Jet Propulsion Laboratory, California Institute of Technology, Pasadena, CA 91109, USA; California Institute of Technology, Pasadena, CA 91125, USA.*
"JPL / Caltech KID Development for Far-IR / mm-Wave Spectroscopy."

3:30 pm **M4.4.** Eduard F.C. Driessen on behalf of the NIKA2 collaboration, *Institut de Radioastronomie Millimétrique (IRAM), 300 rue de la Piscine Domaine Universitaire St Martin d'Heres 38406, France.*
"The NIKA2 instrument at the IRAM 30m-telescope: state-of-the-art KID performance in the mm range."

3:50 pm **M4.5.** Hiroshi Matsuo [1], Hajime Ezawa [1], Hitoshi Kiuchi [1], Mareki Honma [1], Yasuhiro Murata [2], *[1] National Astronomical Observatory of Japan, Mitaka, Tokyo 181-8588, Japan; [2] Institute of Space and Astronautical Science, Japan Aerospace Exploration Agency, Sagamihara, Kanagawa 252-5210, Japan.*
"Prospects of Terahertz Intensity Interferometry."

4:10 pm Coffee Break

Session M5: **Optical Design, Systems, and Components I**
Chair: *Maria Alonso-delPino*

4:40 pm **M5.1.** Irmantas Kašalynas [1], Heiko Richter [2], Simonas Indrišiūnas [1], Ignas Grigelionis [1], Linas Minkevičius [1], Gediminas Račiukaitis [1], Heinz-Wilhelm Hübers [2], *[1] Center for Physical Sciences and Technology, Saulėtekio Ave. 3, LT-10257 Vilnius, Lithuania; [2] German Aerospace Center (DLR), Institute of Optical Sensor Systems, Rutherfordstr. 2, 12489 Berlin, Germany.*
"Development of diffractive optics for 4.7 THz frequency."

5:00 pm **M5.2.** T. Matsumura [1], S. Hanany [2], H. Imada [4], H. Ishino [4], N. Katayama [1], Y. Kobayashi [5], K. Komatsu [4], K. Konishi [6,7], M. Kuwata-Gonokami [6], S. Nakamura [8], H. Sakurai [6], Y. Sakurai [1], R. Takaku [8] , S. Utsunomiya [1], Q. Wen [2], K. Young [2], J. Yumoto [6,7], *[1] Kavli Institute for The Physics and Mathematics of The Universe (WPI), The University of Tokyo, Kashiwa, 277- 8583, Japan; [2] University of Minnesota, Twin Cities, Minneapolis, 55455, USA; [3] Institute of Space and Astronautical Science (ISAS), Japan Aerospace Exploration Agency (JAXA), Sagami-hara, Kanagawa,252-5210, Japan; [4] Okayama University, Okayama, 700-0082, Japan; [5] The Institute for Solid State Physics, The University of Tokyo, Kashiwa, 277-8583, Japan; [6] Institute for Photon Science and Technology, The University of Tokyo, Tokyo, 113-8654, Japan; [7] Department of Physics, The University of Tokyo, Tokyo, 113-8654, Japan; [8] Yokohama National University, Yokohama, 240-8501, Japan.*
"Development of the broadband anti-reflection coating on sapphire using a sub-wavelength structure for a CMB polarization experiment."

5:20 pm **M5.3.** H. Imada [1], T. Matsumura [2], R. Takaku [3], G. Patanchon [4], H. Ishino [5], Y. Sakurai [2], K. Komatsu [5], N. Katayama 2, *[1] Institute of Space and Astronautical Science (ISAS), Japan Aerospace Exploration Agency (JAXA), Sagamihara, Kanagawa, 252-5210, Japan; [2] Kavli Institute for the Physics and Mathematics of the Universe, The University of Tokyo, Kashiwa, Japan; [3] Yokohama National University, Yokohama, Kanagawa, Japan; [4] Paris Diderot University, Paris, France; [5] Okayama University, Okayama, Okayama, Japan.*
"Instrumentally induced spurious polarization of a multi-layer half wave plate for a CMB polarization observation."

5:40 pm **M5.4.** Mikko Kotiranta [1], Axel Murk [1], Karl Jacob [1], Hyunjoo Kim [1], Paul Hartogh [2], [1] *University of Bern, Bern, 3012, Switzerland;* [2] *Max Planck Institute for Solar System Research, Göttingen, 37077, Germany.*
"Optical Design of the Submillimeter Wave Instrument on JUICE."

March 27, 2018 (Tuesday)

Session T1: **Invited II**
Chair: *Matt Bradford*

9:00 am **T1.1.** Dariusz C. Lis, *LERMA, Observatoire de Paris, PSL Research University, CNRS, Sorbonne Universités, UPMC Univ. Paris 06, F-75014, Paris, France.*
"Future prospects for the high-resolution space far- infrared spectroscopy."

Session T2: **SMM Instruments**
Chair: *Victor Belitsky*

9:30 am **T2.1.** Raquel R. Monje, Ken B. Cooper, Robert J. Dengler, Corey J. Cochrane, Stephen L. Durden, Adrian Tang, and Mathieu Choukroun, *Jet Propulsion Laboratory, California Institute of Technology, 4800 Oak Grove Dr., Pasadena, CA 91109, USA.*
"A 95 GHz FMCW thermal-noise-limited radar: sensitivity and range-Doppler measurements."

9:50 am **T2.2.** Deacon J. Nemchick [1], Brian J. Drouin [1], Adrian J. Tang [2], Yanghyo Kim [2], Gabriel Virbila [3], and Mau-Chung Frank Chang [3], [1] *Laboratory Studies and Atmospheric Observation, Jet Propulsion Laboratory, Pasadena, CA 91109, USA;* [2] *Submillimeter Wave Advanced Technology, Jet Propulsion Laboratory, Pasadena, CA 91109, USA;* [3] *Department of Electrical Engineering, University of California at Los Angeles, Los Angeles, CA 90025 USA.*
"A CMOS-Based 90 – 105 GHz Pulsed-Echo Spectrometer: New Approaches for Highly-Mobile and Low-Power in situ Chemical Detections."

10:10 am **T2.3.** Richard J. Roy, Ken Cooper, Matthew Lebsock, Jose V. Siles, Luis Millán, Raquel Rodriguez Monje, Robert Dengler, *Jet Propulsion Laboratory, California Institute of Technology, Pasadena, CA 91109, USA.*
"Differential absorption radar near the 183 GHz water absorption line for inside-cloud humidity profiling."

10:30 am **T2.4.** Eric W. Bryerton and Jeffrey L. Hesler, *Virginia Diodes, Inc., Charlottesville, VA 22902, USA.*
"An Integrated G-Band 4-Channel Direct Detection Radiometer for the TROPICS Mission."

10:50 pm Coffee Break

Session T3: **Superconducting Heterodyne Detectors I**
Chair: *Gregory Goltsman*

11:20 am **T3.1.** Daniel Cunnane [1], Narendra Acharya [2], Wenura K. Withanage [2], Xiaoxing Xi [2], Jonathan Kawamura [1], and Boris S. Karasik [1], *[1] Jet Propulsion Laboratory, Pasadena, CA 91109, USA; [2] Temple University, Philadelphia, PA 19122, USA.*
"MgB_2 Hot Electron Bolometers for Array Receivers."

11:40 am **T3.2.** Sergey Cherednichenko, Evgenii Novoselov, Usman Ul-Haq, Narendra Acharya, *Terahertz and Millimetre Wave Laboratory, Department of Microtechnology and Nanoscience, Chalmers University of Technology, SE-412 96 Gothenburg, Sweden.*
"Prospects with low noise and wide bandwidth MgB_2 HEB THz mixers."

12:00 pm **T3.3.** Sheng-Cai Shi [1,2], Jing Li [1,2], Wen Zhang [1,2], Wei Miao [1,2], Zhen-Hui Lin [1,2,3], Jin-Ping Yang [1,2], Dong Liu [1,2], Wen-Ying Duan [1,2], Zheng Wang [1,2], Zheng Lou [1,2], Qing Shi [1,2,4], Zhi Li [1,2,4], Kang-Ming Zhou [1,2], Ming Yao [1,2,3], Jie Hu [1,2,3], Yun Ren [1,2], and Qi-Jun Yao [1,2], *[1] Purple Mountain Observatory, Chinese Academy of Sciences, Nanjing, China; [2] Key Laboratory of Radio Astronomy, Chinese Academy of Sciences, Nanjing, China; [3] University of Chinese Academy of Sciences, Beijing 100049, China; [4] University of Science and Technology of China, Hefei 230026, China.*
"Development of THz Superconducting Mixers and Detectors for DATE5."

12:20 pm **T3.4.** J. R. G. Silva [1,2], D. Hayton [1,2], B. Mirzaei [4], W. Laauwen [1], Y. Gan [1,2], A. Young [5], C. Kulesa [5], C. Walker [5], J. R. Gao [1,4], *[1] SRON Netherlands Institute for Space Research, Groningen/Utrecht, The Netherlands;[2] Kapteyn Astronomical Institute, University of Groningen, 9747 AD, Groningen, The Netherlands; [3] Jet Propulsion Laboratory, California Institute of Technology, Pasadena, CA 91109, USA;[4] Kavli Institute of NanoScience, Delft University of Technology, Delft, The Netherlands;[5] Steward Observatory, 933 N Cherry Ave., Rm N204, University of Arizona, Tucson, AZ 85721, USA.*
"Preliminary design study of a 4×2 HEB array at 4.7 THz for GUSTO."

12:40 pm Lunch Break

Session T4: **Invited III**
Chair: *Imran Mehdi*

2:00 pm **T4.1.** Paul F. Goldsmith, *Jet Propulsion Laboratory, California Institute of Technology, Pasadena, CA 91109, USA.*
"The Renaissance of Submillimeter Astronomical Spectroscopy."

2:30 pm **T4.2.** Paul Hartogh and the SWI team, *Max Planck Institute for Solar System Research, Göttingen, 37077, Germany.*
"The Submillimeter Wave Instrument on JUICE."

Session T5: **LNAs and Backends**
Chair: *Omid Noroozian*

3:00 pm **T5.1.** Peter Day, Byeong Ho Eom, Henry G. Leduc, *Jet Propulsion Laboratory, California Institute of Technology, Pasadena, CA 91109, USA.*
"Traveling-wave Superconducting Parametric Amplifiers."

3:20 pm **T5.2.** Yanghyo Kim [1,2], Yan Zhang [2], Adrian Tang [1,2], Theodore Reck [1], and Mau-Chung Frank Chang [2], *[1] Jet Propulsion Laboratory, Pasadena, CA 91109, USA; [2] University of California, Los Angeles, Los Angeles, CA 90095, USA.*
"A 1.5 W 3 GHz Back-End Processor in 65-nm CMOS for Sub-millimeter-wave Heterodyne Receiver Arrays."

3:40 pm

Poster Session

Chairs: *Daniel Cunnane and Andrey Khudchenko*

(Coffee will be available during the Session)

PA1. Olivier Auriacombe [1], M. Henry [1], B. N. Ellison [1], M. Jarret [1], J. Charlton [2], S. Parkes [3], I. Rule [4], *[1] Millimetre Wave Technology Group, Chilton, UK; [2] JCR Systems Ltd., South Gloucestershire, UK; [3] STAR Dundee Ltd., Dundee, UK; [4] Met Office., Exeter, Devon, UK.*
"HYper Spectral Microwave Sounder (HYMS)."

PA2. Bertrand Thomas [1], M. Berrada [1], J. Ceru [1], F. Villa [2], A. Walber [1], P. Yagoubov [3], *[1] Radiometer Physics GmbH, Werner-von-Siemens Strasse 4, 53340 Meckenheim, Germany; [2] INAF / IASF Bologna via Gobetti, 101, 40129, Bologna, Italy; [3] European Southern Observatory, Karl-Schwarzschild-Str. 2, 85748 Garching, Germany.*
"A dual-polarized sideband separating Schottky based receiver for ALMA Band 2+3 Warm Cartridge Assembly."

PA3. Ronald Hesper [1], A. Khudchenko [1], M.F Lindemulder [2], M.E. Bekema [1], L.H.R. de Haan-Stijkel [1], J. Barkhof [1], J. Adema [1], and A.M. Baryshev [1], *[1] Kapteyn Astronomical Institute, University of Groningen, 9747 AD Groningen, The Netherlands; [2] KVI - Center for Advanced Radiation Technology, 9747 AA Groningen, The Netherlands.*
"A Deployable 600-720GHz ALMA-Type Sideband-Separating Receiver Cartridge".

PA4. Sean Bryan, Philip Mauskopf, and Christopher Groppi, *Arizona State University, Tempe, AZ 85281, USA.*
"A Low-Power Compact Millimeter-wave Radiometer for a Weather Cubesat".

PA5. Weidong Hu [1], Shi Chen [1], Wenlong Zhang [1], Fen Yue [1], and Leo P. Ligthart [2], *[1] Beijing Institute of Technology, Beijing Key Laboratory of Millimeter wave and Terahertz Technique, Beijing, 100081, China; [2] Delft University of Technology, Faculty of Electrical Engineering, Delft, 2600 AA, Netherlands.*
"The 220GHz Terahertz Cloud Radar System for Atmosphere Observation".

PB6. Jonathan R. Hoh [2], Paul F. Goldsmith (PI) [1], Jose V. Siles [1], Adrian J. Tang [1], Christopher E. Groppi [2], and Jeremy Whitton [2], *[1] Jet Propulsion Laboratory, California Institute of Technology, Pasadena CA 91109, USA; 2 Arizona State University School of Earth and Space Exploration, Tempe AZ 85287, USA.*
"Technology Development for Long Wavelength Satellites".

PD7. Alexey Pavolotsky, Vincent Desmaris, Victor Belitsky, *Chalmers University of Technology, Group for Advanced Receiver Development, Department of Space, Earth and Environment, Göteborg, 412 96, Sweden;*
"Nb/Al-AlN/Nb superconducting tunnel junctions: fabrication process and characterization results".

PD8. Jing Li [1,2], Zhen-Hui Lin [1,2,3], Jin-Ping Yang [1,2], Dong Liu [1,2], Wen-Ying Duan [1,2], Zheng Lou [1,2], Qing Shi [1,2,4], Zhi Li [1,2,4], Wen Zhang [1,2], Wei Miao [1,2], Qi-Jun Yao [1,2], and S.C. Shi [1,2], [1] *Purple Mountain Observatory, Chinese Academy of Sciences, Nanjing, China;* [2] *Key Laboratory of Radio Astronomy, Chinese Academy of Sciences, Nanjing, China;* [3] *University of Chinese Academy of Sciences, Beijing 100049, China;* [4] *University of Science and Technology of China, Hefei 230026, China.*
"Development of the Terahertz Superconducting Imaging Array (TeSIA)".

PD9. Ivan Tretyakov, N. Kaurova, B. M. Voronov, and G. N, Goltsman, *Moscow State University of Education, Moscow, 119991. Russia.*
"About effect of the temperature operating conditions on the Noise Temperature and Noise Bandwidth of the Terahertz Range NbN Hot-Electron Bolometers".

PD10. Nicholas J. Rommelfanger, Bruce Bumble, and Boris S. Karasik, *Jet Propulsion Laboratory, California Institute of Technology, Pasadena, CA 91109, USA.*
"Bandwidth Measurements of NbN HEB Devices with GaN Buffer Layers".

PD11. M. Merker, A. Schmid, C. Bohn, K. Ilin, M. Siegel, *Karlsruhe Institute of Technology, Institute of Micro- and Nanoelectronic Systems, Karlsruhe, 76135, Germany.*
"NbN/AlN/NbN technology on sapphire substrates for SIS based THz receivers".

PD12. Shibo Shu[1], M. Calvo [2,3], J. Goupy [2,3], A. Monfardini [2,3], and E.F.C. Driessen[1], [1] *Institut de RadioAstronomie Millimétrique, 300 rue de la Piscine, 38406 Saint Martin d'Hères, France;* [2] *Université Grenoble Alpes, 621 avenue Centrale, 38400 Saint Martin d'Hères, France;* [3] *Institut Néel, CNRS, BP 166, 38042 Grenoble, France.*
"Optical Response of Lumped-Element Kinetic-Inductance-Detectors".

PD13. Akira Kawakami [1], Yoshihisa Irimajiri [1], Taro Yamashita [1,2], Satoshi Ochiai [1], Yoshinori Uzawa [1], [1] *National Institute of Information and Communications Technology (NICT), Kobe, Hyogo, 651-2492, Japan;* [2] *PRESTO, Japan Science and Technology Agency, 4-1-8 Honcho, Kawaguchi, Saitama 332-0012, Japan.*
"Extension of IF band of THz hot electron bolometer mixer using magnetic thin film".

PD14. Hajime Ezawa [1], Hiroshi Matsuo [1], Masahiro Ukibe [2], Go Fujii [2], and Shigetomo Shiki [2], [1] *National Astronomical Observatory of Japan, Tokyo 181-8588, Japan;* [2] *National Institute of Advanced Industrial Science and Technology, Ibaraki 305-8568, Japan.*
"SIS Photon Detectors for Terahertz Astronomy".

PD15. T. Kojima [1], M. Kroug [1], A. Gonzalez [1], K. Uemizu [1], K. Kaneko [1], A. Miyachi [1], Y. Kozuki [2,1], W. Shan [1], and S. Asayama [1], [1] *National Astronomical Observatory of Japan, Mitaka, 181-8588, Japan;* [2] *Osaka Prefecture University, Sakai, 599-8531, Japan.*
"Development of a waveguide SIS mixer in the 275-500 GHz band and dual band evaluation system".

PD16. Andrey Khudchenko, R. Hesper, J. Barkhof, and A.M. Baryshev, *NOVA/Kapteyn Astronomical Institute, University of Groningen, Groningen, The Netherlands.*
"Contribution of IF Chain Reflections in the Image Rejection Ratio of 2SB SIS receivers".

PD17. J. Zdanevičius [1], K. Ikamas [1], M. Bauer [2,3], J. Matukas [1], A. Lisauskas [1,2], H. Richter [4], H.-W. Hübers [4,5], and H. G. Roskos [2], [1] *Institute of Applied Electrodynamics and Telecommunications, Vilnius University, Vilnius, 10257, Lithuania;* [2] *Physikalisches Institut, Goethe-University Frankfurt am Main, Frankfurt am Main, 60438, Germany;* [3] *Center for Materials Characterization and Testing, Fraunhofer ITWM, Kaiserslautern, 67663, Germany;* [4] *Institute of Optical Sensor Systems, German Aerospace Center (DLR), 12489 Berlin, Germany;* [5] *Humboldt-Universität zu Berlin, Department of Physics, 12489 Berlin, Germany.*
"TeraFET detector for measuring power fluctuations of radiation from a 4.75-THz QCL".

PD18. Daniel Cunnane [1], Narendra Acharya [2], Wenura K. Withanage [2], Xiaoxing Xi [2], and Boris S. Karasik [1], [1] *Jet Propulsion Laboratory, California Institute of Technology, Pasadena, CA 91109, USA;* [2] *Temple University, Philadelphia, PA 19122, USA.*
"THz performance of MgB_2 HEB mixer with non-uniform thickness profile".

PD19. Changyun Yoo [1], M. Huang [1], J. Kawamura [2], B. S. Karasik [2], L. Pfeiffer [3], and M. S. Sherwin [1], [1] *Physics Department and Institute for Terahertz Science and Technology, University of California at Santa Barbara, Santa Barbara, CA 93106, USA;* [2] *Jet Propulsion Laboratory, California Institute of Technology, Pasadena, CA 91109, USA;* [3] *Department of Electrical Engineering, Princeton University, Princeton, NJ 08544, USA.*
"A Tunable Antenna-coupled Intersubband Terahertz Detector".

PD20. Roger O'Brient [1,2], James Bock [2,1], Hien Nguyen [1], Bryan Steinbach [2], Anthony Turner [1], Albert Wandui [2], Jonas Zmuidzinas [2,1], [1] *Jet Propulsion Laboratory, California Institute of Technology Pasadena, CA 91109, USA;* [2] *California Institute of Technology, Pasadena, CA 91125, USA.*
"TKIDs for Cosmic Microwave Background Studies". -- Not presented

PD21. Boris Karasik, *Jet Propulsion Laboratory, California Institute of Technology Pasadena, CA 91109, USA.*
"An array scalable far-IR detector with NEP < 10^{-20} $W/Hz^{1/2}$".

PD22. Matthias Kroug [1], T. Kojima [1], Y. Fujii [1], K. Ohtawara [1], A. Miyachi [1], and Y. Uzawa [2], [1] *National Astronomical Observatory of Japan, Mitaka, 181-8588, Japan;* [2] *National Institute of Information and Communications Technology, Koganei, 184-0015, Japan.*
"Noise Performance of ALMA Band10 Receivers Employing High-Jc SIS Mixers".

PD23. Ghassan Yassin [1], Boon-Kok Tan [1], Andre Hector [1], Sumedh Mahashabde [1], Andrey Khudchenko [2], Andrey M. Baryshev [2], Ronald Hesper [2], and Valery P. Koshelets [3], [1] *Department of Physics (Astrophysics), University of Oxford, Keble Road, Oxford OX1 3RH, UK;* [2] *University of Groningen, Kapteyn Astronomical Institute, 9747 AD, Groningen, The Netherlands;* [3] *Kotel'nikov Institute of Radio Engineering and Electronics RAS, 125009 Moscow, Russia.*
"Investigation of the performance of an SIS mixer with Nb-AlN-NbN tunnel junctions in the 780-950 GHz frequency band".

PD24. Omid Noroozian [1, 2, 3], P. K. Day [4], H.G. Leduc [4], D. Woody [5], J. Zmuidzinas [5], A. R. Kerr [1], J. G. Mangum [1], M. Cyberey [3], A. W. Lichtenberger [3], [1] *National Radio Astronomy Observatory, Charlottesville, VA 22903, USA;* [2] *NASA Goddard Space Flight Center, 20771, Greenbelt, MD, USA;* [3] *University of Virginia, Charlottesville, VA 22903, USA;* [4] *Jet Propulsion Laboratory, California Institute of Technology, Pasadena, CA 91109, USA; 5 California Institute of Technology, Pasadena, CA 91125, USA.*
"Superconducting Parametric Amplifiers: the Next Big Thing in (Sub)Millimeter-wave Receivers".

PD25. Jake A. Connors [1], Paul K. Grimes [2], Raymond Blundell [2], Young Jae Shin [1], and Philip Kim [1], [1] *Department of Physics, Harvard University, Cambridge, MA 02138, USA;* [2] *Smithsonian Astrophysical Observatory, Cambridge, MA 02138, USA.*
"Graphene Field Effect Transistors for Microwave and mm-Wave Applications".

PD26. Lorenza Ferrari [1], S. J. C. Yates [1], M. Eggens [1], A. M. Baryshev [2], and J. J. A. Baselmans [3,4], [1] *SRON Netherlands Institute for Space Research, Groningen, 9747AD, The Netherlands;* [2] *Kapteyn Institute, University of Groningen, Groningen, 9747 AD The Netherlands;* [3] *SRON Netherlands Institute for Space Research, Utrecht, 3584CA, The Netherlands;* [4] *Terahertz Sensing Group, Delft University of Technology, Delft 2628CD, The Netherlands.*
"MKID large format array testbed".

PL27. Artem A. Kuzmin, Steffen Doerner, Stefan Singer, Ilya Charaev, Konstantin S. Ilin, Stefan H. Wuensch, and Michael Siegel, *Institute of Micro- and Nanolectronic Systems of Karlsruhe Institute of Technology, Karlsruhe, Germany.*
"TES readout using an NbN-based Microwave Kinetic-Inductance Nanowire Galvanometer at 4.2 K".

PL28. Takafumi Kojima [1], Y. Uzawa [2,1], and W. Shan [1], [1] *National Astronomical Observatory of Japan, Mitaka, 181-8588, Japan;* [2] *National Institute of Information and Communications Technology, Koganei, Tokyo 184-0015, Japan.*
"Observation of positive gain by a combination of quasiparticle SIS up and down frequency converters".

PL29. Peter J. Sobis [1], Niklas Wadefalk [2], Joel Schleeh [2], Christina Emrich [1], Mats Lindgren [1], and Anders Emrich [1], [1] *Omnisys Instruments AB, Västra Frölunda, SE-421 32, Sweden;* [2] *Low Noise Factory AB, Göteborg, SE-412 63, Sweden.*
"Low Power Cryogenic Rad-hard LNAs for Space".

PL30. Boon-Kok Tan and Ghassan Yassin, *Department of Physics (Astrophysics), University of Oxford, Keble Road, Oxford OX1 3RH, UK.*
"The travelling wave parametric amplifiers - Design considerations". -- **Not presented**

PM31. Yoshihisa Irimajiri [1], Alvaro Gonzalez [2], Satoshi Ochiai [1], Akira Kawakami [1], and Yoshinori Uzawa [1], [1] *National Institute of Information and Communications Technology (NICT), Koganei, Tokyo, 184-0015 Japan;* [2] *National Astronomical Observatory of Japan (NAOJ), Mitaka, Tokyo, 181-8588, Japan.*
"Beam pattern measurements of a quasi-optical HEB mixer at 2 THz".

PM32. Junhan Kim and Daniel P. Marrone, *Department of Astronomy and Steward Observatory, University of Arizona, 933 N. Cherry Avenue, Tucson, AZ 85721, USA.*
"Tilted beam measurement of VLBI receiver for the South Pole Telescope".

PM33. Grigoriy M. Bubnov [1,2], Vyacheslav F. Vdovin [1,2], Dmitriy S. Martynov [2] and Igor I. Zinchenko [1], [1] *Institute of Applied Physics RAS, N.Novgorod, 603950, Russia;* [2] *Nizhniy Novgorod State Technical University, N.Novgorod, 603950, Russia.*
"A refined method of the data processing for astroclimate measurements at millimeter waves".

PM34. Yuki Sakurai [1], T. Matsumura [1], N. Katayama [1], H. Kanai [2], T. Iida [3], [1] *Kavli Institute for The Physics and Mathematics of The Universe (WPI), The University of Tokyo, Kashiwa, Japan;* [2] *Yokohama National University, Yokohama, Kanagawa, Japan;* [3] *ispace inc., Japan.*
"Development of a cryogenic remote sensing thermometer for CMB polarization experiments".

PM35. Jacob W. Kooi [1], Rodrigo A. Reeves [2], Arthur W. Lichtenberger [3], Theodore J. Reck [1], Andy K. Fung [1], Sander Weinreb [4], James. W. Lamb [4], Rohit S. Gawande [1], Kieran. A. Cleary [4], and Goutam Chattopadhyay [1], [1] *Jet Propulsion Laboratory, California Institute of Technology, Pasadena, CA 91109, USA;* [2] *CePIA, Astronomy Department, Universidad de Concepcion, Chile;* [3] *University of Virginia, Charlottesville, VA 22904, USA;* [4] *California Institute of Technology, Pasadena, CA 91125, USA.*
"A Programmable Waveguide Calibration Load".

PM36. Axel Murk [1], Karl Jacob [1], David Marote [2], [1] *University of Bern, Institute of Applied Physics, Sidlerstr. 5 Bern, 3012, Switzerland;* [2] *Airbus Defence and Space SAU (ASE), Avenida de Aragon 404, 28022 Madrid, Spain.*
"Characterization of low-loss reflectors for spaceborne microwave radiometers".

PO37. Hao-Tian Zhu [1,2,4,5], Dong Liu [2], Jie Hu [2], Sheng Li [2], Sheng-Cai Shi [2], and Quan Xue [1,3,4], [1] *State Key Laboratory of Millimeter Waves, City University of Hong Kong, Hong Kong SAR, China;* [2] *Purple Mountain Observatory, Key Laboratory of Radio Astronomy, Chinese Academy of Sciences, Nanjing, 210008, China;* [3] *School of Electronic and Information Engineering, South China University of Technology, Guangzhou, 510640, China;* [4] *Shenzhen Key Lab of Millimeter-Wave and Wideband Wireless Communications, Shenzhen Research Institute, City University of Hong Kong, Shenzhen, 518057, China;* [5] *Poly-Grames, Ecole Polytechnique de Montreal, Montreal, QC, H3T 1J4, Canada.*
"Flexible, Thermal Isolating and Low-loss Rectangular Dielectric Waveguide for THz Superconducting Receivers".

PO38. C. Jarufe [1], R. Rodriguez [2], V. Tapia [2], P. Astudillo [2], D. Monasterio [2], R. Molina [1], F. P. Mena [1], N. Reyes [1], and R. Bronfman [2], [1] *Electrical Engineering Department, Universidad de Chile, Av. Tupper 2007, Santiago, Chile;* [2] *Astronomy Department, Universidad de Chile, Camino el Observatorio 1515, Santiago, Chile.*
"Compact optimized slot antenna for mm (and sub-mm?) applications". -- Not presented

PO39. Kristina Davis [1], Chris Groppi [1], Jose Siles [2], Imran Mehdi [2], Jon Kawamura [2], Craig Kulesa [3], and Phil Mauskopf [1], *Arizona State University, Tempe, AZ 85282, USA;* [2] *Jet Propulsion Laboratory, California Institute of Technology, Pasadena, CA 91109, USA;* [3] *University of Arizona, Tucson, AZ 85721, USA.*
"THz Suborbital Payload Optical System Pointing Stability for Target Studies of Compact Sources".

PO40. Jun Ren, Yijing Deng, Zhenguo Jiang, Md. Itrat Bin Shams, Patrick Fay, and Lei Liu, *University of Notre Dame, Notre Dame, IN 46556, USA.*
"Tunable and Reconfigurable Terahertz Devices Based on Photo-Induced Electromagnetic Band Gap Structures".

PO41. Peter J. Sobis [1], Vladimir Drakinskiy [2], Arvid Hammar [1], Johanna Hanning [2], Anders Emrich [1], Elena Saenz [3], and Jan Stake [2], [1] *Omnisys Instruments AB, Västra Frölunda, SE-421 32, Sweden;* [2] *Chalmers University of Technology, Göteborg, SE-412 96, Sweden;* [3] *European Space Agency – ESA/ESTEC, Noordwijk, PO Box 299, NL-2200 AG, The Netherlands.*
"Membrane Integrated Asymmetric Dual E-plane Probe Ortho Mode Transducer at 424 GHz".

PO42. Y. J. Deng, M. I. B. Shams, J. Ren, P. Fay, and L. Liu, *Department of Electrical Engineering, University of Notre Dame, Notre Dame, IN 46556, USA.*
"High performance Photo-Induced Substrate-Integrated-Waveguide for tunable and reconfigurable THz circuits".

PO43. Duccio Delfini [1], Martina Wiedner [1], Massimiliano Casaletti [2], Julien Serrazin [2], François Joint [1], Thibaut Vacelet [1], and Yan Delorme [1], [1] *LERMA, Observatoire de Paris, PSL Research University, CNRS, Sorbonne Universités, UPMC Univ. Paris 06, F-75014, Paris, France;* [2] *Sorbonne Universités, UPMC Univ Paris 06, UR2, L2E, F-75005, Paris, France.*
"Development of a transmit-array for heterodyne receiver".

PO44. Daniel Koller, Jeffrey L. Helser, Eric W. Bryerton, *Virginia Diodes, Inc., Charlottesville, VA, 22903, USA.*
"WM380 (675-700GHz) Band-Pass Filters in Milled Split-Block Construction".

PO45. Hawal Rashid, V. Desmaris, and V. Belitsky, *Chalmers University of Technology, Gothenburg, 412 96, Sweden.*
"Wideband Passive Circuits for Sideband Separating Receivers".

PO46. Jeremy D. Whitton [1], Philip D. Mauskopf [1], Paul F. Goldsmith [2], Kristina K. Davis [1], and Christopher E. Groppi [1], [1] *Arizona State University, Tempe, AZ 85281, USA;* [2] *Jet Propulsion Laboratories, California Institute of Technology, Pasadena, CA 91109, USA.*
"Prototype of a Dielectrically Embedded Mesh Lens".

PO47. Paul K. Grimes [1], Scott N. Paine [1], Ramprasad Rao [2], Tirupati K. Sridharan [1], and Lingzhen Zeng [1], [1] *Harvard-Smithsonian Center for Astrophysics, Cambridge, MA 02138, USA;* [2] *Institute of Astronomy and Astrophysics, Academia Sinica, Submillimeter Array Observatory, Hilo, HI 96720, USA;*
"Tolerancing of the Submillimeter Array Optics using Physical Optics Simulations".

PO48. Fabien Defrance [1], G. Chattopadhyay [2], J. Connors [3], S. Golwala [1], M. I. Hollister [4], C. Jung-Kubiak [2] , E. Padilla [5], S. Radford [3], J. Sayers [1], E. C. Tong [3], and H. Yoshida [1], [1] *California Institute of Technology, Pasadena, CA 91125, USA;* [2] *Jet Propulsion Laboratories, California Institute of Technology, Pasadena, CA 91109, USA;* [3] *Harvard–Smithsonian Center for Astrophysics, Cambridge, MA 02138, USA;* [4] *Fermi National Accelerator Laboratory, Batavia, IL 60510, USA;* [5] *Cal State San Bernardino, San Bernardino, CA 92407, USA.*
"Wafer-bonded Antireflection Layers for Silicon optics".

PS49. Diego Moro-Melgar, O. Cojocari, I. Oprea, M. Rickes & M. Hoefle, *ACST GmbH, Hanau, 63457, Germany.*
"Single-Chip 150 GHz Doubler with more than 150 mW Output Power based on Flip-Chip Schottky Diodes".

PS50. Carlos G. Pérez-Moreno and Jesús Grajal, *Information Processing and Telecommunications Center, Universidad Politécnica de Madrid, E.T.S.I. Telecomunicación, Av. Complutense 30, 28040 Madrid, Spain.*
"Broadband frequency triplers for radar applications in the THz range".

PS51. Deacon J. Nemchick [1], Brian J. Drouin [1], Adrian J. Tang [2], Yanghyo Kim [2], Gabriel Virbila [3], and Mau-Chung Frank Chang [3], [1] *Laboratory Studies and Atmospheric Observations, Jet Propulsion Laboratory, California Institute of Technology, Pasadena, CA 91109, USA;* [2] *Submillimeter Wave Advanced Technology, Jet Propulsion Laboratory, California Institute of Technology, Pasadena, CA 91109, USA;* [3] *Department of Electrical Engineering, University of California at Los Angeles, Los Angeles, CA 90025 USA.*
"Pure-Rotational Molecular Spectroscopy with a Low-Power CMOS-Based W-Band Transmitter".

PS52. François Joint [1,2], Gregory Gay [1], Thibaut Vacelet [1], L. Li [3], E. Linfield [3], Yan Delorme [1], Raffaele Colombelli [2], [1] *LERMA, Laboratoire d'Études du Rayonnement et de la Matière en Astrophysique et Atmosphères, Observatoire de Paris, Paris, France;* [2] *Centre de Nanosciences et de Nanotechnologies, CNRS, Univ. Paris-Sud, Université Paris-Saclay, C2N – Orsay, 91405 Orsay cedex, France;* [3] *School of Electronic and Electrical Engineering, University of Leeds, Leeds LS2 9JT, UK.*
"Low power consumption quantum cascade lasers at 2.7 THz for compact and sensitive heterodyne detectors".

PT53. Peter J. Sobis [1], Vladimir Drakinskiy [2], Tony Pellikka [1], Slavko Dejanovic [1], Anders Emrich [1], and Jan Stake [2], [1] *Omnisys Instruments AB, Västra Frölunda, SE-421 32, Sweden;* [2] *Chalmers University of Technology, Göteborg, SE-412 96, Sweden.*
"Discrete GaAs Schottky beamlead mixer diodes for space-borne receiver applications".

PT54. Christopher E. Groppi [1], Paul F. Goldsmith [2], Jose V. Siles [2], Philip Mauskopf [1], Christopher K. Walker [3], Daniel C. Jacobs [1], Adrian Tang [2], and Paul Scowen [1], [1] *Arizona State University, Tempe, AZ 85287, USA;* [2] *Jet Propulsion Laboratory, California Institute of Technology, Pasadena, CA 91109, USA;* [3] *University of Arizona, Tucson, AZ 85721, USA.*

"CubeSat Interferometry for THz Astrophysics, Planetary Science and Earth Observing".

PT55. Jonathan Hoh [1], Christopher Groppi [1], Choonsup Lee [2], Robert Lin [2], Philip Mauskopf [1], Phil Putman [3], and Jose Siles [2], [1] *Arizona State University School of Earth and Space Exploration, Tempe AZ 85287, USA;* [2] *Jet Propulsion Laboratory, California Institute of Technology, Pasadena, CA 91109, USA;* [3] *Sierra Lobo Inc., Fremont OH 43429, USA.*
"Integrated Schottky Receiver for Small Satellite Deployment".

PT56. Jeanne Treuttel [1,2], E. Schlecht [1], C. Lee [1], J-V. Siles [1], D. Hayton [1], R.Lin [1], and I. Mehdi [1], [1] *Jet Propulsion Laboratory, California Institute of Technology, Pasadena, CA 91109, USA;* [2] *Observatory of Paris, LERMA, Paris, 75014, France.*
"All Solid-State Receiver Designs at 2 THz for Atmospheric Sounding".

PD57. Faouzi Boussaha, Samir Beldi, Christine Chaumont, Piercarlo Bonifacio, *Galaxies, Etoiles, Physique et Instrumentation (GEPI), Observatoire de Paris, 77 avenue Denfert-Rochereau, 75014 Paris – France.*
"Development of Parallel Capacitor Based Kinetic Inductance Detectors (KIDs)."

PD58. Gert de Lange [1], Peter Roelfsema [1,2], Martin Giard [3], Francisco Najarro [4], KeesWafelbakker 1, Brian Jackson [1], Lee Armus [5], Marc Audard [6], Matt Bradford [7], Yasuo Doi [8], Matt Griffin [9], Frank Helmich [1,2], Inga Kamp [2], Franz Kerschbaum [10], Oliver Krause [11], Bengt Larsson [12], Suzanne Madden [13], David Naylor [14], Takahashi Onaka [8], Luigi Spinoglio [15], Floris van der Tak [1,2], [1] *SRON Netherlands Institute for Space Research, Groningen, The Netherlands;* [2] *Kapteyn Astronomical Institute, Groningen, The Netherlands;* [3] *IRAP, Toulouse, France;* [4] *CAB-INTA, Madrid, Spain;* [5] *IPAC/Caltech, USA;* [6] *ISDC, Geneva, Switzerland;* [7] *JPL/Caltech, USA;* [8] *University of Tokyo, Tokyo, Japan;* [9] *University of Cardiff, Cardiff, United Kingdom;* [10] *University of Vienna, Vienna, Austria;* [11] *MPIA, Heidelberg, Germany;* [12] *Stockholm University, Stockholm, Sweden;* [13] *CEA, Saclay, France;* [14] *University of Lethbridge, Lethbridge, Canada;* [15] *IAPS-INAF, Rome, Italy.*
"The SAFARI grating spectrometer for SPICA."

PD59. Narendra Acharya, Usman Ul-Haq, Sergey Cherednichenko, *Terahertz and Millimetre Wave Laboratory, Department of Microtechnology and Nanoscience, Chalmers University of Technology, SE-412 96 Gothenburg, Sweden.*
"Wide bandwidth measurements of microwave and millimeter wave impedance in MgB_2 HEB mixers."

5:30 pm Break

6:00 pm Banquet

March 28, 2018 (Wednesday)

Session W1: **Invited IV**
Chair: *Andrey Baryshev*

9:00 am **W1.1.** Martina C. Wiedner [1], Andrey Baryshev [2], Victor Belitsky [3], Yan Delorme [1], Vincent Desmaris [3], Anna Di Giorgio [4], Brian Ellison [5], Juan-Daniel Gallego [6], Maryvonne Gerin [1], Paul Goldsmith [11], Christophe Goldstein [7], Frank Helmich [8], Fabrice Herpin [9], Jean-Michel Huet [10], Willem Jellema [8,2], Jean-Michel Krieg [1], Philippe Laporte [10], André

Laurens [7], Imran Mehdi [11], Gary Melnick [12], Benjamin Quertier [9], René Plume [13], Christophe Risacher [14], Russel Shipman [8], OST STDT and NASA Goddard Engineering Team [15], [1] *LERMA, Observatoire de Paris, ENS, PSL Research University, CNRS, Sorbonne Universités, UPMC Univ. Paris 06, avenue de l'Observatoire, 75014 Paris, France;* [2] *Kapteyn Astronomical Institute, University of Groningen, P.O. Box 800, 9700 AV, Groningen, NL;* [3] *Group for Advanced Receiver Development, Chalmers University of Technology, Gothenburg, SE 41296, Sweden;* [4] *Istituto Fisica Spazio Interplanetario INAF, via Fosso del Cavaliere 100, 00133 Roma, Italy;* [5] *Rutherford Appleton Laboratory, Space Department, Harwell Campus, Didcot, OX11 0QX, UK;* [6] *Observatorio de Yebes, CDT-IGN, Apdo. 148, Guadalajara 19080, Spain;* [7] *CNES, 18 Avenue Edouard Belin, 31400 Toulouse, France;* [8] *SRON Netherlands Institute for Space Research, Landleven 12, 9747 AD Groningen, The Netherlands & Kapteyn Astronomical Institute, University of Groningen, Groningen, The Netherlands;* [9] *Laboratoire d'Astrophysique de Bordeaux, Pessac, 33615, France;* [10] *GEPI, Observatoire de Paris, PSL Research University, CNRS, Paris,75014, France;* [11] *Jet Propulsion Laboratory, California Institute of Technology, 4800 Oak Grove Drive, Pasadena CA 91109, USA;* [12] *Harvard-Smithsonian Center for Astrophysics, 60 Garden Street, Cambridge, MA 02138, USA;* [13] *University of Calgary, T2N1N4, Canada;* [14] *Max-Planck-Institut für Radioastronomie, Auf dem Hügel 69, 53121, Bonn, Germany;* [15] *NASA Goddard Space Flight Center, 8800 Greenbelt Rd., Greenbelt, MD 20771, USA*
"HEterodyne Receiver for OST (HERO)."

Session W2: **Large Systems and Applications**
Chair: *Jonathan Kawamura*

9:30 am **W2.1.** A. Smirnov [1], E. Golubev [1], M. Arhipov [1], V. Pishnov [1], T. Kosmovich [1], E. Filina [1], A. Baryshev [1,2], Thijs de Graauw [1], S. Pilipenko [1], S. Likhachev [1] and N. Kardashev [1] on behalf of the Millimetron team, [1] *Astro Space Center of P.N. Lebedev Physical Institute, Moscow, Russia;* [2] *University of Groningen, Kapteyn Astronomical Institute, Groningen, The Netherlands.*
"Millimetron Space Observatory – Large-Aperture and Cooled Space Telescope."

9:50 am **W2.2.** Ken B. Cooper, Raquel R. Monje, Robert J. Dengler, Corey J. Cochrane, Stephen L. Durden, Adrian Tang, and Mathieu Choukroun, *Jet Propulsion Laboratory, California Institute of Technology, 4800 Oak Grove Drive, Pasadena CA 91109, USA.*
"The Gas and Ice Spectrometer/Radar (GAISR): A Millimeter/Submillimeter-Wave Multi-Functional Instrument for Cometary Jet Observations."

10:10 am **W2.3.** Sean Bryan [1], Amanda Clarke [1], Loÿc Vanderkluysen [2], Christopher Groppi [1], Scott Paine [3], Daniel W. Bliss [1], James Aberle [1], and Philip Mauskopf [1], [1] *Arizona State University, Tempe, AZ, 85281, USA;* [2] *Drexel University, Philadelphia, PA, 19104, USA;* [3] *Smithsonian Astrophysical Observatory, Cambridge, MA, 02138, USA.*
"Measuring Volcanic Eruption Dynamics with WAMS, a Millimeter-wave Radar and Imager."

10:30 am **W2.4.** A. M. Baryshev [1], R. Hesper [1], A. Khudchenko [1], K. Rudakov [1,2], [1] *NOVA/Kapteyn Astronomical Institute, University of Groningen, Groningen, The Netherlands;* [2] *Institute of Radioengineering and Electronics, Moscow, Russia.*
"High Frequency High Spectral Resolution Focal Plane Arrays for ATLAST."

10:50 am Coffee Break

Session W3: **Sources and Local Oscillator Systems II**
Chair: *Jeffrey Hesler*

11:20 am **W3.1.** Jose V. Siles, Ken B. Cooper, Choonsup Lee, Robert Lin, Goutam Chattopadhyay, and Imran Mehdi, *Jet Propulsion Laboratory, California Institute of Technology, 4800 Oak Grove Drive, Pasadena CA 91109, USA.*
"Next Generation of Room-Temperature Broadband Frequency Multiplied LO Sources with 10 times Higher Output Power in the 100 GHz – 1.9 THz Range."

11:40 am **W3.2.** Christopher A. Curwen [1], John L. Reno [2], Tatsuo Itoh [1], Benjamin S. Williams [1], [1] *Department of Electrical Engineering, University of California, Los Angeles, CA 90095, USA;* [2] *Sandia National Laboratories, Center of Integrated Nanotechnologies, MS 1303, Albuquerque, NM 97185, USA.*
"Broadband metasurface external-cavity QC-lasers."

12:00 pm **W3.3.** Oleg Cojocari, Diego Moro-Melgar, and Ion Oprea, *ACST GmbH, Hanau, 63457, Germany.*
"Diamond-Substrate Schottky Diodes for high-power MM-wave Multipliers."

12:20 pm **W3.4.** Diego Moro-Melgar, O. Cojocari, I. Oprea, M. Hoefle & M. Rickes, *ACST GmbH, Hanau, 63457, Germany.*
"High Power Flip-Chip Schottky Diodes based 275-305 GHz Transceiver for FMCW-Radar."

12:40 pm Lunch Break

Session W4: **Invited V**
Chair: *Jose Siles*

2:00 pm **W4.1.** Christopher Walker, *University of Arizona, 933 N. Cherry Ave., Tucson, AZ 85721, USA.*
"Terahertz Astronomy from Near Space and Beyond."

Session W5: **Superconducting Heterodyne Detectors II**
Chair: *Jenna Kloosterman*

2:30 pm **W5.1.** Wenlei Shan, Shohei Ezaki, Shinichiro Asayama, Takashi Noguchi, and Satoru Iguchi, *National Astronomical Observatory of Japan, Tokyo, 181-8588, Japan.*
"Planar-integration of Arrayed SIS Receiver Frontends."

2:50 pm **W5.2.** Jacob W. Kooi, Darren J. Hayton, Bruce Bumble, Rick Leduc, Pierre Echternach, Anders Skalare, Jonathan Kawamura, Goutam Chattopadhyay, and Imran Medhi, *Jet Propulsion Laboratory, California Institute of Technology, 4800 Oak Grove Drive, Pasadena CA 91109, USA.*
"Submillimeter and Terahertz Receiver Technology for the Detection of Water Isotopes on Cometary Bodies."

3:10 pm **W5.3.** John D. Garrett [1], Jamie Leech [1], Brian Ellison [2], and Ghassan Yassin [1], [1] *University of Oxford, Oxford, OX1 3RH, UK;* [2] *Rutherford Appleton Laboratory, Didcot, OX11 0QX, UK.*
"A 1x4 Focal Plane Array Using 230 GHz SIS Mixers."

3:30 pm **W5.4.** Denis Sych [1,2,3], Michael Shcherbatenko [3,4], Michael Elezov [3], Gregory N. Goltsman [3,5],

[1] *P.N. Lebedev Physical Institute of the Russian Academy of Sciences, 119991 Moscow, Russia;* [2] *Russian Quantum Center, 143025 Moscow, Russia;* [3] *Moscow State Pedagogical University, 119992, Moscow, Russia;* [4] *Moscow Institute of Physics and Technology, Moscow Oblast, Russia;* [5] *Zavoisky Physical-Technical Institute of the Russian Academy of Sciences, 420029, Kazan, Russia.*
"Towards the improvement of the heterodyne receiver sensitivity beyond the quantum noise limit."

3:50 pm **W5.5.** C. Edward Tong [1], Paul Grimes [1], Lingzhen Zeng [1], Wei-Chun Lu [2], Tse-Jun Chen [2], Yen-Pn Chang [2], and Ming-Jye Wang [2], [1] *Harvard-Smithsonian Center for Astrophysics, Cambridge, MA 02138, USA;* [2] *ASIAA, Taipei, 10617, Taiwan ROC*
"Development of SIS Receivers with Ultra-wide Instantaneous Bandwidth for wSMA."

Session W6: **Optical Design, Systems, and Components II**
Chair: *Christopher Groppi*

4:10 pm Coffee Break

4:40 pm **W6.1.** Stephen J. C. Yates [1], Kristina K. Davis [2], Willem Jellema [3,4], Jochem J. A. Baselmans [4,5], Lorenza Ferrari [1], Ronald Hesper [3], and Andrey M. Baryshev [3], [1] *SRON Netherlands Institute for Space Research, Groningen, 9747AD, The Netherlands;* [2] *Arizona State University, 781 Terrace Rd., Tempe, AZ, USA;* [3] *Kapteyn Institute, University of Groningen, Groningen, 9747AD, The Netherlands;* [4] *SRON Netherlands Institute for Space Research, Utrecht, 3584CA, The Netherlands;* [5] *Terahertz Sensing Group, Delft University of Technology, Delft 2628CD, The Netherlands.*
"Vector Beam Pattern measurements of an 850 GHz wide field Microwave Kinetic Inductance Detector camera."

5:00 pm **W6.2.** Maria Alonso-delPino, Cecile Jung-Kubiak, Theodore Reck, Choonsup Lee, and Goutam Chattopadhyay, *Jet Propulsion Laboratory, California Institute of Technology, 4800 Oak Grove Drive, Pasadena CA 91109, USA.*
"Integrated Micro-Lens Antennas for THz Heterodyne Receivers."

5:20 pm **W6.3.** Jeffrey Hesler, *Virginia Diodes Inc., Charlottesville, VA 22902, USA.*
"Sensitivity limits and design of THz absorption spectrometers."

5:40 pm **W6.4.** Thierry Wiertz [1], S. Carpentier [1], Y. Pennec [1], J-M Niot [1], W. Zhang [2], Q.J. Yao [2], K. Zhang [2], J. Li [2], S.C. Shi [2], [1] *Air Liquide Advanced Technologies, Sassenage, 38360, France;* [2] *Purple Mountain Observatory, Nanjing, 210008, China.*
"HiPTC: a compact, efficient and low vibration cooler for Terahertz detection."

6:00 pm Wrap-up and Farewell

6:10 pm Symposium ends

2018 29th IEEE International Symposium on Space Terahertz Technology (ISSTT 2018)

March 26-28, 2018, Pasadena, California, USA

Technical Papers

Session M1: **Sources and Local Oscillator Systems I**

4.7 THz local oscillator for GUSTO

Ali Khalatpour[1], Qing Hu[1], and John L. Reno[2]

[1]*Massachusetts Institute of Technology (MIT), Cambridge, Massachusetts 02139, USA*
[2]*Center for Integrated Nanotechnology, Sandia National Laboratories, Albuquerque, New Mexico 87123, USA*
Contact: akhalat@mit.edu

Abstract— NASA has initiated a science mission (GUSTO) that will measure emissions from the interstellar medium. This data will help scientists determine the life cycle of interstellar gas in our Milky Way galaxy to understand the dynamics and gas flow in the vicinity of the center of our galaxy. In particular, GUSTO requires a local oscillator in its heterodyne receiver to detect natural Oxygen line at 4.744 THz. Considering the practical aspects, GUSTO demands a local oscillator with high power (>1 mw), tight beam pattern, and single mode of operation. Quantum Cascade Lasers (QCL) with 3rd-order distributed feedback (3rd DFB) grating on metal-metal waveguides can deliver high power and tight beam pattern. However, to achieve single mode operation at 4.744 THz, the lateral dimension of such waveguides should be much smaller than the wavelength (~20 um in the semiconductor). For such narrow lasers in metal-metal waveguides, the confinement factor gets unfavorably small and consequently decreases the output power and negatively affects the beam pattern. Lack of confinement also decreases the maximum lasing temperature and complicates the fabrication process. The second challenge is to tune the frequency of lasing and hit the target frequency within 3GHz. Fabrication fluctuation can readily produce up to ~300 GHz uncertainty in the lasing frequency. In this paper, we report a strategy to suppress higher lateral modes and still achieve high power and tight beam pattern though novel waveguide design. We also discuss a strategy to achieve exact frequency (within 3 GHz) for third order DFB's through systematic variation of laser geometry.

Compact and efficient 4.7-THz local oscillator with a GaAs/AlAs quantum-cascade laser

H.-W. Hübers[1,2*], T. Hagelschuer[1], H. Richter[1], M. Wienold[1], L. Schrottke[3], X. Lü[3], B. Röben[3], K. Biermann[3], and H. T. Grahn[3]

[1]*German Aerospace Center (DLR), Institute of Optical Sensor Systems, Rutherfordstr. 2, 12489 Berlin, Germany*
[2]*Humboldt-Universität zu Berlin, Department of Physics, Newtonstr. 15, 12489 Berlin, Germany*
[3]*Paul-Drude-Institut für Festkörperelektronik, Leibniz-Institut im Forschungsverbund Berlin e. V., Hausvogteiplatz 5–7, 10117 Berlin, Germany*
[*]Contact: heinz-wilhelm.huebers@dlr.de

Abstract— A major challenge for heterodyne receivers operating above approximately 3 THz is the local oscillator (LO), which has to be a compact source requiring high output power and low electrical input power. The atomic oxygen fine structure transition at 4.7 THz is of particular interest for astronomy as well as for atmospheric research. Since almost four years, a 4.7-THz LO, which is based on a quantum-cascade laser (QCL), is in operation on SOFIA, the Stratospheric Observatory for Infrared Astronomy. It is part of the heterodyne receivers GREAT and upGREAT (German REceiver for Astronomy at Terahertz frequencies) [1]. While the laser itself is only a few mm long, the mechanical cryocooler, which is necessary for operation of the QCL, determines the mass and power budget of the LO. The LO of GREAT/upGREAT has a mass of about 40 kg and a power consumption of about 400 W. For application in a space mission, an LO with the same performance has to be significantly more compact and less power consuming than for airborne instruments. We will report on the development of a compact, easy-to-use source, which employs a QCL operating in a very compact, low-input-power linear Stirling cooler. The QCL, which is based on a hybrid design using the GaAs/AlAs material system [2], has been developed for high output power and low electrical pump power. Efficient carrier injection is achieved by resonant longitudinal-optical phonon scattering. The QCL operates on a single mode at 4.75 THz exhibiting more than three times higher wall plug efficiency than GaAs/$Al_{0.25}Ga_{0.75}As$ QCLs with an almost identical design. At the same time, the threshold current density is significantly reduced. The cryocooler with the QCL weighs less than 4 kg. The QCL provides up to 7.5 mW output power at an operating temperature of 42 K in an almost Gaussian-shaped beam profile. The results indicate that a compact LO based on a GaAs/AlAs QCL is feasible for spaceborne applications.

[1] H. Richter, M. Wienold, L. Schrottke, K. Biermann, H. T. Grahn, and H.-W. Hübers, 4.7-THz local oscillator for the GREAT heterodyne spectrometer on SOFIA, IEEE Trans. Terahertz Sci. Technol. **5**, 539 (2015).
[2] L. Schrottke, X. Lü, G. Rozas, K. Biermann, and H. T. Grahn, Terahertz GaAs/AlAs quantum-cascade lasers, Appl. Phys. Lett. **108**, 102102 (2016).

Spatial Power Combining Amplifier (SPCA) for W-Band Radar in Earth and Planetary Science

Mark B. Taylor*, Lorene A. Samoska, Jose E. Velazco, Robert H. Lin, Andy Fung, Alejandro Peralta

Jet Propulsion Laboratory, California Institute of Technology, Pasadena, CA, 91109, USA
*Contact: mark.taylor@jpl.nasa.gov

Abstract**—**We are developing a high efficiency, compact W-Band (94 GHz) spatial power combining amplifier (SPCA) in an effort to demonstrate a new type of millimeter-wave high-power amplifier for use in earth science cloud radar and planetary landing radar. The SPCA concept utilizes multiple monolithic millimeter-wave integrated circuit (MMIC) amplifiers, which are combined spatially within a resonant microwave cavity. In the SPCA device, an input signal enters the input cavity where it sets up a resonant cavity mode; in this case the TM_{110} cavity mode. The signal power is then equally split between multiple MMIC amplifiers, and, after amplification, is spatially recombined within a second output cavity. The cavity is then coupled to WR10 waveguide by means of an iris for suitable mode conversion and matching. This combining scheme is scalable, allowing for additional MMICs to be added in parallel to increase power without compromising gain or efficiency. Additionally, this method of combining provides a factor of 2-5 times smaller footprint as compared to standard waveguide-based power-combining packages.

This effort builds upon previous work in which we developed and tested the SPCA concept at S-band (2.4 GHz) and X-band (8.5 GHz) frequencies [1]. In this paper we present the design and results of a W-band (94 GHz) version first prototype demonstration utilizing state-of-the-art GaN HEMT MMIC amplifiers, each capable of 1 Watt saturated output power at W-band, targeting a combined power output of ~1.5 Watts. This new SPCA design was scaled by a factor of ~40 from the previous S-band version. For the next phase of development, we plan to implement a 6 MMIC combiner targeting 5 Watts of output power.

The first stage of the design involved EM simulations to determine cavity dimensions and optimize the microstrip antenna coupling, center frequency, thru-power, and bandwidth. The structure of the cavity-mode power combiner was then designed using CAD software to accommodate the EM and tolerance criteria established through EM simulations. Testing of the two-way combiner block using thru-lines indicates bandwidth and matching compare well with the simulated results. The power combiner operates between 93-107 GHz (a 3dB bandwidth of 14 GHz) as measured using microstrip thru-lines in place of the MMICs, with greater than 10 dB return loss. Our latest power-combined results will be presented at the conference.

[1] Jose Velazco, Mark Taylor, "Spatial Power Combining Amplifier for Ground and Flight Applications," *Interplanetary Network Progress Reports* 42-207, Nov. 2016. https://ipnpr.jpl.nasa.gov/progress_report/42-207/207B.pdf

Prototype 4.7 THz array local oscillator for GUSTO

B. Mirzaei[1,*], J. R. G. Silva[2], D. Hayton[3], W. Laauwen[2], Y. Gan[2,4], Q. Hu[5], C. Groppi[6], and J. R. Gao[1,2]
[1] Kavli Institute of Nanoscience, Delft University of Technology, Lorentzweg 1, 2628 CJ, Delft, The Netherlands
[2] SRON Netherlands Institute for Space Research, Sorbonnelaan 2, 3584 CA Utrecht, The Netherlands
[3]Jet Propulsion Laboratory, California Institute of Technology, Pasadena, CA 91109, USA
[4]Kapteyn Astronomical Institute, University of Groningen, 9747 AD, Groningen, The Netherlands
[5]Massachusetts Institute of Technology (MIT), Cambridge, Massachusetts 02139, USA
[6]School of Earth and Space Exploration, Arizona State University, AZ, USA
*Contact: B.Mirzaei@tudelft.nl

***Abstract*— We present an 8-beam local oscillator (LO) for the astronomically significant [OI] line at 4.7 THz. The beams are generated using a quantum cascade laser (QCL) in combination with a Fourier phase grating. The grating is fully characterized using a third order distributed feedback (DFB) QCL with a single mode emission at 4.7 THz as the input. The measured diffraction efficiency of 74.3 % is in an excellent agreement with the calculated result of 75.4 % using a 3D simulation. We show that the power distribution among the diffracted beams is uniform enough for pumping an array receiver. To validate the grating bandwidth, we apply a far-infrared (FIR) gas laser emission at 5.3 THz as the input and find a very similar performance in terms of efficiency, power distribution and spatial configuration of the diffracted beams. Both results represent the highest operating frequencies of THz phase gratings reported in the literature. By injecting one of the eight diffracted 4.7 THz beams into a superconducting hot electron bolometer (HEB) mixer, we find that the coupled power, taking the optical loss into account, is in consistency with the QCL power value. This paper has been published in Optics Express 25(24), 29587 (2017), and is submitted to ISSTT 2018 as a direct way of exposure to the most relevant community.**

I. Introduction

Heterodyne spectroscopy provides valuable scientific information for modern astrophysics, in particular, the process of star formation and the life cycle of interstellar clouds in our own and nearby galaxies. It has the capability to measure molecular and atomic fine structure lines in the super-terahertz frequency region (> 1 THz) [1] including ionized nitrogen [NII] at 1.4 THz, ionized carbon [CII] at 1.9 THz, hydrogen deuteride [HD] at 2.6 THz and neutral oxygen [OI] at 4.7 THz, each containing unique source of information. One important observation is to map particular THz lines in star forming areas within our Milky Way at very high spectral resolution. Such large-scale surveys with velocity-resolved imaging of key fine structure lines require multi-pixel heterodyne receivers [2] with the state-of-the-art mixer performance in order to increase the observation speed and thus the observing efficiency of the telescope. A single heterodyne receiver with extremely high spectral resolving power of 106-107, consists typically of a mixing detector, a local oscillator (LO), a low noise GHz cryogenic amplifier and a backend GHz spectrometer. Because of the complexity of the receiver and especially the power dissipation of low noise cryogenic amplifiers, the state-of-the-art heterodyne receiver arrays typically have in the order of 10 pixels. Furthermore, very few instruments using multi-pixel receivers have been demonstrated at THz frequencies in the literature.

A multi-pixel heterodyne receiver therefore requires multiple parallel receivers; each pumped with its own LO beam. In general, advanced fabrication technology allows making relatively uniform array of mixers, so that (as will be shown later in this paper) within a certain variation of the LO power they still deliver receiver sensitivities with negligible differences. For generating multiple THz LO beams two approaches have been reported: solid-state waveguide multiple beam generators based on frequency multipliers driven by a common microwave source [3,4], and reflective phase gratings for generating multiple beams by diffracting a single THz source [5,6]. A phase grating is a periodic arrangement of a unit cell with a specific surface morphology for phase manipulation of the incident coherent radiation. Such a surface profile determines the intensity distribution among the diffracted beams, their angular configuration, and the diffraction efficiency. The latter is defined as the ratio of the total power of the desired diffraction orders to the power of the incoming beam. We will call it efficiency in the rest of the paper.

The choice of the LO technologies depends strongly on the frequency. The neutral oxygen [OI] line at 4.7 THz mostly probes warm natural gas heated by massive newly formed stars [7,8]. As such, it is useful for probing the star formation process, especially when the emission is spectrally resolved. However, because the technology at this frequency is not sufficiently mature yet, it has been relatively less studied so far. Since waveguide based multi-beam LOs have not been demonstrated at such a high frequency, the QCL [9] is the only practical solid-state source at 4.7 THz, which can provide multiple beams in combination with a phase grating. At present, given the power level and beam pattern of the current generation of THz QCLs suitable for use as an LO, such as third order DFB QCLs [10], they can typically only drive a single pixel superconducting hot electron bolometer (HEB) mixer. Progress, however, is rapid enough to overcome these issues soon with the next generation of 4.7 THz QCLs. One could think of using an array of QCLs instead. However, it is in general very challenging to produce multiple lasers with the same frequency. Furthermore, phase or frequency locking of multiple QCLs can be quite complicated.

No detailed study of a phase grating-based LO has so far been published beyond 1.4 THz [11], in which a FIR gas laser was applied. There are two challenges for such a system at higher frequencies. One is the realization of the phase grating since its surface finesse increases by the frequency and becomes more difficult to fabricate. The other is the 4.7 THz QCL with non-Gaussian beam and insufficient output power. Furthermore, the wave fronts of THz QCLs, which are not fully understood [12], might also affect the grating performance.

In this work we report an 8-pixel Fourier phase grating developed for operation at 4.7 THz together with its full characterization using a third order DFB QCL. In addition to the diffraction beam pattern and efficiency measurements, the power uniformity issue of such a LO array is discussed. We present an experiment of optically pumping an HEB mixer using one of the diffracted beams. The effects of the number of applied Fourier coefficients (FC) on the efficiency and also on the manufacturability of the grating are studied as well. We also demonstrate the grating bandwidth advantage using a FIR gas laser emission at 5.3 THz as the input.

II. Grating Optimization and Realization

A phase grating is formed by 2D repetition of a unit cell with a surface profile determined by the Fourier series based on a limited number of FCs, which modulates the phase of the incoming beam [6]. In addition to providing the required diffraction pattern and high efficiency, the unit cell profile has to have a minimum radius of curvature (MRC) meeting the manufacturing capability, beyond which the machine is not able to faithfully pattern. This minimum radius of curvature is limited by the diameter of available ball-end milling tools for fabrication via Computer Numerically Controlled (CNC) machining. Thus an optimization process needs to be performed. There are two main parameters for such an optimization. First is the number of applied FCs, which is proportional to the efficiency and inversely proportional to the MRC; second is the unit cell size, which is proportional to the MRC and inversely proportional to the divergence of the diffraction pattern.

The FCs for deriving the 2D profiles for beam multiplexing by 4 and 2 are calculated in MATLAB and optimised using a genetic algorithm to ensure the maximum achievable efficiency. They are then orthogonally superimposed to make the surface topology of the unit cell of an 8-pixel grating. The rectangular uniform arrangement of the diffracted beams requires the power to be distributed only among the 8 products of the diffraction orders of m = ±1 and n = ±3.

We use 13 FCs to simulate a 3x3 cm2 grating, consisting of 25 times the unit cell of 1.2x1.2 mm2 in two orthogonal directions, for operation at 4.7 THz with 25° incident angle with respect to the normal. A 3D drawing of 16 unit cells of the phase grating is shown in Fig. 1(a). We only show a part of the whole grating (225 unit cells), otherwise it would not allow a detailed view of the surface structure. Note that for clarity, the height is out of scale.

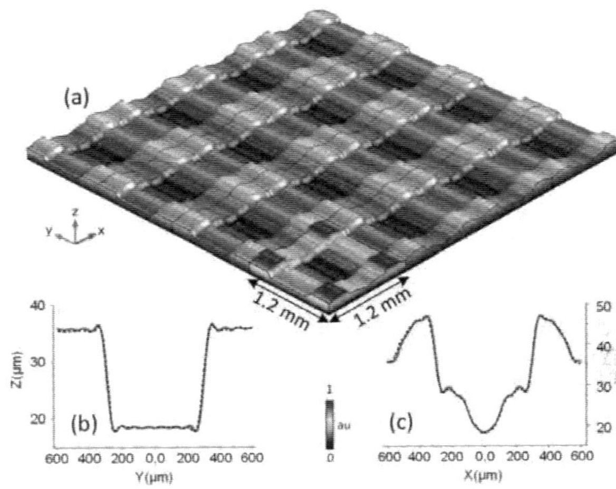

Fig. 1. (a) 3D drawing of 16 unit cells of the phase grating. A single unit cell is shown in colour. The height is out of scale for clarity, (b), (c) The calculated and manufactured 2D cross-section profiles of a unit cell. Z shows the height, and X and Y are the two lateral directions. The dashed black and solid red curves are the manufactured and calculated results respectively.

Based on the grating equation [13], $\sin\beta=\sin\alpha+m\lambda/\Lambda$, an angle of ~ 6° is expected between two adjacent diffracted beams, where α and β are the angles of incidence and the m'th (±1 and ±3) diffraction order respectively and Λ is the grating periodicity (equal to the unit cell size). The desired angle is chosen mainly based on the beam configuration of an existing HEB mixer array and its distance to the grating. This design results in an MRC of 112 μm, which is larger than the 90-μm radius of the ball end mill (available at the time we started) used with the KERN EVO micro-milling machine (at Arizona State University) that is employed for the grating manufacturing. In other words, the available machine is able to pattern all the fine surface structures. An aluminium plate (Aloca QC-10 Mold Alloy) is chosen as the grating material because of its very high reflectivity [14] and manufacturability. The manufactured 2D cross sections of a unit cell are measured by a Dektak XT30 stylus profiler and plotted in Figs. 1(b) and 1(c) together with the calculated ones. They show very good matches with deviations in amplitude of < 1 μm.

III. Quantum Cascade Laser and the Experimental Setup

As the input source, we use a third order DFB QCL with an GaAs/AlGaAs active region based on four-well resonant phonon depopulation scheme [15]. It consists of 21 periods over a length of about 0.55 mm with 17-µm width and 10-µm thickness. Multiple lasers with a 7.5 GHz frequency spacing are grouped together on a single chip with a linear frequency coverage that allows targeting of a specific frequency. The active region was grown at Sandia National Laboratories by molecular beam epitaxy (MBE) while the QCLs were designed and fabricated at MIT. We choose a QCL with a single mode emission at 4.756 GHz mainly because of a relatively low water absorption in air at this frequency. The QCL was operated at a temperature of ~ 7 K using a pulse tube cooler. No power measurement was done for this specific QCL. However, based on the measured power of a similar laser on a different chip [16], we expect it to be around 0.25 mW.

A room temperature pyro-electric detector with an aperture diameter of 2 mm, mounted on a X-Y translational stage, is used for all the beam pattern measurements in this work. The QCL beam profile is measured in front of the cryostat window at a ~ 3 cm distance from the laser and plotted in Fig. 2(a).

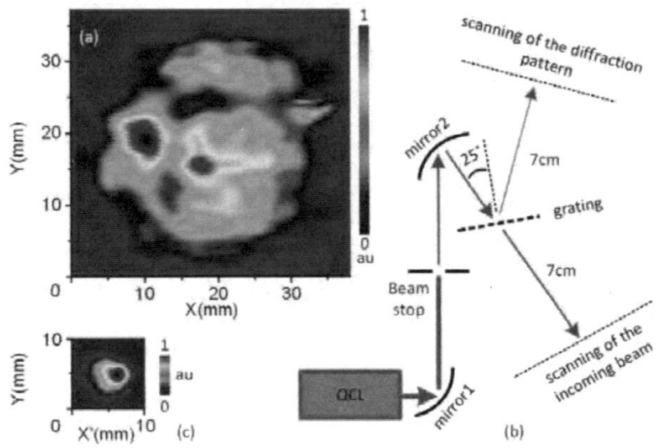

Fig. 2. (a) Quantum cascade laser beam pattern measured in front of the cryostat window, (b) Optical setup for collimation and filtering the laser beam. the line thickness is an indication of the intensity, (c) incoming collimated beam to the grating

The QCL beam is highly divergent with a strong deviation from a Gaussian profile. This QCL has the beam pattern that is worse than the others studied previously in our group [12,16]. Directly applying such a beam to the grating leads to the same order of divergence for the individual diffracted beams, which is more than their angular separation. This means that they are merged and would not be resolved from each other. Therefore we have introduced a spatial mode filter shown schematically in Fig. 2(b). Two parabolic mirrors are used for beam collimation. An iris beam stop in between the mirrors reduces the beam size by partially passing the left bright spot (and blocking the rest). This is the optimum setup we can make for this purpose regarding to the geometrical constraints. The resulted collimated beam shown in Fig. 2(c) then illuminates the grating at 25° incident angle with respect to the normal. For the reason to be explained in the next section, we measure the filtered beam at 7 cm away from the grating position when it is removed. The aperture size of the beam stop is adjusted to maximize the passing QCL power whilst still resolving separated diffracted beams at a reasonable distance from the grating. By integrating the intensity of the beam patterns and considering about 1.4 and 0.5 dB absorption losses by the cryostat UHMW-PE window and the air respectively, we find that about 78% of the total QCL power in the whole beam pattern is blocked by this setup. This large power waste already explains the crucial role of the laser beam pattern. If the latter was more concentrated in a small area, a much smaller (or no) portion would be needed to be removed.

IV. MEASUREMENT AND SIMULATION RESULTS

Since the grating unit cell has two different orthogonal cross-sections, it results in two different diffraction beam patterns (either 4 by 2 or 2 by 4) when illuminated from each orthogonal direction [11]. The beam patterns measured at a distance of 7 cm from the grating using the setup shown in Fig. 2(b), for both cases are plotted in Fig. 3. The high-resolution patterns have been achieved by fine scanning with a step size of 0.1 mm. We measure the incoming beam in the same distance from the (removed) grating to cancel out the effect of the air absorption in the efficiency calculations later on. As expected, all diffracted beams copy the incoming beam profile, but horizontally flipped. The predicted ~ 6° angular distance between each two adjacent beams is experimentally confirmed. The unwanted higher diffraction orders are visible around the desired ones with much weaker intensities. This gives a good visual feeling of the efficiency. Since both diffraction patterns show a similar behaviour in terms of power and spatial distribution, we focus on the result shown on the left side (4x2) in the rest of the paper.

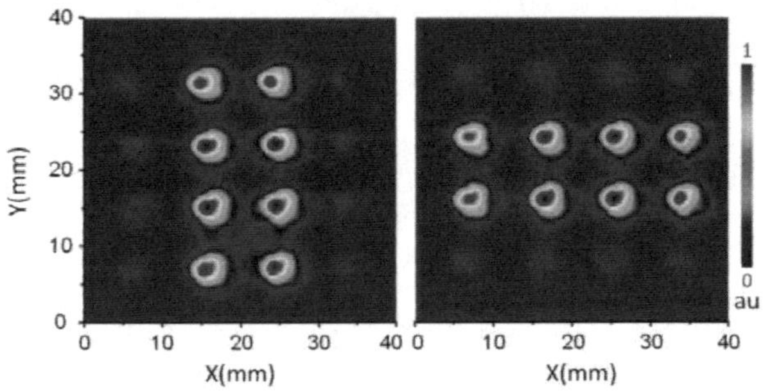

Fig. 3. Measured diffraction patterns of the grating at 4.7 THz using a quantum cascade laser with 25° of incidence in two different orthogonal directions.

For estimating the efficiency, we take the ratio between the integrated intensities of all the desired diffracted beams and the one of the incoming beam. It should be mentioned that the noise floor of the room temperature detector, which can affect the efficiency analysis, is removed before the integration. We find an efficiency of 74.3%, which is close to 75.4% predicted by a 3D simulation using the RF module of COMSOL. The latter is done by importing the grating surface profile and simulating a periodic port with periodic boundary conditions. Extracting the S parameters of the port then gives the efficiency. The maximal grating efficiency is dependent on the diffraction pattern. So the 75.4 % efficiency is associated with our prerequisite 4x2 rectangular beam configuration, otherwise the efficiency can be different.

We analyse the uniformity of the power distribution among the diffracted beams by applying the same method of integrating the intensity. The fractions of the incoming beam power distributed to the diffracted beams shown on the left side of the Fig. 3 from top to bottom of the left column are 9 %, 10.4 %, 9.5 %, and 8.6 % and on the right column are 8.6 %, 10.3 %, 9.3 % and 8.4 %. The power in the individual beams varies within ~ 21 % around its average value. The 3D simulation also shows that the intensities of the four central beams are slightly higher to the same extent.

In order to confirm that such a power variation among the diffracted beams may not deteriorate the performance of an HEB array, in which all the HEBs require the same LO power, we use our existing experimental data to evaluate the effect of the LO power tolerance to the sensitivity of a single mixer. This particular measurement was done at 4.3 THz (very close to our frequency of interest). The receiver noise temperature (T_{rec}) of the HEB mixer was measured as a function of LO power, reflected by the HEB current (inversely proportional to the LO power) at a constant bias voltage of 0.8 mV. The result is shown in the inset of Fig. 4 and suggests the optimum current of ~ 38 μA, in which the T_{rec} becomes minimal [17]. The correspondence between the current and the LO power can be estimated from the pumped current-voltage (I-V) characteristics by using the isothermal technique [18]. The multiple I-V curves, measured at different LO powers, are plotted in Fig. 4, where the top curve (in brown) is recorded without LO (superconducting state) and the bottom (in red) with a sufficiently high LO power to fully pump the device (to the resistive state). By estimating all the LO powers and their corresponding T_{rec}s we find that the change in the T_{rec} is less than 4 % when the LO power varies within about 21 %. We therefore conclude that despite the power variation (~ 21 %), the multi-beam LO made by our phase grating can pump a uniform HEB array without affecting the uniformity of the T_{rec}.

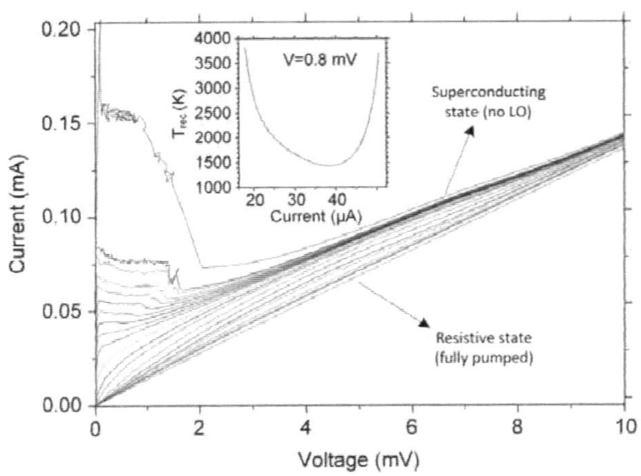

Fig. 4. The current-voltage characteristics of the HEB mixer at different levels of the LO power, which is zero in the top curve (in brown) and sufficient to fully pump the device in the bottom curve (in red). Inset: Measured receiver noise temperature (Trec) of a hot electron bolometer (HEB) mixer versus the current at a bias voltage of 0.8 mV. The current is inversely proportional to the absorbed local oscillator (LO) power.

We study the effect of the number of applied FCs on the grating efficiency and manufacturability using the same 3D simulation. The results, which are plotted in Fig. 5(a), show that decreasing this factor from 13 to 6 degrades the efficiency from 75.4 to 69 % (by ~ 6%) while increases the MRC by ~ 274 μm (~ 300 %). As explained before, the MRC determines the manufacturing requirement, which is the size of the ball end mill in our case. This is an interesting and useful finding since one can ease the manufacturing considerably by sacrificing a few percent of the efficiency. 3D surface profiles of the resulting unit cells from applying 6 and 13 FCs are plotted in Figs. 5(b) and 5(c), respectively, illustrating the effect on the surface fines and challenge of the manufacturing.

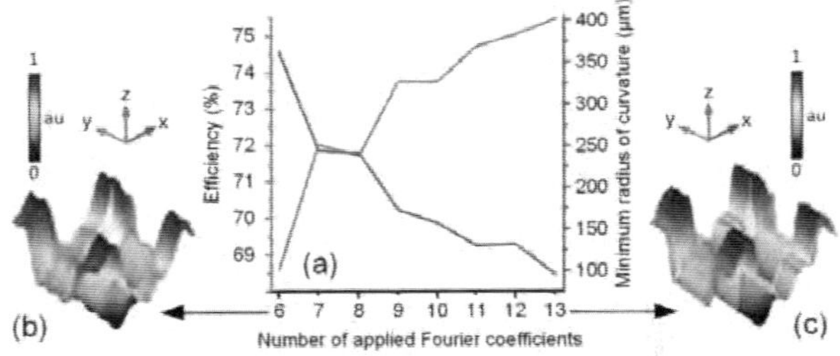

Fig. 5. (a) Grating efficiency and the surface minimum radius of curvature (MRC), versus the number of applied Fourier coefficients (FCs) in design, (b), (c) 3D profiles of the unit cells made by 6 and 13 FCs, respectively.

V. Grating Operation Bandwidth

The bandwidth of a phase grating is an interesting property, which has so far been less explored. In theory, it can perform well over a wide frequency band by properly changing the incident angle. A grating designed for the frequency f with an incident angle θ with respect to the normal works with the same efficiency as long as the value of $f \times \cos(\theta)$ is fixed. In our case, if we consider an applicable range of 0 to 45 degrees for the incident angle, the grating can function optimally from 4.3 to 6 THz.

Due to the limited available THz sources, we verify the performance of the 4.7 THz grating using a FIR gas laser line at 5.3 THz. The measured incident beam and the resulting diffraction pattern at 15 cm away from the grating are plotted in Fig. 6.

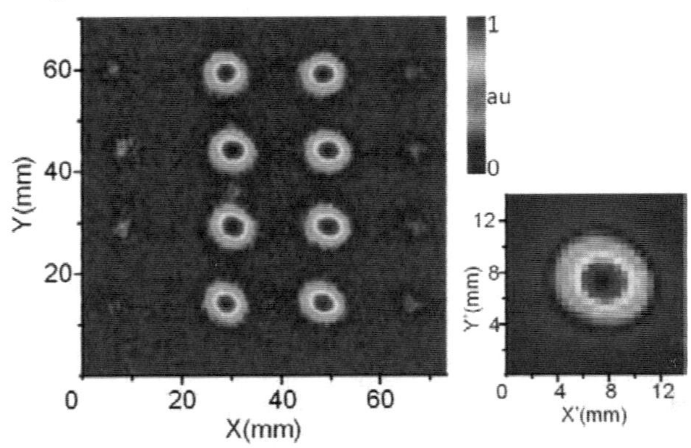

Fig. 6. Diffraction beam pattern measured at a distance of 15 cm from the grating (left) and the incident beam (right) at 5.3 THz measured using a FIR gas laser.

In this particular measurement, the incident angle is chosen to be 37° (as opposed to 25° for 4.7 THz). The diffracted beams are in the same order of spatial and power distributions as the case of 4.7 THz described earlier in this paper. The derived efficiency by applying the same analysing method is 74%, being almost equal to the 4.7 THz case. Our measurement at 5.3 THz, being 0.6 THz higher than the targeted frequency, confirms that the grating bandwidth should be more than 13%. The broadband grating characteristic offers a potential for other applications including multiple imaging at different frequencies using a single grating [19].

VI. Pumping a superconducting HEB mixer

To further quantify the performance of the 4.7 THz grating, we perform an experiment, in which we apply one of the eight diffracted beams (originated from the QCL) as the LO to pump a superconducting HEB mixer. We have to change the optical setup in order to redirect the diffracted beams towards a suitable place for putting and manually fine positioning a bulky liquid helium cryostat, in which the HEB is mounted. By doing so and considering the very limited geometrical flexibility, we could illuminate the grating with only about 9 μW. An additional beam stop is introduced after the grating to pick only one, out of the eight diffracted beams. Then it is coupled into a quasi-optical (lens-spiral antenna) HEB mixer using a focusing lens. We succeed in coupling ~ 50 nW to the HEB mixer, which is derived from the pumped I-V characteristics using the isothermal technique [18]. By taking into account the total optical loss of ~ 32.6 dB [20] in the entire system, we find that ~ 91 μW of the QCL power is used (we assume that the QCL has a 0.25 mW output power). We think this power fraction is reasonable by attributing the difference to the partial mismatch between the non-Gaussian LO beam and beam of the HEB mixer, and also to the uncertainty in the actual QCL output power.

Let us now estimate the required QCL power for a practical 8-pixel array with spiral antenna coupled HEB mixers. To do so, we consider the system as the following; a) each mixer needs a pumping power of 220 nW; b) Si lens for the mixer has an antireflection (AR) coating at 4.7 THz; c) a 3 μm thick Mylar beam splitter is used for a heterodyne measurement; d) the setup is in vacuum; e) the focusing lens is made of silicon with double sided AR coating at 4.7 THz; f) The QCL has an ideal Gaussian beam. This system has an optical loss of ~ 26.2 dB [21], implying that a power of ~ 92 μW from the QCL is needed in this case. However, the fraction of the power in the beam in the fundamental Gaussian mode can be much less (e.g. 10 %) in practice, which increases the required QCL output power considerably (e.g. 0.9 mW). It is worth to mention that producing THz QCLs with either higher power or improved beam profile seems to be feasible [22,23] considering the rapid progress being made in the community.

Conclusions

We report for the first time a Fourier phase grating for generating eight local oscillator beams using a single mode 4.7 THz DFB quantum cascade laser as the input source. We characterized the grating performance by measuring the diffraction pattern and deriving the diffraction efficiency of 74.3%, which is in an excellent agreement to the 3D simulation result. We demonstrated that the power uniformity of such a multi beam LO is sufficient, leading to uniform sensitivities of the array mixers. The effects of applying different number of Fourier coefficients in simulation, on the efficiency and manufacturing is discussed. The broad-band characteristic of the grating is verified by its evaluation at 5.3 THz, where we found the same performance as the case for 4.7 THz. Finally we analysed the power issue by coupling one of the diffracted beams generated by the QCL-grating combination to a superconducting hot electron bolometer mixer. We estimate that an 8-pixel array receiver, in which each mixer needs a power of 220 nW, requires from 92 μW to 0.9 mW for the QCL power if its Gaussian component has a power fraction from 100 to 10 %.

Our work demonstrates the principle of a novel array local oscillator technology, which is demanded for future space missions. It is a prototype LO unit for a NASA science mission GUSTO (Galactic/X-galactic Ultra-duration-balloon Spectroscopic/Stratospheric Terahertz Observatory), which is scheduled to launch in 2021 from Antarctica and map three THz fine structured lines including [OI] using an 8-pixel receiver array.

Acknowledgments

One of the authors (B. M.) acknowledges the support and encouragement from Leo Kouwenhoven. We also acknowledge useful discussions with Chris Walker, Kobus Kuipers, Paul Urbach, and Akira Endo. The authors would like to thank Yuchen Luo for making his grating codes available for this work, Matt Underhill for fabricating the gratings and David Thoen for helping with the surface profile measurements. EU RadioNet and TU Delft Space Institute are acknowledged for their supports. The work at MIT was supported by NASA. The work at Sandia was performed, in part, at the Center for Integrated Nanotechnologies, an Office of Science User Facility operated for the U.S. Department of Energy (DOE) Office of Science. Sandia National Laboratories is a multimission laboratory managed and operated by National Technology and Engineering Solutions of Sandia, LLC., a wholly owned subsidiary of Honeywell International, Inc., for the U.S. Department of Energy's National Nuclear Security Administration.

References

[1] C. Kulesa, "Terahertz spectroscopy for astronomy: from comets to cosmology," IEEE Trans. THz Sci. Tech. 1, 232-240 (2011).

[2] C. Walker, Terahertz Astronomy (CRC, Taylor & Francis Group, 2016), Chap. 5&6.

[3] J. V. Siles, R. H. Lin, C. Lee, E. Schlecht, A. Maestrini, P. Bruneau, A. Peralta, J. Kloosterman, J. Kawamura and I. Mehdi, "Development of high-power multi-pixel LO sources at 1.47 THz and 1.9 THz for astrophysics: present and future," in Proceedings of the 26th International Symposium on Space Terahertz Technology (Cambridge, MA, 2015), pp. 40–42.

[4] C. Groppi, C. Walker, C. Kulesa, D. Golish, J. Kloosterman, S. Weinreb, G. Jones, J. Barden, H. Mani, T. Kuiper, J. Kooi, A. Lichtenberger, T. Cecil, G. Narayanan, P. Pu'tz and A. Hedden, "SuperCam: A 64 pixel heterodyne array receiver for the 350 GHz atmospheric window," in

Proceedings of the 20th International Symposium on Space Terahertz Technology (Charlottesville, 2009), pp. 90–96.

[5] J. A. Murphy, C. O. Sullivan, N. Trappe, W. Lanigan, R. Colgan and S. Wittington, "Modal Analysis of the Quasi-Optical Performance of Phase Gratings," Int. J. Inf and Mill Waves **20**, 1469–1486 (1999).

[6] U. U. Garf and S. Heyminck, "Fourier gratings as submillimeter beam splitters," IEEE Trans. Antennas Propag. **49**, 542–546 (2001).

[7] A. G. G. M. Tielens and D. Hollenbach, "Photodissociation regions. I. basic model. II. A model for the Orion photodissociation Region" ApJ. **291**, 722–754 (1985).

[8] S. Leurini, F. Wyrowski, H. Wiesemeyer, A. Gusdorf, R. Gusten, K. M. Menten, M. Gerin, F. Levrier, H. W. Hubers, K. Jacobs, O. Ricken and H. Richter, "Spectroscopically resolved far-IR observations of the massive star-forming region G5.89-0.39," A&A. **584**, A70 (2015).

[9] B. S. Williams, "Terahertz quantum-cascade lasers," Nat. Photon. **1**, 517–525 (2007).

[10] M. I. Amanti, G. Scalari, F. Castellano, M. Beck and J. Faist, "Low divergence Terahertz photonic-wire laser," Opt. Express **18**, 6390–6395 (2010).

[11] B. Mirzaei, J. R. G. Silva, Y. C. Luo, X. X. Liu, L. Wei, D. J. Hayton, J. R. Gao and C. Groppi, "Efficiency of multi-beam Fourier phase gratings at 1.4 THz," Opt. Express **25**, 6581-6588 (2017).

[12] M. Cui, J. N. Hovenier, Y. Ren, N. Vercruyssen, J. R. Gao, T. Y. Kao, Q. Hu and J. L. Reno, "Beam and phase distributions of a terahertz quantum cascade wire laser", Appl. Phys. Lett. **102**, 111113 (2013).

[13] W. C. Elmore and M. A. Heald, *Physics of waves* (Dover Publications, Inc., New York, 1969).

[14] H. Ehrenreich, H. R. Philipp and B. Segall, "Optical properties of aluminum," Phys. Rev. **132**, 1918–1928 (1963).

[15] B. S. Williams, S. Kumar, Q. Hu and J. L. Reno, "Resonant phonon terahertz quantum-cascade laser operating at 2.1 THz ($\lambda \approx 141$ μm)," Elec. Lett. **40**, 431–433 (2004).

[16] J. L. Kloosterman, D. J. Hayton, Y. Ren, T. Y. Kao, J. N. Hovenier, J. R. Gao, T. M. Klapwijk, Q. Hu, C. K. Walker and J. L. Reno, "Hot electron bolometer heterodyne receiver with a 4.7-THz quantum cascade laser as a local oscillator," Appl. Phys. Lett. **102**, 011123 (2013).

[17] W. Zhang, P. Khosropanah, J. R. Gao, T. Bansal, T. M. Klapwijk, W. Miao and S. C. Shi, "Noise temperature and beam pattern of an NbN hotelectron bolometer mixer at 5.25 THz," J. Appl. Phys. **108**, 093102 (2010).

[18] H. Ekström, B. S. Karasik, E. Kollberg and K. S. Yngvesson, "Conversion gain and noise of niobium superconducting hot-electron-mixers," IEEE Trans Microw. Theory Tech. **43**, 938–947 (1995).

[19] L. P. Boivin, "Multiple Imaging Using Various Types of Simple Phase Gratings," Appl. Opt. **11**, 1782-1792 (1972).

[20] The quoted optical loss is contributed by the HEB-antenna coupling (0.5 dB), the spiral antenna (3 dB, for a linearly polarized Gaussian beam), the Si lens reflection loss (1.2 dB), the QMC heat filter (0.4 dB), the HEB cryostat UHMW-PE window (1.4 dB), the HDPE focusing lens (1.2 dB), the air loss (1 dB), the phase grating (10 dB), the beam stop (in combination with two mirrors) (12.5 dB), and the UHMW-PE window of the QCL cryostat (1.4 dB).

[21] The quoted optical loss is contributed by the HEB-antenna coupling (0.5 dB), the spiral antenna (3 dB, for a linearly polarized Gaussian beam), the AR coated Si lens reflection loss (0.15 dB), the QMC heat filter (0.4 dB), the HEB cryostat UHMW-PE window (1.4 dB), the Si focusing lens with AR coating on both sides (0.3 dB), the phase grating (10 dB), the beam splitter (9 dB), and the UHMW-PE window of the QCL cryostat (1.4 dB).

[22] T. Y. Kao, X. Cai, A. W. M. Lee, J. Reno and Q. Hu, "Antenna coupled photonic wire lasers," Opt. Express **23**, 17091-17100 (2015).

[23] A. Khalatpour, J. L. Reno, N. P. Kherani and Q. Hu, "Unidirectional photonic wire laser," Nat. Photon. **11**, 555-559 (2017).

Session M2: **SMM and THz Receivers**

1080-1280GHz Schottky Receiver for JUICE-SWI with 1600-2600K DSB Receiver Noise Temperature

A. Maestrini[1*], L. Gatilova[1+2], J. Treuttel[1], Y. Jin[2], A. Cavanna[2], D. Moro Melgar[1], T. Vacelet[1], A. Féret[1], S. Caroopen[1], G. Gay[1], F. Dauplay[1], J-M. Krieg[1], P. De Maagt[3], C. Goldstein[4]

[1]*LERMA, Observatoire de Paris, PSL Research University, CNRS, UMR 8112, Sorbonne Universités, UPMC Univ. Paris 06, F-75014 Paris, France*
[2]*C2N-Marcoussis, Route de Nozay, F-91460 Marcoussis, France*
[3]*ESA-ESTEC, Keplerlaan 1, PO Box 299, NL-2200 AG Noordwijk, The Netherlands*
[4]*Centre National d'Etudes Spatiales, 18 avenue Edouard Belin, F-31401 Toulouse cedex 9, France*
[*]Contact: alain.maestrini@obspm.fr

Abstract—The Sub-millimeter Wave Instrument (SWI) for ESA Jupiter Icy Moons Explorer (JUICE) will be the first planetary instrument to feature 1200GHz and 600GHz heterodyne receivers. SWI will investigate the temperature structure, composition and dynamics of Jupiter's stratosphere and troposphere, and the exospheres and surfaces of the icy moons.

This paper will present the current design and status of SWI 1080-1280GHz Schottky receiver frontend, which is developed at LERMA-Observatoire de Paris in partnership with C2N-Marcoussis (formerly LPN). The receiver specifications and goal were respectively T_rec_DSB≤4000K and T_rec_DSB≤3000K. A double side band receiver noise temperature of T_rec_DSB=1600K has been recorded at 1114GHz at an ambient temperature of 150K, with T_rec_DSB ranging from 1600K to 2570K across the entire band, with an average of 1950K and a standard deviation of 220K. At 301K ambient temperature, T_rec_DSB is ranging from 3280K to 6320K across the entire band, with an average of 3840K. This is believed to be the best performance of a Schottky receiver in this frequency range.

Noise performance of a balanced waveguide NbN HEB mixer utilizing a GaN buffer-layer at 1.3 THz

S. Krause*, D. Meledin, V. Desmaris, A. Pavolotsky and V. Belitsky
[1]Group for Advanced Receiver Development, Chalmers University of Technology, SE-41296, Gothenburg, Sweden
*Contact: sascha.krause@chalmers.se, phone +46-73-303 36 72

***Abstract*— We report on the initial measurement results of a balanced waveguide phonon cooled NbN mixer employing a 5.5 µm thin GaN membrane, which was operated at frequencies around 1.3 THz. The uncorrected DSB noise temperature amounts to approximately 750 K at 1 GHz IF and increases to only 900 K at 4 GHz IF, which was deduced from the standard Y-factor measurement technique. The recorded IF spectrum from 0.5 GHz to 8 GHz suggests a measured noise bandwidth of approximately 7 GHz owing to the employment of a GaN buffer-layer, which promotes the single crystal growth of NbN films and provides high phonon transparency, thus lowering the phonon escape time.**

We emphasize with the implementation of a waveguide balanced receiver scheme and using NbN/GaN mixers the possibility to extend the operational IF range of phonon cooled NbN HEBs, yet providing low noise performance.

Index terms – **Hot Electron Bolometer mixer, THz receiver, NbN thin film**

I. Introduction

Hot electron bolometers (HEB) based on ultra-thin NbN films for low-noise heterodyne THz receivers have been the technology of choice for high-resolution spectroscopy for more than two decades. In recent years, the complexity of such receivers has increased by using balanced waveguide schemes [1], [2] or going to multi-pixel arrays operating at frequencies as high as up to 4.7 THz [3]. Superconducting NbN mixers are superior to Schottky diodes as THz mixer in terms of noise temperature and LO power consumption requirements. They do not experience, in principle, an upper RF frequency limit, and do not require an applied magnetic field unlike THz SIS mixers. However, phonon cooled HEBs show a deficit in their IF bandwidth, which is governed by intrinsic relaxation processes of electrons under non-equilibrium conditions, namely the electron-phonon interaction in the superconducting film and the phonon escape from the film to the substrate. The latter is believed to limit the operational IF bandwidth in these NbN film based devices to only 3-4 GHz [1], [4], [5], [6]. This restricts the efficient use of observation time as well as complicates the study of far distant objects, which are associated with increased spectral line broadening. Two promising approaches to enlarge the IF bandwidth have been pursued in recent years. With advances in the growth of thin layers of the high critical temperature (T_c) superconductor MgB_2, and utilizing its very short electron-phonon and phonon escape time, it was possible to demonstrate an extended noise bandwidth in a quasi-optical mixer setup operating at 1.63 THz [7]. However, the reported noise temperature of those MgB_2 based devices [7], [8] is still higher than these made of NbN, and they demand more LO power, which limits their prospective use in future, especially multi-pixel applications. Other studies focused on the improvement of NbN's film quality by exploring different buffer-layers [9], [10], [11] and reducing the phonon escape time, which is believed to limit the IF bandwidth. It has been shown that hexagonal GaN promotes the epitaxial growth of NbN ultra-thin films and enhances the phonon escape significantly compared to commonly used Si substrates due to the close acoustic match to NbN and a low defect interface [12], [13], [14]. Nonetheless, the noise performance of such THz mixers that employ a GaN buffer-layer has not been demonstrated before the work presented here.

II. Experiment

A. Receiver design

The balanced waveguide receiver scheme offers major advantages over the single-ended configuration at the cost of higher complexity. The separation of the RF and LO path in conjunction with a waveguide RF hybrid makes a potentially lossy beam splitter obsolete, utilizes all available LO power and provides more flexibility when designing receiver arrays. Moreover, amplitude modulations and thermal instabilities of the LO source are suppressed in the balanced mode [15] and higher receiver stability is achieved [1], which allows for prolonged integration times of astronomical observations.

The implemented layout consists of a waveguide RF quadrature hybrid and two NbN-on-GaN HEB mixers, which were integrated into a full-metal mixer block in back-piece configuration, as illustrated in Fig. 1. Details on the fabrication of the quadrature RF hybrid and waveguide components with cross-section of 90 µm x 180 µm can be found in [16], [17]. The HEB mixers have been designed and optimized in the full-wave

3D simulator *Ansys HFSS* and are, according to simulations, able to operate in the 1.0 THz to 1.5 THz frequency band [18]. The individual IF outputs of the HEBs are combined in-phase by using a compact superconducting multi-section Wilkinson combiner [19], which was followed by a cryogenic SiGe low noise amplifier (LNA) [20]. The IF chain was complemented with a warm broadband LNA and the read-out was performed with a spectrum analyzer.

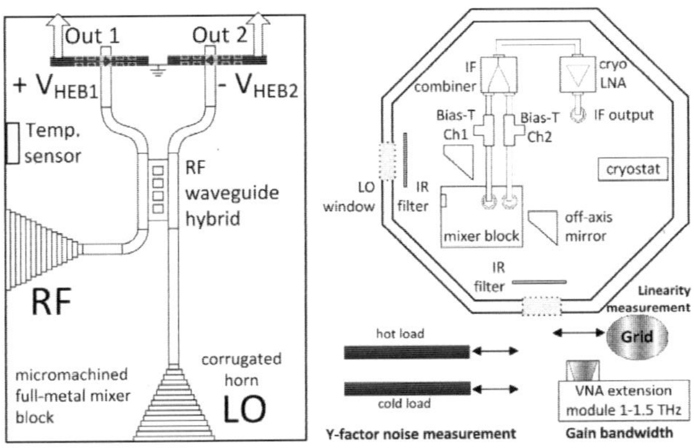

Fig. 1 Left: Schematic of the micro machined full-metal mixer block in back-piece configuration comprising of the RF hybrid and two NbN-on-GaN mixer chips. Right: Arrangement of the mixer block inside the wet cryostat. The receiver noise temperature is deduced from a Y-factor measurement on a hot and cold load.

The HEB mixer chips were processed on a thinned Si substrate onto which a 5.5 µm thick GaN buffer-layer was grown. The NbN film with approximate thickness of 4.5 nm was deposited by DC magnetron sputtering in a reactive N_2/Ar atmosphere at elevated temperatures. A detailed description of the deposition method can be found in [12]. The T_c of 12.8 K of the NbN film was deduced prior to processing by performing a resistance versus temperature measurement in a liquid helium Dewar.

The GaN beam has been defined by using photo-lithography and dry-etching techniques in chlorine chemistry, which is highly selective to the underlying Si substrate and provides a smooth surface and high process controllability. The patterning of the Si from the back into a Π-shaped structure as seen in Fig. 2 facilitated the mounting of the mixer and greatly improved the alignment accuracy [21], as the Fig. 2(right) suggests. The rigidity that is added with employing such Si frame makes it possible to use wire bonds for electrical contacting, which is considered a space-compatible technology. Details on the fabrication of such GaN beams can be found in [22].

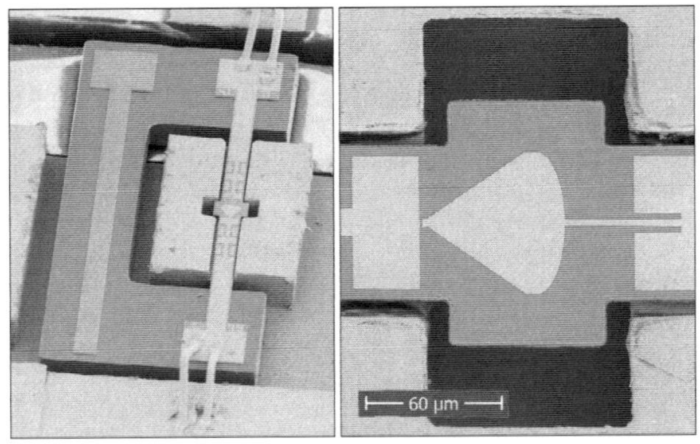

Fig. 2 Left: SEM images of the HEB mixer with bonding wire for DC bias and IF extraction. The Π-shaped Si frame, which hosts the suspended GaN membrane of 5.5 µm thickness, is a great aid for handling and alignment of the chips. Right: Shows the well-aligned mixer chips inside the waveguide opening. The HEB bridge is located between narrow part of the E-probe and the RF choke structure, which prevents leakage of the THz signal through the IF stripline.

III. Results and Discussion

A. Current-voltage characteristics

The IVC provides valuable information on the actual operation state of the HEB mixer, and will depend on its physical temperature and the amount of applied LO power, i.e. the pumping of the mixer. The bath temperature, measured directly at the mixer block, was 4.6 K and remained constant throughout all performed measurements.

Under zero bias conditions and without applied LO power, the HEB mixers are in the superconducting state and reveal their critical current (I_c), which can be read to 150 µA, as seen in Fig. 3. More importantly, both HEB mixers feature nearly the same I_c, as well as normal state resistance of 125 Ω, which allows to fully utilize the advantages of the balanced receiver scheme.

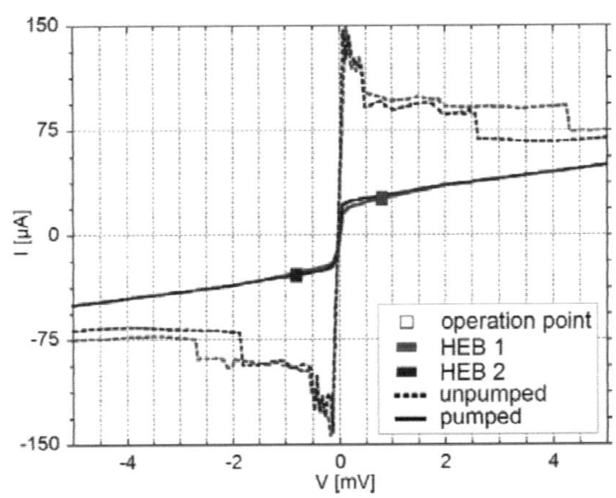

Fig. 3 Current-voltage characteristic (IVC) of both HEB mixers in the balanced receiver scheme. The operation points, which were used for the noise and gain bandwidth measurement are indicated with □. The I_c of both HEBs is nearly identical and measured to be 150 µA.

B. *Noise measurement*

The LO source was tunable in the range from 1.25-1.39 THz, however, the maximum output power of 9 μW [23] was only available in the frequency range between 1.28-1.29 THz and restricted the optimal pumping of the HEB mixers to these frequencies.

The receiver noise temperature was measured utilizing the Y-factor technique, which is based on two receiver output power measurements of black body emitters that are held at different physical temperatures (T_{hot}=296 K and T_{cold}=78 K). The IF power spectrum of the receiver was recorded with a spectrum analyzer for a hot load and a cold load in the frequency range from 0.5 GHz to 8 GHz and was used to calculate the DSB noise temperature of the receiver (T_{rec}).

The optimal bias voltage, yielding lowest noise performance was found to be in the range between 0.6-0.8 mV. The uncorrected DSB noise temperature is displayed in Fig. 4 as red curve, and suggests 750 K at 1 GHz IF and 900 K at 4 GHz IF, respectively. The measured noise temperature experienced a steeper than theoretically expected roll-off above approximately 5.5-6 GHz, which is most likely caused by the deteriorating noise performance of the used cryogenic LNA as indicated by the green curve in Fig. 4. However, its effect on the receiver noise roll-off can easily be removed, as measurement data on the LNA noise temperature was available and the total receiver conversion gain was deduced from a U-factor measurement, which is described in more detail in [24]. The blue curve in Fig. 4 shows the receiver noise temperature with removed frequency dependent LNA contribution. The noise bandwidth at which the noise temperature has increased to twice its value extrapolated to zero frequency, amounts to f_n=7 GHz.

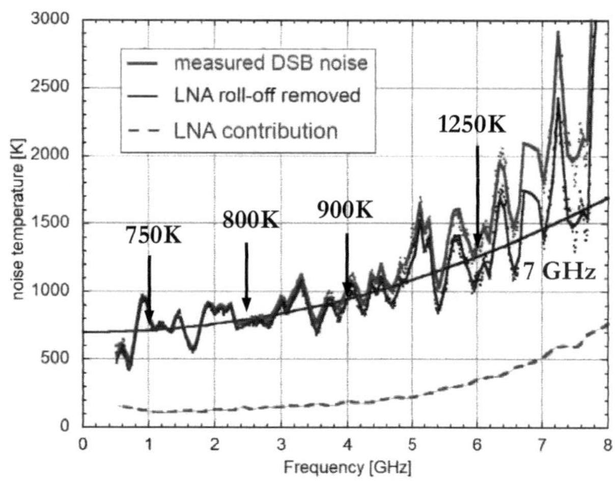

Fig. 4 DSB receiver noise temperature as measured (red) and without frequency dependent LNA contribution (blue). The green curve presents the corrected HEB mixer noise temperature. The fit suggest a nose bandwidth of f_n=7 GHz.

IV. Conclusion

A waveguide balanced NbN-on-GaN HEB receiver operating at 1.3 THz LO frequency has been presented. The uncorrected DSB receiver noise temperature was as low as 750 K at 1 GHz IF and 900 K at 4 GHz IF, respectively, and results in a noise bandwidth of 7 GHz, which is a substantially higher than the typically achieved 4 GHz for NbN-on-Si substrates HEB mixers. This improvement is ascribed to the employment of a GaN buffer-layer, which enhances the phonon escape of the epitaxially grown NbN film to the substrate. The contribution of the HEB mixer on the receiver noise temperature has been estimated to be 300 K or ~5 hf/k by taking into account losses in the optical path and waveguide components as well as the receiver's conversion gain, and is comparable to the sensitivity of state-of-the-art receivers.

The measurement results strongly promote the use of a GaN buffer-layer in future NbN based low noise THz heterodyne instruments with demands on an extended IF bandwidth, but also show its suitability to be implemented in balanced waveguide receiver schemes.

References

[1] D. Meledin, A. Pavolotsky, V. Desmaris, I. Lapkin, C. Risacher, V. Perez, D. Henke, O. Nyström, E. Sundin, D. Dochev, M. Pantaleev, M. Fredrixon, M. Strandberg, B. Voronov, G. Gol'tsman and V. Belitsky, "A 1.3-THz Balanced Waveguide HEB Mixer for the APEX Telescope," *IEEE Transactions on Microwave Theory and Techniques, vol. 57, no. 1,* 2009.

[2] F. Boussaha, J. Kawamura, J. Stern and C. Jung-Kubiak, "2.7 THz Balanced Waveguide HEB mixer," *IEEE TRANSACTIONS ON TERAHERTZ SCIENCE AND TECHNOLOGY,* vol. 4, no. 5, pp. 545-551, 2014.

[3] C. Risacher, R. Güsten, J. Stutzki, H.-W. Hübers, D. Büchel, U. Graf, S. Heyminck, C. Honingh, K. Jacobs, B. Klein, T. Klein, C. Leinz, P. Pütz, N. Reyes, O. Ricken, H.-J. Wunsch, P. Fusco and S. Rosner, "First Supra-THz Heterodyne Array Receivers for Astronomy With the SOFIA Observatory," *IEEE Transactions on Terahertz Science and Technology,* vol. 6, no. 2, pp. 199-211, 2015.

[4] D. Büchel, P. Pütz, K. Jacobs, M. Schultz, U. Graf, C. Risacher, H. Richter, O. Ricken, H. W. Hübers, R. Gusten, C. E. Honingh and J. Stutzki, "4.7-THz superconducting hot electron bolometer waveguide mixer," *IEEE Transactions on Terahertz Science Technology,* vol. 5, no. 2, pp. 207-214, 2015.

[5] P. Pütz, D. Büchel, K. Jacobs, M. Schultz, C. E. Honingh and J. Stutzki, "Waveguide hot electron bolometer mixer development for upGREAT," in *39th International Conference on Infrared, Millimeter, and Terahertz waves (IRMMW-THz),* Tucson, AZ, 2014.

[6] S. Cherednichenko, V. Drakinskiy, T. Berg, P. Khosropanah and E. Kollberg, "Hot-electron bolometer terahertz mixers for the Herschel Space Observatory," *Review of Scientific Instruments,* vol. 79, p. 034501, 2008.

[7] E. Novoselov and S. Cherednichenko, "Gain and Noise in THz MgB2 Hot-Electron Bolometer Mixers With a 30-K Critical Temperature," *IEEE Transactions on Terahertz Science and Technology,* vol. 7, no. 6, pp. 704-710, 2017.

[8] D. Cunnane, J. Kawamura, M. Wolak, N. Acharya, X. Xi and B. Karasik, "Optimization of Parameters of MgB2 Hot-Electron Bolometers," *IEEE Trans. on Appl. Supercond.,* vol. 27, no. 4, p. 2300405, 2017.

[9] D. Dochev, V. Desmaris, A. Pavolotsky, D. Meledin, Z. Lai, A. Henry, E. Janzen, E. Pippel, J. Woltersdorf and V. Belitsky, "Growth and

characterization of epitaxial ultra-thin NbN films on 3C-SiC/Si substrates for terahertz applications," *Supercond. Sci. Technol.*, 2011.

[10] J. Gao, M. Hajenius, F. Tichelaar, T. Klapwijk, B. Voronov, E. Grishin, G. Gol'tsman, C. Zorman and M. Mehregany, "Monocrystalline NbN nanofilms on a 3C-SiC/Si substrate," *Applied Physics Letters*, vol. 91, p. 062504, 2007.

[11] D. Meledin, E. Tong, R. Blundell, N. Kaurova, K. Smirnov, B. Voronov and G. Gol'tsman, "Study of the IF Bandwidth of NbN HEB Mixers," *Applied Superconductivity*, vol. 13, no. 2, pp. 164-167, 2003.

[12] S. Krause, D. Meledin, V. Desmaris, A. Pavolotsky, V. Belitsky, M. Rudzinski and E. Pippel, "Epitaxial growth of ulta-thin NbN films on AlxGa1-xN buffer-layers," *Supercond. Sci. Technol.*, vol. 27, no. 6, p. 065009, Apr 2014.

[13] S. Krause, V. Afanas'ev, V. Desmaris, D. Meledin, A. Pavolotsky, V. Belitsky, A. Lubenschenko, A. Batrakov, M. Rudzinski and E. Pippel, "Ambient temperature growth of mono- and polycrystalline NbN nanofilms and their surface and composition analysis," *Applied Superconductivity*, 2016.

[14] S. Krause, V. Mityashkin, S. Antipov, G. Gol'tsman, D. Meledin, V. Desmaris, V. Belitsky and M. Rudzinski, "Reduction of Phonon Escape Time for NbN Hot Electron Bolometers by Using a GaN Buffer Layers," *IEEE Transactions on Terahertz Science and Technology*, vol. 7, pp. 53-59, 2017.

[15] J. Kooi, R. Chamberlin, R. Monje, B. Force, D. Miller and T. Phillips, "Balanced Receiver Technology Development for the Caltech Submillimeter Observatory," *IEEE Transactions on THz Sci. and Techn.*, vol. 2, no. 1, pp. 71-84, 2012.

[16] V. Desmaris, D. Meledin, A. Pavolotsky, R. Monje and V. Belitsky, "All-metal micromachining for the fabrication of sub-millimetre and THz waveguide components and circuits," *Journal of Micromechanics and Microengineering*, vol. 18, no. 9, p. 095004, 2008.

[17] V. Desmaris, D. Meledin, D. Dochev, A. Pavolotsky and V. Belitsky, "Terahertz components packaging using integrated waveguide technology," *IEEE MTT-S International Microwave Workshop on Millimeter Wave Integration Technologies*, no. 2011, pp. 81-84, 2011.

[18] S. Krause, D. Meledin, V. Desmaris and V. Belitsky, "Design of a wideband balanced waveguide HEB mixer employing a GaN buffer-layer for the 1-1.5 THz band," in *28TH INTERNATIONAL SYMPOSIUM ON SPACE TERAHERTZ TECHNOLOGY*, Cologne, Germany, 2017.

[19] H. Rashid, D. Meledin, V. Desmaris and V. Belitsky, "Passive Circuits for Wideband Receiver Applications," in *Swedish microwave days*, 2016.

[20] K. Dawson, *SiGe low noise amplifier CITLF4*, Pasadena, California: California Institute of Technlogy, Caltech, Department of Electrical Engineering, 2015.

[21] D. Dochev, V. Desmaris, A. Pavolotsky, D. Meledin and V. Belitsky, "A Technology Demonstrator for 1.6-2 THz Waveguide HEB Receiver with a Novel Mixer Layout," *Journal of Infrared, Millimeter and Terahertz Waves*, vol. 32, p. 451, 2011.

[22] S. Krause, V. Desmaris, A. Pavolotsky, D. Meledin and V. Belitsky, "Suspended GaN beams and membranes on Si as a platform for waveguide-based THz applications," *Journal of Micromechanics and Microengineering*, 2018.

[23] V. D. Inc., *1251-1389 GHz x72 Frequency Extension Module - User Guide*, Virginia Diodes Inc., 2004.

[24] S. Krause, D. Meledin, V. Desmaris, A. Pavolotsky and V. Belitsky, "Noise and IF gain bandwidth of a balanced waveguide NbN/GaN hot electron bolometer mixer operating at 1.3 THz," *IEEE Transactions on Terahertz Science and Technology*, vol. 8, no. 3, pp. 365-371, 2018.

A Fully Integrated Heterodyne Receiver Based on a Hot Electron Bolometer Mixer and a Quantum Cascade Laser

W. Miao[1,2], H. Gao[1,2], Z. Lou[1,2], J. Hu[1,2], W. Zhang[1,2], Y. Ren[1,2], K.M. Zhou[1,2], S.C. Shi[1,2,*], H. Li[3], J.C. Cao[3], and Y. Delorme[4]

[1]*Purple Mountain Observatory, CAS, Nanjing, 210008, China*

[2]*Key Lab of Radio Astronomy, CAS, Nanjing, 210008, China*

[3]*Shanghai Institute of Microsystem and Information, CAS, Shanghai, 200050, China*

[4]*Observatoire de Paris, Paris, 75014, France*

*Contact: scshi@mail.pmo.ac.cn

***Abstract*—We report on a very compact heterodyne receiver by integrating a superconducting NbN hot electron bolometer (HEB) mixer and a semi-insulating surface-plasmon quantum cascade laser (QCL) operating at 2.5 THz as local oscillator in a single block. To ensure effective pumping of the HEB mixer, the QCL's beam is collimated with a parabolic mirror that is integrated in the block too. After the collimation, the beam coupling efficiency between the HEB mixer and the QCL reaches approximately 0.9 and the superconducting HEB mixer can be easily pumped by the QCL. We then measured the receiver noise temperature of the integrated HEB/QCL receiver with a vacuum experimental setup. The lowest uncorrected receiver noise temperature is found to be about 750 K at 2.5 THz.**

I. Introduction

It is well known that the terahertz (THz) frequency region is featured by a diversity and large number of fine structure and molecular lines. They are very important tracers for studying the energetics and kinematics of astronomical objects such as stars and planetary systems. To detect these spectral lines with high frequency resolution, a sensitive and compact heterodyne receiver is essentially required. At frequencies up to 1.9 THz, heterodyne receivers based on superconducting mixers [1, 2] and solid-state multiplier-chain local oscillator (LO) sources have been realized for many space and ground-based telescope applications [3, 4]. At higher frequencies, the development of heterodyne receivers is seriously limited by the availability of suitable LO sources although superconducting hot electron bolometer (HEB) mixers have shown a superior sensitivity up to 5.3 THz [5]. Recently, THz quantum cascade lasers (QCLs) have been developed as a new LO source at frequencies above 1.9 THz, with high output power and good power stability [6]. Much progress has been made in using a THz QCL as an LO for superconducting HEB receivers, such as QCL's frequency and amplitude stabilization [7-10], and gas phase spectroscopy [11]. However, due to the high power dissipation of the QCL, most of them use an additional 4-K cryostat for the cooling of the QCL. It makes the receiver system fairly complicated. To simplify the receiver system, Richter et al. integrated the HEB mixer and the QCL on the different cold stages of a pulse-tube (PT) cooler [12]. Such integrated receiver, however, requires a THz QCL of good performance at relatively high temperatures. Recently, we have also demonstrated the integration of a HEB mixer and a 2.7 THz QCL on the same 4-K stage of a cryostat [13]. However, the HEB mixer and the 2.7 THz QCL are still mounted in two separate blocks and the noise performance of the integrated receiver is largely limited by optical losses. In this paper, we report on the demonstration of a fully integrated heterodyne receiver with a superconducting HEB mixer and a THz QCL in a block. The noise temperature of the integrated receiver is characterized using a vacuum experimental setup.

II. HEB/QCL Integrated Receiver

Fig. 1 (a) shows the photograph of the integrated receiver, in which a superconducting niobium nitride (NbN) HEB mixer and a semi-insulating surface-plasmon QCL operating at 2.5 THz are integrated in a block. The superconducting NbN HEB mixer used in this work was fabricated at Laboratoire d'Etudes du Rayonnement et de la Matiere en Astrophysique (LERMA). The HEB mixer consists of a 0.2 μm long and 2 μm wide NbN microbridge (connected to a log spiral antenna). The mixer has a normal resistance of 90 Ω and a critical temperature of 9.7 K.

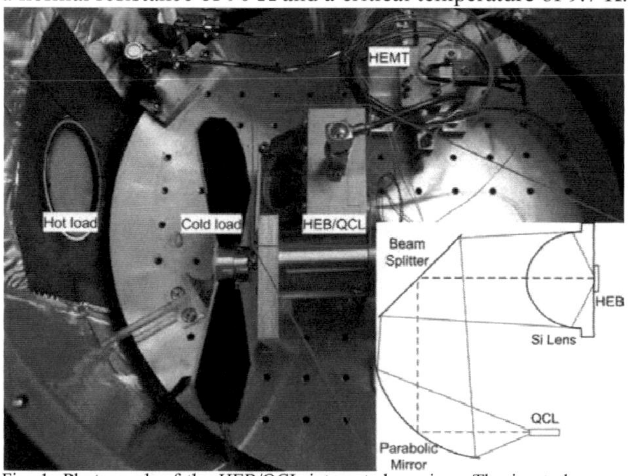

Fig. 1. Photograph of the HEB/QCL integrated receiver. The inset shows a schematic diagram of the coupling of the LO radiation from the QCL to the superconducting HEB mixer inside the block.

The QCL used here was fabricated at Shanghai Institute of Microsystem and Information (SIMIT), based on a bound-to-continuum (BTC) active region design and a semi-insulating surface-plasmon waveguide. The beam profile of the semi-insulating surface-plasmon QCL is known to be still divergent due to the sub wavelength lateral dimension of the cavity and the coherent emission from the facets and sides [13]. So we used a parabolic mirror to collimate the radiation of the semi-insulating surface-plasmon QCL (see the inset of Fig. 1). The parabolic mirror has a diameter of 20 mm and a focal length of 12 mm. After the parabolic mirror, the radiation is coupled to the HEB mixer with a 6 μm thick beam splitter.

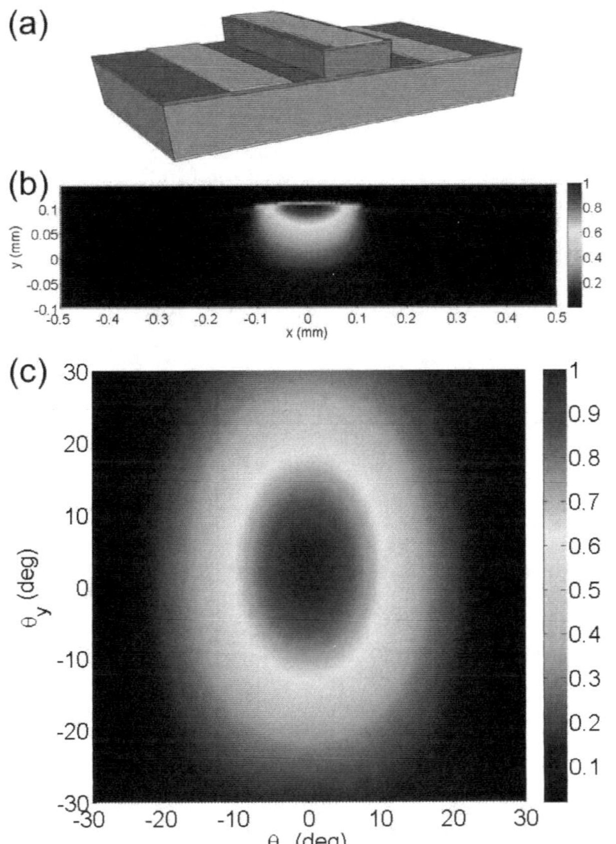

Fig. 2. (a) Schematic view of the semi-insulating surface-plasmon QCL. (b) Electric field intensity distribution at the front facet of the QCL. The electric field is simulated by using COMSOL. (c) Calculated beam pattern of the 2.5 THz semi-insulating surface-plasmon QCL after the parabolic mirror.

Fig. 2 (a) shows a schematic view of the semi-insulating surface-plasmon QCL. Fig. 2 (b) and (c) show the simulated electric field intensity distribution at the front facet of the semi-insulating surface-plasmon QCL and the calculated far field beam pattern of the QCL after the parabolic mirror. Here the far field beam pattern of the QCL is calculated from the electric field intensity distribution at the front facet by using spatial Fourier transform and physical optics propagation. It can be seen that the calculated far field beam pattern after the parabolic mirror is quite symmetric with a small angle offset in the vertical direction. We then evaluated the beam coupling efficiency between the superconducting HEB mixer and the QCL from the convolution of their far field beam patterns. We found that after the collimation, the beam coupling efficiency is as high as 0.9 (not including the reflection loss of the beam splitter).

III. Measurement Results and Discussions

With the collimated beam, the superconducting NbN HEB mixer can be easily pumped by the semi-insulating surface-plasmon QCL. Fig. 3 shows the measured current-voltage (I-V) curves of the superconducting HEB mixer. We then measured the noise performance of the HEB/QCL integrated receiver by using a vacuum experimental setup (see Fig. 1). In the setup, a 295 K blackbody inside the liquid helium cryostat is used as hot load and a 4.2 K blackbody on a chopper is used as cold load. The two loads can be selected by rotating the chopper, controlled by a Phytron stepper motor. Fig. 4 shows the measured mixer's intermediate frequency (IF) output power (responding to the hot and cold loads) and the receiver noise temperature as a function of the bias voltage. We can find that the lowest uncorrected receiver noise temperature is about 750 K at 2.5 THz.

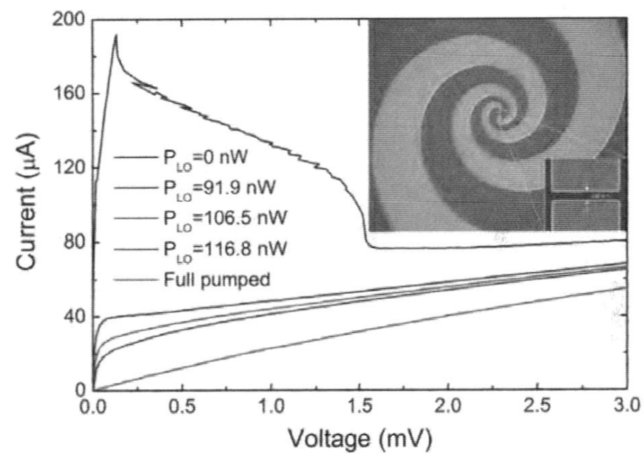

Fig. 3. Measured I-V curves of the superconducting HEB mixer. The inset shows a SEM micrograph of the superconducting HEB mixer.

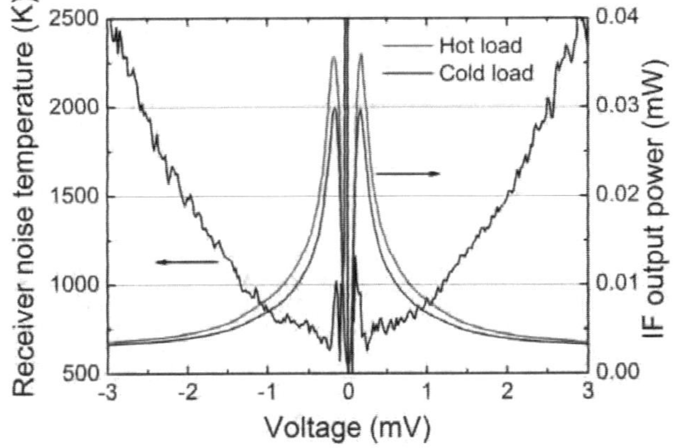

Fig. 4. Measured IF output power (responding to the hot and cold loads) and the resulting receiver noise temperatures as a function of the bias voltage.

CONCLUSIONS

We have demonstrated a compact and sensitive heterodyne receiver by integrating a superconducting NbN HEB mixer and a low power consumption QCL emitting at 2.5 THz as LO in a block. We have found that the superconducting HEB mixer can be easily pumped by the QCL with its beam collimated with a parabolic mirror. We have characterized the noise performance of the integrated HEB/QCL receiver using a vacuum experimental setup. The measured double sideband receiver noise temperature is as low as 750 K at 2.5 THz.

ACKNOWLEDGMENT

We acknowledge W.Y. Duan, K. Zhang, and Q.J. Yao of the Purple Mountain Observatory for their helpful assistance in the measurement. This work was supported by the Chinese Academy of Sciences (CAS) under Grant XDB23020200, the NSFC under Grant 11473075, and the CAS Joint Key Lab for Radio Astronomy.

REFERENCES

[1] J.R. Tucker and M.J. Feldman, "Quantum detection at millimeter wavelengths," *Rev. Mod. Phys.*, vol. 57, pp. 1055-1113, 1985.

[2] E.M. Gershenzon, G.N. Gol'tsman, I.G. Gogidze, Y.P. Gusev, A.I. Elant'ev, B.S. Karasik, and A.D. Semenov, "Millimeter and submillimeter range mixer based on electronic heating of superconducting films in the resistive state," *Sov. Phys. Supercond.*, vol. 3, pp. 1582-1597, 1990.

[3] Th. de Graauw et al., "The Herschel-heterodyne instrument for the far infrared (HIFI)," *Aston. & Astrophys.*, vol. 518, L6, 2010.

[4] M. Tarenghi, "The Atacama large millimeter/submillimeter array: overview & status," *Astrophys. Space Sci.*, vol. 313, pp. 1-7, 2008.

[5] W. Zhang, P. Khosropanah, J.R. Gao, E.L. Kollberg, K.S. Yngvesson, T. Bansal, R. Barends, and T.M. Klapwijk, "Quantum noise in a terahertz hot electron bolometer mixer," *Appl. Phys. Lett.*, vol. 96, p. 111113, 2010.

[6] E. Kohler, A. Tredicucci, F. Beltram, H.E. Beere, E.H. Linfield, A.G. Davies, D.A. Ritchie, R.C. Iotti, and F. Rossi, "Terahertz semiconductor heterostructure laser," *Nature*, vol. 417, pp. 156-159, 2002.

[7] H. Richter, S.G. Pavlov, A.D. Semenov, L. Mahler, A. Tredicucci, H.E. Beere, D.A. Ritchie, and H.W. Hubers, "Submegahertz frequency stabilization of a terahertz quantum cascade laser to a molecular absorption line," *Appl. Phys. Lett.*, vol. 96, p. 071112, 2010.

[8] Y. Ren, J.N. Hovenier, M. Cui, D.J. Hayton, J.R. Gao, T.M. Klapwijk, S.C. Shi, T.K. Kao, Q. Hu, and J.L. Reno, "Frequency locking of single mode 3.5 THz quantum cascade lasers using a gas cell," *Appl. Phys. Lett.*, vol. 100, p. 041111, 2012.

[9] Y. Ren, D.J. Hayton, J.N. Hovenier, M. Cui, J.R. Gao, T.M. Klapwijk, S.C. Shi, T.K. Kao, Q. Hu, and J.L. Reno, "Frequency and amplitude stabilized terahertz quantum cascade laser as local oscillator," *Appl. Phys. Lett.*, vol. 101, p. 101111, 2012.

[10] J.L. Kloosterman, D.J. Hayton, Y. Ren, T.Y. Kao, J.N. Hovenier, J.R. Gao, T.M. Klapwijk, Q. Hu, C.K. Walker, and J.L. Reno, "Hot electron bolometer heterodyne receiver with a 4.7 THz quantum cascade laser as a local oscillator," *Appl. Phys. Lett.*, vol. 102, p. 011123, 2013.

[11] Y. Ren, J.N. Hovenier, R. Higgins, J.R. Gao, T.M. Klapwijk, S.C. Shi, A. Bell, B. Klein, B.S. Williams, S. Kumar, Q. Hu, and J.L. Reno, "Terahertz heterodyne spectrometer using a quantum cascade laser," *Appl. Phys. Lett.*, vol. 97, p. 161105, 2010.

[12] H. Richter, A.D. Semenov, S.G. Pavlov, L. Mahler, A. Tredicucci, H.E. Beere, D.A. Ritchie, K.S. Il'in, M. Siegel, and H.W. Hubers, "Terahertz heterodyne receiver with quantum cascade laser and hot electron bolometer mixer in a pulse tube cooler," *Appl. Phys. Lett.*, vol. 93, p. 141108, 2008.

[13] W. Miao, Z. Lou, G.Y. Xu, J. Hu, S.L. Li, W. Zhang, K.M. Zhou, Q.J. Yao, K. Zhang, W.Y. Duan, S.C. Shi, R. Colombelli, H.E. Beere, and D.A. Ritchie, "Demonstration of a fully integrated superconducting receiver with a 2.7 THz quantum cascade laser," *Opt. Exp.*, vol. 23, no. 4, pp. 4453-4458, 2015.

A MEMS-based, Dicke-switched radiometer at 560GHz

Theodore Reck*, Cecile Jung-Kubiak[1], Maria Alonso-delPino[1], Goutam Chattopadhyay[1]

[1]Jet Propulsion Laboratory, Pasadena, 91109, USA
*Contact: theodore.reck@jpl.nasa.gov

Abstract—Future planetary missions require terahertz spectrometers and radiometers to be lower-power and lower-mass to fly on smaller platforms or allow more instruments to be deployed on a single platform. While significant advances have been made in reducing the power and mass of the receivers, LO sources and backend spectrometers, to date, no advances have been made to the calibration system above 200 GHz. Presently, calibration is performed with quasi-optical flip mirror redirecting the feed to a free-space absorber. Although well proven, this approach is problematic for two reasons: 1) it relies on mechanical motion, which presents significant risk to the instrument, and 2) the free-space optics are physically large compared to the waveguide-based components of the receiver.

This work demonstrates a MEMS-based waveguide single-pole double-throw switch to enable compact Dicke-switched calibration of a 560 GHz receiver. This device uses large-deflection, silicon micromachined comb-drive actuators to move a septum across the waveguide to direct the signal path. The MEMS device is packaged within a metal split-block housing for compatibility with standard waveguide components. The packaged switch demonstrates low-loss (<2 dB) and high-isolation (>20dB) to maximize sensitivity and avoid contamination of the received signal by the calibration source. A Dicke-switched Schottky-diode based receiver operating at 560GHz is presented and the implications of standing waves on the calibration accuracy are discussed.

Session M3: **Invited I**

Results of SMILES and the plan of follow-on THz mission, SMILES-2

Satoshi Ochiai*

National Institute of Information and Communications Technology, Koganei, Tokyo 184-8795, Japan
*Contact: ochiai@nict.go.jp

Abstract—The Superconducting Submillimeter-Wave Limb-Emission Sounder (SMILES) measured the Earth's atmosphere by receiving submillimeter limb-emission spectra with SIS receivers at 625 and 650 GHz on the International Space Station from October 2009 to April 2010. The SMILES observations provided new knowledge of the middle atmosphere. For example, the diurnal variation of the stratospheric ozone was observationally shown for the first time. The evidence of the diurnal variation has impacted on long-term ozone trend analysis that uses several ozone data obtained at different local time depending on the measurement methods. SMILES also demonstrated wind measurement by submillimeter limb sounding. It had been challenging to detect the small Doppler shift due to wind, which is less than a hundredth of the line width of molecular emission in the stratosphere. The successful wind measurement inspired new projects on wind measurement by submillimeter limb sounding.

SMILES-2 is a proposed mission for comprehensive observation of the Earth's atmosphere with SIS and HEB receivers. The main objectives of SMILES-2 are to understand diurnal variations and true steady background of the atmosphere above the tropopause. Frequency bands including 2 THz channel are chosen for SMILES-2 so that wind and temperature in all the altitude range from the stratosphere to lower thermosphere can be measured with high precision, and also various minor species in the atmosphere can be observed.

In this talk, I present some of the scientific results of SMILES, as well as a plan of future comprehensive atmospheric observation by submillimeter limb sounding from space.

Session M4: **Direct Detectors and Instruments**

Single Photon Detection of 1.5THz Radiation with the Quantum Capacitance Detector

P.M. Echternach*[1], B.J. Pepper[1], T. Reck[1], and C.M. Bradford [1]

[1]*Jet Propulsion Laboratory, California Institute of Technology, Pasadena, 91109, USA*

*Contact: pierre.m.echternach@jpl.nasa.gov

Abstract— The Quantum Capacitance Detector (QCD) is a new high-sensitivity direct detector under development for low background applications such as far-infrared spectroscopy from a cold space telescope. The QCD has demonstrated an optically-measured noise equivalent power of $2x10^{-20}$ W $Hz^{-1/2}$ at 1.5THz, making it among the most sensitive far-IR detectors systems ever demonstrated, and meeting the requirements for spaceborne spectroscopy. Under these low-background conditions, the photon arrival rate is of the order of 100Hz making it possible to detect individual photons, provided the detector has enough speed. In this work we describe a new fast readout technique for the QCD that enabled single photon detection and counting at 1.5THz. Single photon detection and counting of single photons was demonstrated between 100Hz and 10 kHz. The QCD also demonstrates high absorption efficiency: both the photon arrival rate in counting mode, and the statistics of the shot noise in the non-counting mode indicate a total photon absorption and detection rate which is within a few percent of that expected for the experimental setup. Our measurements provide further confidence in the QCD as a detector approach for future ultrasensitive far-IR instrumentation.

Enabling Technologies for Photon-Counting Spectroscopy with the Origins Space Telescope (OST) in the Far/Mid-Infrared Region

O. Noroozian*[1, 2, 3], E. M. Barrentine[2], Jochem J.A. Baselmans[4], A. D. Brown[2], M. Cyberey[3], L. A. Hess[2], A. W. Lichtenberger[3], T. R. Stevenson[2], E. J. Wollack[2], S. H. Moseley[2]

[1]*National Radio Astronomy Observatory, 22903, Charlottesville, VA, USA*
[2]*NASA Goddard Space Flight Center, 20771, Greenbelt, MD, USA*
[3]*University of Virginia, 22903, Charlottesville, VA, USA*
[4]*Netherlands Institute for Space Research (SRON), Utrecht, Netherlands*

*Contact: o.noroozian@gmail.com

Abstract— Photon-counting detectors are highly desirable for reaching the ~ 10^-20 W/ √ Hz sensitivity permitted by the Origins Space Telescope (OST). They address the most difficult technology challenge for the OST: to capture and detect every single photon. The OST, a mid- to far-infrared cryogenic observatory concept, is being developed for consideration as a major NASA mission for the next Decadal Survey. An objective this facility is to allow rapid spectroscopic surveys of the high redshift universe at 420 – 800 μm, using arrays of integrated spectrometers with moderate resolutions ($R = \lambda/\Delta\lambda \sim 1000$), to create a powerful new data set for exploring galaxy evolution and the growth of structure in the universe. A second objective is to perform higher resolution (R ~ 100,000) spectroscopic surveys at 20–300 μm, a uniquely powerful tool for exploring the evolution of protoplanetary disks. Finally, the OST aims to do sensitive mid-infrared (5–40 μm) spectroscopy of thermal emission from rocky planets in the habitable zone using the transit method. These OST science objectives are exciting, but they are impossible to reach without a significant leap forward in detector technology.

We are developing unique Kinetic Inductance Detectors (KIDs) that can count photons in the far/mid-IR. Combined with μ-Spec - our Goddard-based on-chip far-IR spectrometer - these detectors will enable the first OST science objective mentioned above. For OST's mid-IR science objectives, a different optical coupling design is needed for the KIDs, but the detector technology is readily transferable. Since KIDs are highly multiplexable in nature their scalability will be a major improvement over current technologies that are severely limited in observing speed due to small numbers of pixels.

To reach the required sensitivities we are developing superconducting resonators made from thin aluminum films on single-crystal silicon substrates. Under the right conditions, small-volume inductors made from these films can become ultra-sensitive to single photons >90 GHz. Understanding the physics of these superconductor-dielectric systems is critical to performance. At GSFC we have achieved a very high quality factor of 0.5 x 10^6 for a 10-nm aluminum resonator at n ~ 1 microwave photon drive power, by far the highest value for such thin films reported in the literature. We measured a residual dark electron density of < 5 /μm^3 and extremely long excitation lifetime of ~ 6.0 ms, both within requirements for photon-counting. Using a detailed model we simulated the detector when illuminated with randomly arriving photon events and show that photon counting with >95% efficiency at 0.5 - 1.0 THz is achievable. To realize an optically coupled device, the challenge is to integrate these resonators with the fabrication process of our on-chip spectrometer, μ-Spec.

I will report on the above developments and plans to continue this project through funding from a recently awarded ROSES-APRA grant and a Roman Technology Fellowship award. I will also put in context work on two related technologies: a) μ-Spec at GSFC and b) wide-bandwidth quantum-limited parametric amplifiers. Together, these three technologies will enable a wide spectrum of science investigations with the OST.

JPL / Caltech KID Development for Far-IR / mm-Wave Spectroscopy

Charles M. (Matt) Bradford[*1] on behalf of the SuperSpec team and the STARFIRE development team

[1]JPL, Caltech, Pasadena, CA, 91126, USA
[*]Contact: bradford@caltech.edu

Abstract—I will provide a brief overview of two projects developing kinetic inductance detectors (KIDs) in the micro-devices laboratory (MDL) at JPL. Both systems use lumped-element resonators in the 100-400 MHz band, and have demonstrated sensitivities below (better than) the photon noise for their intended applications.

The first is the SuperSpec on-chip spectrometer. SuperSpec consists of a filterbank covering the full 190-310 GHz band patterned in superconducting niobium-on-silicon-nitride microstrip, integrated with an array of 300 titanium-nitride (TiN) kinetic inductance detectors (KIDs). Light is coupled onto the 1-micron-wide microstrip with a planar antenna, and the filterbank channels are formed with half-wavelength sections of the microstrip coupled via proximity to both the main trunk line and the TiN KIDs. The filterbank electromagnetic design has been optimized to provide high efficiency at a resolving power of ~300. The active portion of each SuperSpec KIDs is 20 nm film of TiN patterned into a meander with 0.25-micron wide lines. The total inductor volume is 3 cubic microns, the small value is key for high responsitivity and good sensitivity. We have demonstrated a noise equivalent power (NEP) of 4×10^{-18} W $Hz^{-1/2}$ at the operating temperature of 220 mK, and an even lower 8×10^{-19} W $Hz^{-1/2}$ when cooled to below 120 mK. The 1/f knee frequency is at or below 1 Hz for these devices. We are currently preparing for a SuperSpec demonstration at the Large Millimeter Telescope (LMT) in late 2018.

The second is a horn-coupled direct-absorbing aluminum device under development for far-IR applications including the proposed STARFIRE balloon-borne spectrometer covering 240-420 microns at R=500. A direct-drill multi-flare-angle horn couples incident radiation into a circular waveguide which illuminates the inductor / absorber. The inductor / absorber is 30 nm film patterned into a 300-micron diameter circular meander which couples both polarizations. The silicon substrate is etched from the backside to create a quarter-wave backshort. We have carefully characterized noise both dark and under loaded, and response using a cryogenic blackbody and 350-micron bandpass filter. The devices demonstrate optical efficiency as expected from simulations, and a device NEP of 4×10^{-18} W $Hz^{-1/2}$. These devices also have a 1/f knee around 1 Hz. The run of noise and response with applied load constrains the quasiparticle recombination dynamics, and we infer that the current devices have maximum quasiparticle lifetimes of ~20 microseconds, which limits the NEP. We are currently pursuing higher quality aluminum which we expect will provide longer lifetimes and thus lower NEP.

The NIKA2 instrument at the IRAM 30m-telescope: state-of-the-art KID performance in the mm range

E.F.C. Driessen*[1], on behalf of the NIKA2 collaboration

[1]*Institut de Radioastronomie Millimétrique (IRAM), St Martin d'Hères, France*
*Contact: driessen@iram.fr

Abstract—The New IRAM KID Array 2 (NIKA2) is a state-of-the-art mm-wave continuum camera that has been installed on the IRAM 30-m telescope since October 2015. Its focal planes consist of 3 lumped-element kinetic-inductance detector arrays (2896 pixels) cooled down to 100 mK using a closed-cycle dilution refrigerator. These arrays permit simultaneous observation of the 260 GHz and 150 GHz sky with a 6.5 arcmin field of view and 11 and 18 arcsec resolution, respectively, as well as a polarization capability for the 260 GHz band. The total-power commissioning stage of the instrument has recently come to an end, and the instrument is now opened for astronomical observations.

I will describe the general characteristics of the instrument, its performance as determined during the commissioning stage, and finally I will discuss possible future improvements of the performances, mostly from a detector point of view.

Prospects of Terahertz Intensity Interferometry

Hiroshi Matsuo[1*], Hajime Ezawa[1], Hitoshi Kiuchi[1], Mareki Honma[1], Yasuhiro Murata[2]

[1]National Astronomical Observatory of Japan, Mitaka, Tokyo 181-8588, Japan
[2]Institute of Space and Astronautical Science, Japan Aerospace Exploration Agency, Sagamihara, Kanagawa 252-5210, Japan
[*]Contact: h.matsuo@nao.ac.jp

Abstract—Far-infrared and terahertz intensity interferometry is studied for future high angular resolution observation from space. The intensity interferometers were demonstrated by Hanbury-Brown and Twiss in radio and optical wavelengths, but low correlation efficiency, requirements for large dynamic range and missing phase information limited their application. We have proposed that fast photon counting detectors in terahertz frequencies give high correlation efficiencies and delay time measurements using photon bunches, which are used to define complex visibilities for an aperture synthesis imaging. The photon counting detectors serve for high sensitivity observations from space; several orders of magnitude more sensitive than heterodyne receivers.

To show the feasibility of the terahertz intensity interferometry, we have made numerical simulations to compare the performance to heterodyne interferometry. When the noise temperature is the same, intensity interferometers require longer integration time to obtain complex visibilities from delay time measurements. In this respect, the intensity interferometry is not a technology to replace existing interferometers. On the other hand, due to the long coherence time, intensity correlations are stable against atmospheric phase fluctuation. When interferometer observation is made at 1 THz under phase fluctuation larger than 100 μm in path length, amplitude correlations drop exponentially, but intensity correlations stay the same. So, the intensity interferometry can be applied to long baseline interferometers at high terahertz frequencies from ground.

Further advantage is the use of direct detectors and photon counting detectors. Since direct detectors do not measure phase information, their sensitivities are not limited by the receiver quantum limit. Fast photon counting detectors also serve for high dynamic range measurements; such that 1 GHz bandwidth SIS photon detectors with NEP less than 10^{-17} W/Hz$^{0.5}$ can count 10^8 photons/sec from bright terahertz sources.

Based on the simulation studies, a roadmap towards the space terahertz intensity interferometry is discussed. Using fast photon detectors, such as SIS photon detectors, the terahertz intensity interferometry is to be demonstrated in lab to show imaging performance. For ground-based astronomical observations, existing or planned telescopes in Atacama or in Antarctica could be used to demonstrate the intensity interferometry by observing bright compact sources, such as massive star-forming regions and proto-planetary disks. For the space terahertz intensity interferometry, combination of two cryogenic telescopes is proposed to observe compact and bright sources catalogues by IRAS and AKARI missions.

Session M5: **Optical Design, Systems and Components I**

Development of diffractive optics for 4.7 THz frequency

Irmantas Kašalynas[1*], Heiko Richter[2], Simonas Indrišiūnas[1], Ignas Grigelionis[1], Linas Minkevičius[1], Gediminas Račiukaitis[1], Heinz-Wilhelm Hübers[2]

[1] *Center for Physical Sciences and Technology, Saulėtekio Ave. 3, LT-10257 Vilnius, Lithuania*
[2]*German Aerospace Center (DLR), Institute of Optical Sensor Systems, Rutherfordstr. 2, 12489 Berlin, Germany*
[*]Contacts: irmantas.kasalynas@ftmc.lt

Abstract— Direct laser ablation (DLA) is a mask-less technology used for the research and development of optical components of various materials [1]. The typical fabrication process combines a mask production and many technological steps such as a surface preparation, mask alignment, photoresist development, materials deposition, plasma and chemical treatment [2]. In some cases, such an approach is inefficient and time-consuming, especially in a small volume production. The relevance of the DLA technology is verified in the sub-terahertz frequency range demonstrating the terahertz zone plates with increased frequency selectivity [3], the binary zone plate on-chip with the terahertz detector [4], multilevel phase Fresnel lenses on silicon [5], etc.

In this work, four zone plates were developed: one from a 30-μm thick metal foil and three from a 500-um thick Si wafer for focusing of the 4.7 THz radiation. Processed by the DLA technology, all the lenses were of 25 mm diameter and #f/2 *f*-number but differed in the number of phase-levels. The focusing performance was determined by measuring the Gaussian-shaped beam profile of a 4.745 THz quantum cascade laser along the optical axis with a real-time bolometric camera (pixel size of 25 μm). With the binary optics the focusing gain up to 22 dB and the diffraction efficiency of 41 % was achieved. This was almost independent on whether the lens was developed on an optically thin metal foil or on a thick semiconductor substrate. An increased number of the phase-levels till 8 on the silicon zone plate provided the focusing gain and diffraction efficiency values up to 27 dB and 100 %, respectively.

[1] B. Voisiat, et al. "Laser processing for precise fabrication of the THz optics", Proc. SPIE vol. 10091, LAMOM XXII; 100910F (2017); Event: SPIE LASE, 2017, San Francisco, CA, USA
[2] E. D. Walsby, et al., "Multilevel silicon diffractive optics for terahertz waves", J. Vac. Sci. Technol. B 20, 2780 (2002).
[3] L. Minkevičius, et al., "Terahertz zone plates with integrated laser ablated bandpass filters", Electron. Lett. 49, pp. 49-50 (2013).
[4] L. Minkevičius, et al., "On-chip integration of laser-ablated zone plates for detection enhancement of InGaAs bow-tie terahertz detectors", Electron. Lett. 50, p. 1367 (2014).
[5] L. Minkevičius, et al., "Terahertz multilevel phase Fresnel lenses fabricated by laser patterning of silicon", Optics Letters Vol. 42, pp. 1875-1878 (2017)

Prototype demonstration of the broadband anti-reflection coating on sapphire using a sub-wavelength structure

T. Matsumura[1], R. Takaku[2,3], S. Hanany[4], H. Imada[5,6], H. Ishino[7], N. Katayama[1], Y. Kobayashi[8], K. Komatsu[7], K. Konishi[9], M. Kuwata-Gonokami[3], S. Nakamura[2], H. Sakurai[3], Y. Sakurai[1], Q. Wen[4], K. Young[4], J. Yumoto[3,9]

[1]*Kavli Institute for The Physics and Mathematics of The Universe (WPI), The University of Tokyo, Kashiwa, 277-8583, Japan*
[2]*Yokohama National University, Yokohama, 240-8501, Japan*
[3]*Department of Physics, The University of Tokyo, 113-0033, Japan*
[4]*The University of Minnesota, Twin Cities, MN 55455 USA*
[5]*Institute of Space and Astronautical Science (ISAS), Japan Aerospace Exploration Agency (JAXA), Sagamihara, 252-5210, Japan*
[6]*Laboratoire de l'Accélérateur Linéaire, Université Paris-Sud, CNRS/IN2P3, Université Paris-Saclay, Université Paris-Sud, Bâtiment 200, 91898 Orsay, France*
[7]*Okayama University, Okayama, 700-8530, Japan*
[8]*The Institute for Solid State Physics, The University of Tokyo, Tokyo, 277-0882, Japan*
[9]*Institute for Photon Science and Technology, The University of Tokyo, Tokyo, 113-0033, Japan*

*Contact: tomotake.matsumura@ipmu.jp

***Abstract*—We report the development of the broadband anti-reflection (AR) coating on sapphire using a sub-wavelength structure for a cosmic microwave background (CMB) polarization experiment. One of the challenges in the CMB polarization experiment is to disentangle the sources of the polarized emission among the CMB and the galactic emissions from our Galaxy. This leads a need of a cryogenically compatible broadband AR coating on a millimeter wave optical element used in a CMB telescope. We design and fabricate the AR sub-wavelength structure using a laser machining on a sapphire plate, which is a commonly used material as a half-wave plate at millimeter wavelength. We fabricate a prototype sample and evaluate the transmittance of an SWS AR sapphire for the fractional bandwidth of about 1.5. We show the agreement between the measured transmittance and the prediction based on the fabricated shape. The projected transmittance based on the measurement achieves above 90% above 50 GHz. We finally discuss the room of the improvement and the applicability for this technology for a future CMB polarization experiment.**

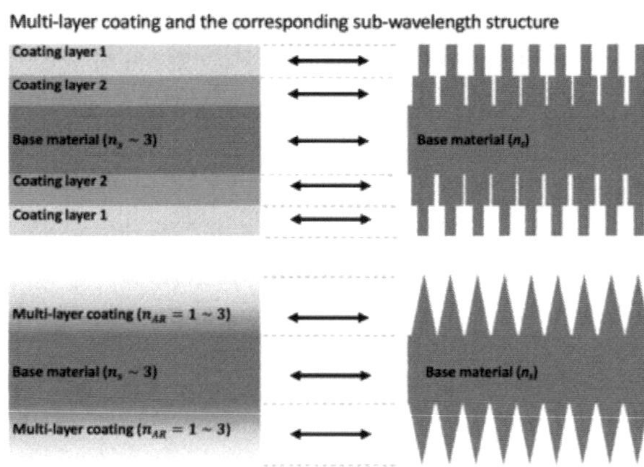

Figure 1 A sketch to show the correspondence between the multi-layer based AR coating and the SWS based AR strategy.

I. Introduction

The measurement of the cosmic microwave background (CMB) has played an important role in the modern cosmology. The recent rapid progress of the technological developments in the CMB experiment brings the community to the next target, testing the hypothesis of the cosmic inflation experimentally. Today, there is a strong enhancement between the scientific motivation and the technological development [1,2].

One of the major challenges in the CMB polarization experiment is to make a distinction between the CMB and the galactic polarized emissions from our own Galaxy. The standard method is to separate the different sources of the emission by relying on the difference of the spectral shape by observing a broad bandwidth. As a result, the recent CMB polarization telescope is in need of covering the broadband frequency coverage, ideally speaking from a centimeter wavelength to sub-millimeter wavelength.

One of the limitations in a broadband observation comes from the available bandwidth of the anti-reflection (AR) coating on an optical element in a CMB telescope. Typically, a CMB telescope employs a filter, a lens, and a half-wave

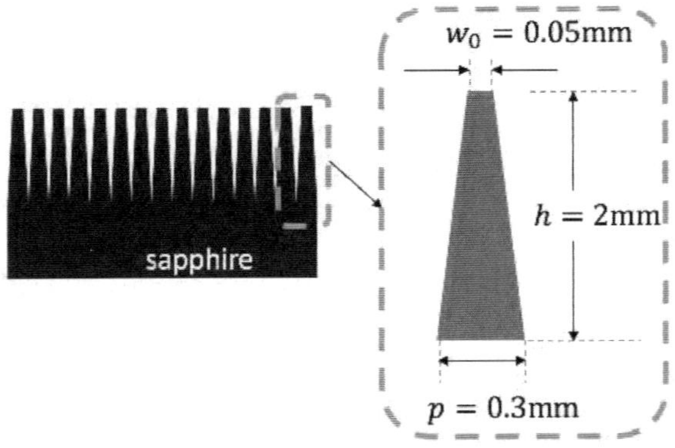

Figure 2 The sketch shows the parameterization of the pyramidal structure using three parameters.

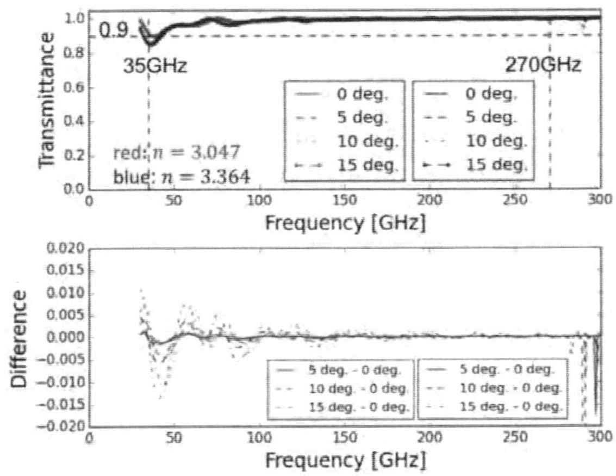

Figure 3 Top panel: The transmittance as a function of the frequency for the geometry in Figure 2. Bottom panel: The differential transmittance of the non-normal incident radiation with respect to the normal incident radiation. The thickness of the sapphire substrate, not including the AR layer, is assumed to be 1 mm in this simulation.

plate. All of these are a transmissive optical element, and thus it suffers from reflection without any AR coating. The typical materials, e.g. silicon, alumina, sapphire, have the index of refraction of about 3, and the resultant reflection without any AR coating is the reflectance of about 50%. Therefore, it is essential to apply a proper AR coating in order to maintain enough throughput of a telescope.

For a narrow bandwidth application, the standard technique to achieve the AR coating is to employ a thin dielectric layer with the index of refraction, $\sqrt{n_s}$. The valuable, n_s, is the index of refraction of the substrate on which we apply the AR coating. The thickness of the thin dielectric layer is $d_{AR} = \lambda_c/4\sqrt{n_s}$, where λ_c is the wavelength to minimize the reflection. This condition achieves a perfect transmission at a single frequency, $\nu_c = c/\lambda_c$, and its harmonics. Transmittance drops when the frequency of an interest departs away from the targeted frequency, ν_c. It achieves the fractional bandwidth of ~30% for above 90% of transmittance [3].

For a broadband coverage, the conventional method is to employ a multi-layer based AR coating. In recent years, there has been a tremendous development effort by using this multi-layer based AR coating on a flat disk and/or a curved lens [4-6]. The multi-layer based AR coating has two potential problems. One limitation is in its availability of materials for the various indices of refractions with a low dielectric loss. Another limitation is a mismatch of the coefficient of the thermal contraction among the materials. The coated layers potentially peel off the layers and eventually lead to the crack of each layer. In the worst case, it breaks the substrate itself.

In order to overcome these effects, another approach has been taken using sub-wavelength structure (SWS) based AR. The SWS, also known as a moth's eye structure, can provide the effective index of refraction between vacuum and a substrate. The desired index can be obtained by tuning the volume fraction between vacuum and the substrate in a unit volume as shown in Figure 1. The direct machining to the optical material can mitigate any issues originated from the differential thermal contraction. This also eliminates the search of a material that matches with a proper index of refraction and low loss.

Multiple methods have been explored to fabricate the SWS, dicing saw, etching, and laser machining [7-12]. In this paper, we address the laser machining approach on a sapphire plate, which is a typical material for a half-wave plate material at millimeter wavelength. The development is aimed to achieve the transmittance above 90% for the frequency range between 35 and 270 GHz. This bandwidth corresponds to the fractional bandwidth of 1.5, which has never been demonstrated.

This work is motivated by the technological development program, which is driven by the requirement from the next generation CMB satellite proposal, LiteBIRD [13,14]. It is a candidate mission proposal for a JAXA/ISAS strategic large class mission category and it is in a conceptual study phase. One of the telescopes of LiteBIRD is designed to cover the frequency range from 35 GHz to 161 GHz and possibly 270 GHz. One of the mission instruments for LiteBIRD is a half-wave plate polarization modulator using a sapphire. We require the design of the AR coating for a sapphire plate using a SWS method to cover 35 GHz to 161 GHz and we set the goal to cover up to 270 GHz.

In this paper, we report the design and the fabrication of the AR SWS on a sapphire plate. We also discuss the design trade-off, the feasibility and the implication of this technology toward a future CMB project.

II. Sample Preparations

A. Design

We design the AR structure on a flat sapphire. The design goal is to cover the frequency range from 35 GHz to 270 GHz, which the LiteBIRD low frequency telescope was originally targeted to cover. We set the first requirement of this development to be transmittance of above 90% to cover this bandwidth. We design the SWS using a second order effective medium theory (EMT) [15]. We start with a simple design, which is parameterized by the three parameters, a pitch, a height, and a top width as shown in the right of Figure 2. Generally speaking, the pitch determines the highest frequency which the AR structure can be functional below the diffraction limit. The combination of the height and the top width determines the AR performance at the low frequency end.

We verify the design performance without relying on the EMT, which is an approximated method. We compute predicted transmittance using a commercially available software, DiffractMod, which is based on Rigorous Coupled-Wave Analysis (RCWA) algorithm [16]. The parameters of a pitch of $p = 300$ μm, a height of $h = 2$mm, a top width of $w_0 = 50\ \mu$m on both sides of the A-cut sapphire result transmittance as shown in the top panel of Figure 3. In this simulation, we assume the thickness of the sapphire to be 1.0 mm. The pitch is determined to avoid the effect of the diffraction. We choose the pitch to be sufficiently finer than the frequency range of interest in order to avoid the effect of diffraction. The choice of the pitch, $p = \frac{c}{n\nu_c} = 300\ \mu$m, results the diffraction limited frequency, ν_*, to be ν_*=294 GHz for $n = n_e$. Also, the index of the refraction is assumed to be $n_o = 3.047$ and $n_e = 3.364$, and the dielectric loss is neglected [17]. This first iteration of the design can achieve transmittance above 90% at the nearly full targeted bandwidth.

We also computed for the incident angle range in $\pm 5, 10,$ and 15 degrees. All the curves essentially overlap on the same curves. The bottom panel of Figure 3 shows the differential transmittance of the non-normal incident radiation with respect to the normal incident radiation. The RMS differential transmittance over the entire frequency range for 15-degree incident radiation is 0.3% for both the ordinary and extraordinary axes. The maximum reduction of transmittance is less than 1.5% from the transmittance of the normal incident radiation at the 40-50 GHz range.

B. Fabrication

We fabricate the SWS AR pattern on a C-cut sapphire by using a 3 W nano-second UV laser machine. We name this sample as the sample A. The top and middle panels of Figure 4 show the fabricated structure, which is observed by a confocal microscope. We have not done any optimization in machining time, and it took about 70 hours to fabricate this sample. There are a number of optimizations which we can address in the machining speed, but this is in progress and beyond the scope of this paper.

We choose a sapphire sample to be a C-cut crystal instead of an A-cut crystal. This is because we want to avoid the degeneracy between the performance of the AR structure and the effect due to the optic axis alignment with respect to the incident polarization angle. As a result, the interpretation is straightforward at the time of the experimental demonstration. The structure was made only one side for the demonstration purpose. The area of the machining is centered with the diameter of 21 mm to minimize the machining time.

The bottom of Figure 4 shows one dimensional slices of the image at two locations as shown with the dashed line in the middle panel of Figure 4. The measured shape parameters are

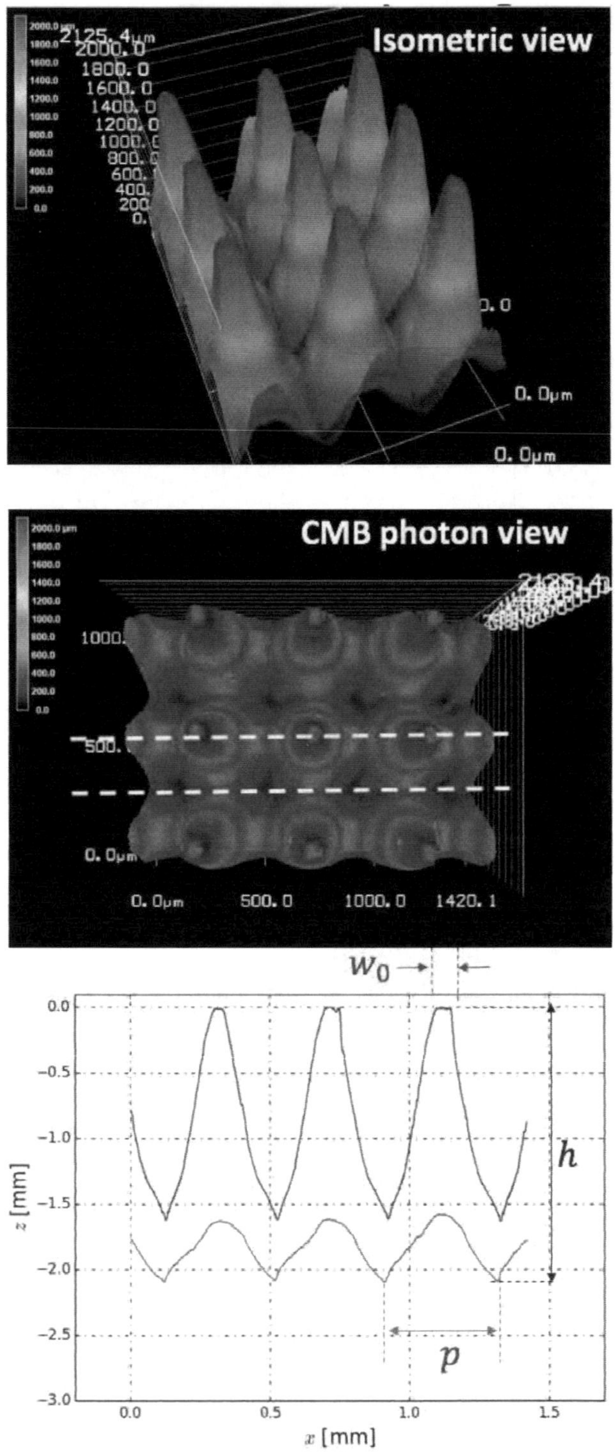

Figure 4 Top and middle: The confocal microscope images of the fabricated sample A. Bottom: The one dimensional sliced profile. The blue profile

corresponds to the slice which crosses the tip of the pyramid. The green profile corresponds to the slice which crosses the valley of the pyramid.

$w_0 = 51.3 \pm 7.2\,\mu\text{m}$, $p = 398.7 \pm 6.1\,\mu\text{m}$, and $h = 2105.0 \pm 21.3\,\mu\text{m}$ while the designed values are $w_0 = 50\,\mu\text{m}$, $p = 400\,\mu\text{m}$, and $h = 2000\,\mu\text{m}$. We picked the pitch

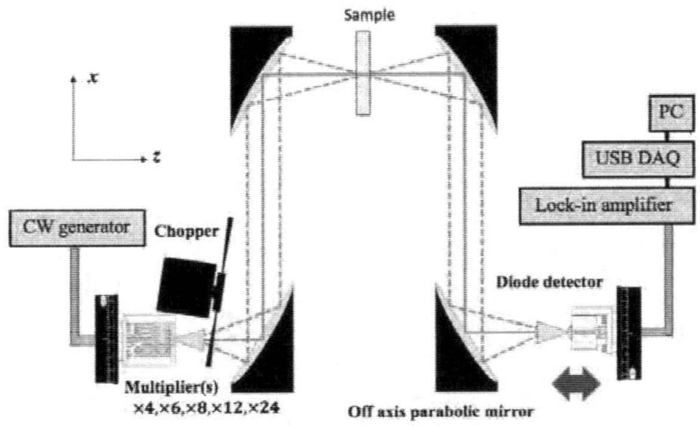

Figure 5 A sketch of the transmittance measurement set up at Kavli IPMU.

to be about 0.4 mm instead of 0.3 mm that is easier to machine due to the lower aspect ratio. The corresponding diffraction limited frequency is $\nu_*(n_o) = 246$ and $\nu_*(n_e) = 223$ GHz for the index of refraction of n_e. The quoted errors are computed by the RMS fluctuation of the measured parameters from the 45 pyramids. We measure the shapes using the confocal microscope at 5 locations of the machined area, center, and 4 corners, and one location contains 9 pyramids.

This error can be interpreted as a measure of the repeatability of the SWS shape.

Due to the nature of a laser machining, the shape does not come out to be exactly the same as the designed shape. Thus, we evaluate the performance by both the millimeter wave transmittance experimentally and the simulation based prediction with the fabricated shape.

III. Experiments

The measurement of transmittance is carried out by the millimeter-wave transmittance measurement setup at Kavli IPMU. Figure 5 shows the sketch of the measurement setup. We use a synthesizer to generate the electromagnetic wave frequency in range between 8 and 18 GHz. The source electromagnetic wave is propagated to a multiplier. We employ 6 different multipliers, Q-band (× 4), V-band (× 4), W-band (× 6), F-band (× 8), G-band (× 12), Y-band (× 24), to cover the frequency range from 35 to 265 GHz. Each transmittance measurement is done at a monochromatic frequency, and we sweep the frequency to cover the entire range. The frequency step of this sweep is chosen to be about 1 GHz to capture most of the spectral feature in transmittance. We normalize the transmitted power by dividing the measured power with the sample by the measured power without the sample. The sample is placed between the second and third mirrors, which is in focus. This is because we want to speed up the iteration of the development by preparing only a small sample, which saves a machining time. At the Gaussian waist, the spot diameter in the measurement bandwidth ranges from 2 mm to 15 mm. The Rayleigh length ranges from 20 mm to 158 mm, which is longer than the sample thickness of about 3 mm. The sample is tilted in the *y-z* plane by 5 degrees in order to minimize the standing wave along the chief ray. We also inserted attenuators to minimize this effect in the optical path.

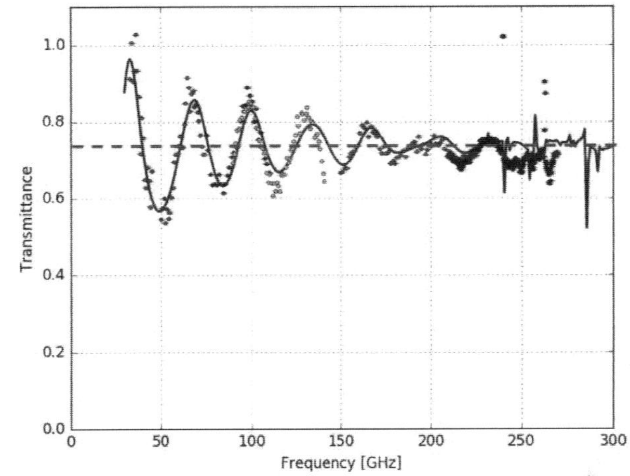

Figure 6 Fig. 6. The data points are the measured transmittance of the sample A and the solid line is the prediction based on the measured shape. The red dashed line is drawing at 0.74, which is the expected mean of transmittance from one side AR sample. The shape spikes in the prediction is due to the effect of diffraction.

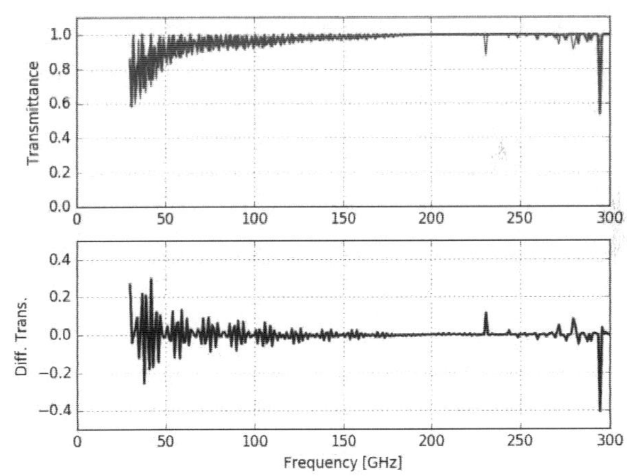

Figure 7 Top panel: The predicted transmittance given the fabricated geometry of the sample A using the DiffractMOD. The green and red curves are transmittance for the ordinary and extra-ordinary axes, T_o and T_e, respectively. Bottom panel: The differential transmittance between the ordinary and the extra-ordinary axes.

IV. Results

The top panel of Figure 6 shows the measured transmittance as a function of frequency. The oscillatory features around the mean of 0.74 is due to the fact that the SWS AR is only applied on one side of the sapphire surface. A set of the colored data points corresponds to the same multiplier source. The predicted curve is over-plotted to the data points. The predicted curve is computed by the

DiffractMOD based on the measured shape from the confocal microscope. The accuracy of the calculating transmittance is below 10^{-4}. There is no free parameter to fit and the data and the prediction agree well. This also shows the validity of the predicted transmittance that is computed by the DiffractMOD given the measured shape data using a confocal microscope.

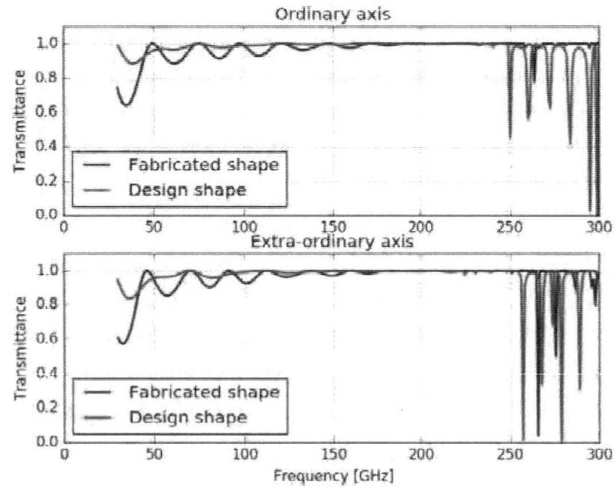

Figure 8 A comparison of the transmittance between the design shape and the fabricated shape.

The top panel of Figure 7 shows the prediction of transmittance when the fabricated AR structure is applied on both sides of the A-cut sapphire. The thickness is assumed to be 28.3 mm, which is assumed to have 3.14 mm×9 for use as an achromatic HWP [18]. Transmittance achieves above 90% for the frequency range above 70 GHz. When the band is averaged over 30% fractional bandwidth, the fast-oscillatory feature is smeared and averaged transmittance achieves above 90% for the frequency range above 50 GHz. The bottom panel of Figure 7 shows differential transmittance, $T_e - T_o$.

V. Discussions

A. Design shape V.S. Fabricated shape

There is a clear difference between the design shape and the fabricated shape when the laser machining is used. This is a unique feature due to the nature of the laser machining as compared to the mechanical machining. Figure 8 shows the comparison of two transmittances between one based on the design shape and one based on the fabricated shape. While the exact shape is not identical, the overall shape which is parametrized by the pitch, height, and top width are close to the design values and the resultant transmittance is qualitatively tracing well.

B. Optimization in shape

We investigate the design optimization in the detailed shape. The shape is parameterized as shown in Figure 9. In addition to the three parameters, the height, the pitch, and the top width, we also introduce the curvature (slope parameter, α) and the bottom width. The curvature is modelled as

$$w(z) = w_0 + \{(p-b) - w_0\}\{1 - \left(\frac{z}{h}\right)^{\alpha}\}.$$

Figure 9 shows the example of the three profiles for $\alpha = 0.5, 1,$ and 1.5. The profile with $\alpha = 1$ corresponds to the straight line.

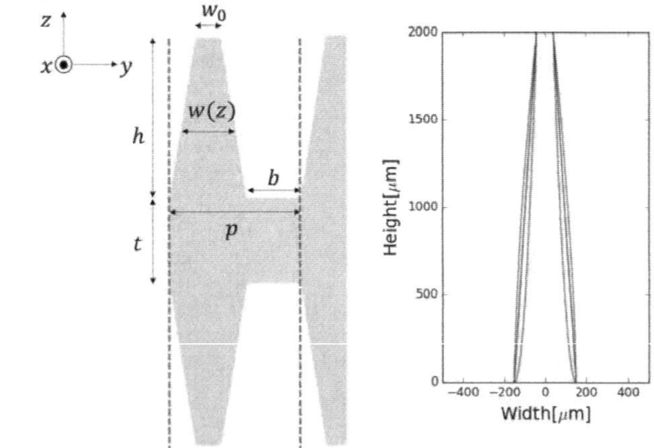

Figure 9 Left shows the parametrization for the optimization of the pyramidal shape. Right panel shows the pyramidal shape profile with the three different slope parameters, α = 0.5, 1, and 1.5.

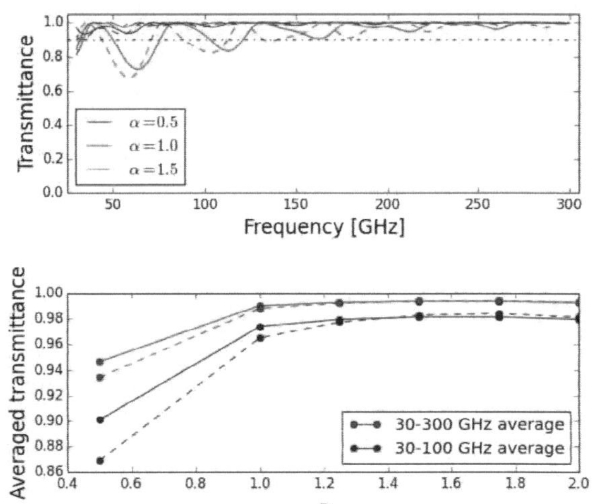

Figure 10 Top: Transmittance for the slope parameters, α =0.5, 1, and 1.5. Bottom: The band averaged transmittance for various slope parameters. For both plots, the solid line is Transmittance for n_o and the dashed line is for n_e.

We have studied the sensitivity of Transmittance to two parameters, the slope parameter, α, and the top width, w_0. The top panel of the Figure 10 shows the difference of transmittance computed by RCWA for three different α values. The bottom panel of the Figure 10 shows the averaged transmittance between 30 and 300 GHz and also between 30 and 100 GHz as a function of the slope parameter, α. The profile with $\alpha > 1$ is more desirable to increase transmittance at low frequency, <~ 100 GHz, as compared to the case with $\alpha < 1$. However, the higher slope parameter eventually reaches to the limit which makes the pyramidal shape to be a rectangular. There is a wide range of an optical slope parameter ranges between 1 and 1.8. The top panel of Figure 11 shows transmittance for various values in the top width. The bottom panel of Figure 11 shows average transmittance as

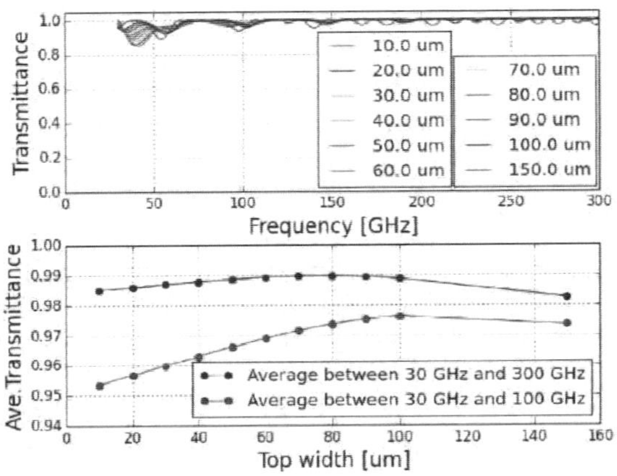

Figure 11 Top: Transmittance is plotted for various top width values. Bottom: Averaged transmittance as a function of the top width.

a function of the top width. We see the optimal top width to be around 80-100 μm depending on the bandwidth of interest. The fabricated sample is $w_0 = 51\ \mu$m, and clearly there is a room for more improvement in transmittance particularly at low frequencies, below 100 GHz, by increasing the top width to be wider.

Based on this study, we explore further optimizations in the fabrication process toward the optimal shape. The laser machining contains various parameters to tune, including the laser pulse width, wavelength, energy, laser optical properties, and scan strategy. There is a huge parameter space to explore and it is beyond this paper to cover the entire optimization process. Yet, we shall describe one example of the fabricated shape which leads to higher transmittance in the low frequency range.

We have constructed a small prototype sample shown in Figure 12. We label this sample as a sample B. Figure 13 is transmittance that is computed by the DiffractMOD for the sample B assuming that this SWS is applied on the both surfaces of the sapphire plate. Clearly, transmittance from the sample B shows higher transmittance in the low frequency range, below 100 GHz, as compare to that from the sample A.

From the confocal microscope data, we can compute the area fraction of a material to the air region along the z-axis for both samples. Based on this data, we can compute the effective index of refraction at each slice in the z-axis using the second order EMT. The top and the bottom panels of Figure 14 show these relationships. We also over-plot the Klopfenstein profile as a reference [19,20], which is an optimal profile for a broadband AR coating particularly at the low frequency side. Based on the study in Figure 10 and 11, transmittance in the low frequency region is higher for $\alpha > 1$ and $w_0 \sim 100\ \mu$m The shape of the sample B is closer match to these conditions as compared to that of the sample A.

C. Beyond transmittance

We started the development of the AR SWS with the design goal which is set by transmittance. High transmittance is very

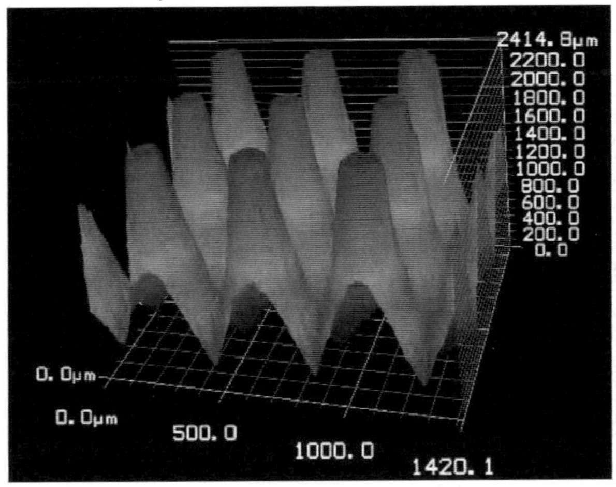

Figure 12 The confocal image of the sample B.

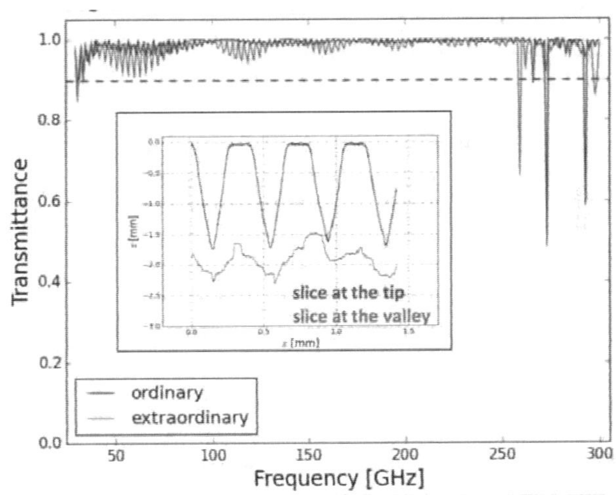

Figure 13 Projected transmittance of the sample B when the AR SWS is applied on the both surfaces of the sapphire plate. The inset shows the sliced shape profile at the tip and the valley.

important in the CMB polarization experiment. The reduction of the total signal power due to low transmittance degrades the signal-to-noise of the experiment and also becomes a source of a systematic effect, such as multiple reflections within a system.

When the AR coating is considered to a sapphire for use as a HWP, there are other parameters to be optimized. Transmittance along the ordinary and extra-ordinary axes results two different transmittances, T_e and T_e, due to the different indices of refraction. Differential transmission can be a source of an instrumental systematic effect, and this is a secondary figure-of-metric to minimize. In our design, we did not take into account this effect, but we can include this in our design requirement.

Projected differential transmission based on the fabricated AR SWS has its own spectral signature as shown in the bottom panel of Figure 7. At the time of design, the modeled pyramidal shape was symmetric about the z-axis. In principle, we can increase the degree of freedom by introducing the

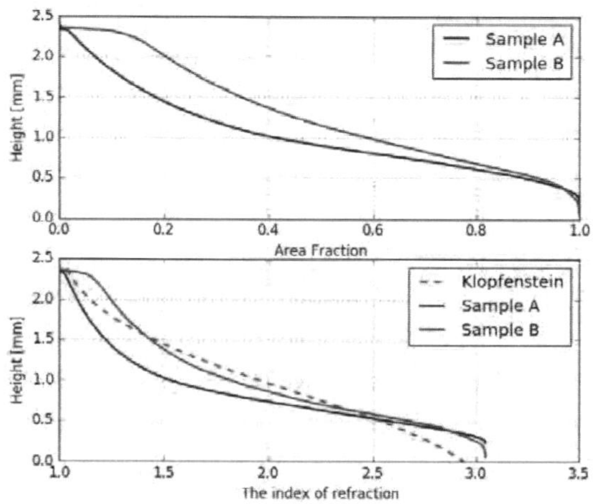

Figure 14 Top: The relationship between the height and the fractional area of the material at each slice along the z-axis. Bottom: The relationship between the height and the index profile which is estimated by the second order EMT.

asymmetry, e.g. rectangular base and top area with, p_x, p_y, and w_{0x}, w_{0y}. This allows the better impedance match for both the ordinary and extraordinary axes if the SWS is patterned with a proper alignment with the crystal axis.

Conclusions

We design the AR on sapphire using the SWS. We fabricate the SWS AR on one side of the C-cut sapphire prototype sample and measure transmittance. Measured transmittance and the predicted performance given the AR shape agree. Prospective transmittance when the AR SWS is applied on both surfaces is calculated. Prospective transmittance achieves above 90% for the frequency range above 70 GHz. When the band is averaged over 30% fractional bandwidth, the oscillatory feature is smeared and averaged transmittance achieves above 90% for the frequency range above 50 GHz. We also study the sensitivity of the AR design parameters and discuss the pros and cons for the further improvement in transmittance.

Acknowledgment

This work was supported by JSPS KAKENHI Grant Number 15H05441, 17H01125, 17K14272. This work was also supported by JSPS Core-to-Core Program, A. This work was also supported by World Premier International Research Center Initiative (WPI), MEXT, Japan.

References

[1] Planck collaborations, "Planck 2015 results. I. Overview of products and scientific results", A&A 594, A1 (2016).
[2] M. H. Abitbol et al., "CMB-S4 Technology Book, First Edition", arXiv:1706.02464.
[3] E Hecht, "Optics", Addison-Wesley.
[4] D. Rosen et al., "Epoxy-based broadband antireflection coating for millimeter-wave optics," Appl. Opt. 52, 8102–8105 (2013).
[5] Y. Inoue et al., "Cryogenic infrared filter made of alumina for use at millimeter wavelength". Appl. Opt., 53(9):1727–1733, Mar 2014.
[6] O. Jeong, A. Lee, C. Raum, and A. Suzuki, "Broadband plasma-sprayed anti-reflection coating for millimeter-wave astrophysics experiments", Journal of Low Temperature Physics, pages 1–6, 2016.
[7] R. Datta, C. D. Munson, M. D. Niemack, J. J. McMahon, J. Britton, E. J. Wollack, J. Beall, M. J. Devlin, J. Fowler, P. Gallardo, J. Hubmayr, K. Irwin, L. Newburgh, J. P. Nibarger, L. Page, M. A. Quijada, B. L. Schmitt, S. T. Staggs, R. Thornton, and L. Zhang, "Large aperture wide-bandwidth antireflection-coated silicon lenses for millimeter wavelengths", Appl. Opt., 52(36):8747–8758, Dec 2013.
[8] T. Matsumura, K. Young, Q. Wen, S. Hanany, H. Ishino, Y. Inoue, M. Hazumi, J. Koch, O. Suttman, and V. Schuetz, "Millimeter-wave broadband antireflection coatings using laser ablation of subwavelength structures", Applied Optics, 55:3502, May 2016.
[9] Karl Young, Qi Wen, Shaul Hanany, Hiroaki Imada, Jürgen Koch, Tomotake Matsumura, Oliver Suttmann, and Viktor Schütz., "Broadband millimeter-wave anti-reflection coatings on silicon using pyramidal sub-wavelength structures", Journal of Applied Physics 121, 213103 (2017).
[10] Viktor Schütz, Karl Young, Tomotake Matsumura, Shaul Hanany, Jürgen Koch, Oliver Suttmann, Ludger Overmeyer, and Qi Wen, "Laser Processing of Sub-Wavelength Structures on Sapphire and Alu- mina for Millimeter Wavelength Broadband Anti-Reflection Coatings", JLMN-Journal of Laser Micro/Nanoengineering Vol. 11, No. 2, 2016.
[11] Fabien Defrance, Cecile Jung-Kubiak, Jack Sayers, Jake Connors, Clare deYoung, Matthew I. Hollister, Hiroshige Yoshida, Goutam Chattopadhyay, Sunil R. Golwala, and Simon J. E. Radford, "1.6:1 bandwidth two-layer antireflection structure for silicon matched to the 190–310 GHz atmospheric window", Applied Optics Vol. 57, Issue 18, pp. 5196-5209 (2018).
[12] Patricio A. Gallardo, Brian J. Koopman, Nicholas F. Cothard, Sarah Marie M. Bruno, German Cortes-Medellin, Galen Marchetti, Kevin H. Miller, Brenna Mockler, Michael D. Niemack, Gordon Stacey, and Edward J. Wollack, "Deep reactive ion etched anti-reflection coatings for sub-millimeter silicon optics", Applied Optics Vol. 56, Issue 10, pp. 2796-2803 (2017).
[13] Ishino et al., "LiteBIRD: lite satellite for the study of B-mode polarization and inflation from cosmic microwave background radiation detection", Proceedings Volume 9904, Space Telescopes and Instrumentation 2016: Optical, Infrared, and Millimeter Wave; 99040X (2016); doi: 10.1117/12.2231995.
[14] Y. Sekimoto et al., "Concept design of the LiteBIRD satellite for CMB B-mode polarization", submitted to SPIE conference proceedings (2018).
[15] R. Bräuer et al., "Design of antireflection gratings with approximate and rigorous methods," Appl. Opt. 33, 7875–7882 (1994).
[16] DiffractMOD from RSoft/Synopsys.
[17] B. R. Johnson, "MAXIPOL: A bolometric, balloon-borne experiment for measuring the polarization anisotropy of the cosmic microwave background radiation", Ph.D. thesis, University of Minnesota.
[18] K. Komatsu et al., "Prototype design and evaluation of the 9 layer achromatic half-wave plate for LiteBIRD low frequency telescope", submitted to the SPIE proceedings 2018.
[19] R. W. Klopfenstein, "A transmission line taper of improved design", Proceedings of the IRE (Volume: 44, Issue: 1, Jan. 1956).
[20] E. B. Grann et al., "Optimal design for antireflective tapered two-dimensional subwavelength grating structures", Vol. 12, No. 2/February 1995/J. Opt. Soc. Am. A.

Instrumentally induced spurious polarization of a multi-layer half wave plate for a CMB polarization observation

Hiroaki Imada[1,2,*], Tomotake Matsumura[3], Ryota Takaku[4], Guillaume Patanchon[5], Hirokazu Ishino[6], Yuki Sakurai[3], Kunimoto Komatsu[6], Nobuhiko Katayama[3]

[1]*Institute of Space and Astronautical Science (ISAS), Japan Aerospace Exploration Agency (JAXA), 3-1-1 Yoshinodai, Chuo-ku, Sagamihara, Kanagawa, 252-5210, Japan*

[2]*Laboratoire de l'Accélérateur Linéaire, Université Paris-Sud, CNRS/IN2P3, Université Paris-Saclay, Université Paris-Sud, Bâtiment 200, 91898 Orsay, France*

[3]*Kavli Institute for the Physics and Mathematics of the Universe (Kavli IPMU), The University of Tokyo Institutes for Advanced Study (UTIAS), The University of Tokyo, Kashiwa, Japan*

[4] *Graduate school of Science, The University of Tokyo, Tokyo, Japan*

[5] *Paris Diderot University, Paris, France*

[6] *Okayama University, Okayama, Japan*

*Contact: imada@lal.in2p3.fr

***Abstract*—Recent cosmic microwave background (CMB) polarization experiments employ a rotating half wave plate (HWP) as a polarization modulator. In order to cover a wide range of a radio frequency, a multi-layer HWP has been studied. In this paper, we set up a numerical model of a 9-layer HWP based on the design for LiteBIRD and estimated the optical properties. Sub-wavelength structure for anti-reflection was also modeled. We used a commercial software based on Rigorous Coupled-Wave Analysis. We have confirmed that spurious polarization by a HWP does not appear for a normal incidence. On the other hand, the spurious polarization can be observed for an oblique incidence. When the incident angle is 10 degrees the magnitude of the spurious polarization is about 10^{-4}, which is comparable to the effect of the gravitational lensing B-mode. We have also confirmed that this design has the potential to achieve a highly efficient polarization modulator.**

I. Introduction

Many inflation theories predict that the evidence of inflation should appear in the cosmic microwave background (CMB) B-mode polarization pattern. Various CMB polarization experiments to search for the B-mode polarization pattern are proposed and some of them have started to observe CMB [1-7]. CMB B-mode polarization is so weak that it is hard to detect the signal. Therefore, recent CMB polarization experiments require three things: a field of view to be sufficiently wide to arrange many detectors on a focal plane, observing bands to be sufficiently broad to remove the effect of foreground sources, and a rotating half wave plate (HWP) to modulate the CMB polarization. If we modulate the signals with a rotating HWP, it is not necessary to take a difference between two detector outputs by adjusting different gains. In addition, a HWP can save the weak signals from $1/f$ noise by shifting them to the higher frequency.

A HWP accurately makes a phase retardation of π radians at a specific wavelength. When a HWP is employed to CMB experiments, the polarization properties of a HWP should be independent of wavelengths and incident angles. An achromatic HWP can be achieved by combining several HWPs but the polarization properties of a HWP vary as a function of the incident angle [8-11]. In the point of view of a CMB polarization experiment, the leakage from unpolarized light to linear polarization, hereafter I-to-P leakage, can contaminate the CMB polarization map.

In this paper, a HWP is assumed to be one that is designed for the Low Frequency Telescope (LFT) on the LiteBIRD [1], a next generation CMB polarization satellite. The HWP for LiteBIRD will be composed of 9 sapphire plates [12] and equipped with a sub-wavelength structure for anti-reflection. We summarize the characterization of a HWP with the Mueller matrix from the point of view of a CMB experiment in Section II. In order to calculate all the Mueller matrix elements as a function of frequency and incident angle, we set up a model for the numerical simulation in Section III. Section IV shows the simulation results and an I-to-P leakage in the results can be found. Finally, we discuss the causes of the I-to-P leakage and whether or not the amount of I-to-P leakage can be acceptable for LiteBIRD.

II. Characterization of HWP

Mueller matrix [13] is one of useful tools to express the character of a polarizing optical element. When we introduce the angle of an optic axis, ρ, with respect to the direction along which a detector is sensitive to light polarized (Fig. 1), an ideal HWP can be expressed as

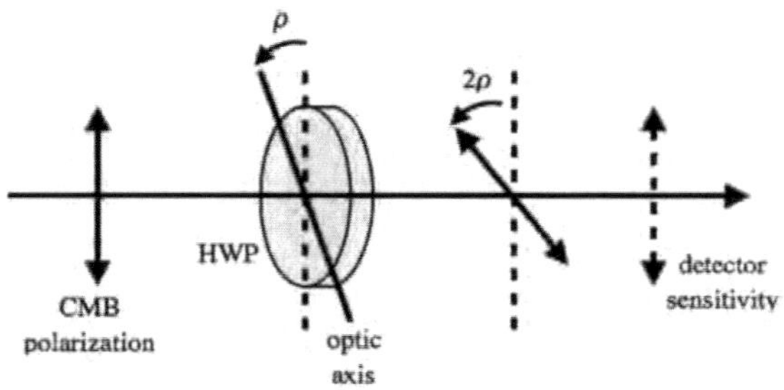

Fig. 1. A diagram showing then incident CMB polarization, HWP optic axis, and detector sensitivity.

If we employ a polarized detector sensitive to the Stokes parameter $Q > 0$, i.e. linear polarization, the ideal HWP allows us to detect CMB polarization at a frequency of 4ρ as follows:

When we consider a more realistic case, each element of the Mueller matrix can have a non-zero component at a frequency of 4ρ due to the combination of fresnel reflection, oblique incidence, and so on. Therefore, it is assumed, in general, that the Mueller matrix should be written with the summation of Fourier harmonics of the HWP rotation angle of ρ as

where X, and Y represent I, Q, U, and V, Θ is an incident angel to a HWP, and $\triangleleft$ is an azimuthal angle. Using A_{XY}, C_{XY}, and $\varphi^{(4)}{}_{XY}$, We can define a modulation efficiency η, for example, as

Since modulated CMB signals appear at a frequency of 4ρ, it is useful to focus on what effects each coefficient C_{XY} provides on the output signals.

We can categorize C_{XY} into 4 groups. First group contains C_{QQ}, C_{QU}, C_{UQ}, and C_{UU}. They are unity for the ideal HWP case and determine the modulation efficiency η and polarization angle through $\varphi^{(4)}{}_{QQ}$, $\varphi^{(4)}{}_{QU}$, $\varphi^{(4)}{}_{UQ}$ and $\varphi^{(4)}{}_{UU}$ for a general HWP case. Second group contains C_{II}, C_{VI}, C_{VQ}, and C_{VU}. They cause the modulation efficiency to decrease because they convert light to circular polarization that is identified as a constant signal independent of the rotation angle ρ for a linear polarized detector. C_{IV}, C_{QV}, C_{UV}, and C_{VV} are sorted into third group. They can be ignored as long as CMB and astronomical sources have Stokes V equal to zero. The last group, which is most important in terms of spurious signal, has C_{IQ}, C_{IU}, C_{QI}, and C_{UI}. They cause spurious polarization from unpolarized light. They should be 10^{-4} or less from our study. When the magnitude of C_{IQ}, C_{IU}, C_{QI}, and C_{UI} are 10^{-4}, we estimate that the contamination from unpolarized light has an effect on the polarimetry of CMB signal, which is comparable to the gravitational lensing effect. Table 1 summarizes the four groups.

III. Simulation

The HWP for the LiteBIRD LFT will be made of 9 sapphire plates with each thickness of 3.14mm. The refractive indices for ordinary and extraordinary rays are assumed to be 3.047 and 3.361, respectively. No dielectric loss is assumed. The optic axes are arranged at 18.5, 37.5, 73.9, 141.5, 73.9, 37.5, 18.5, 22.7 degrees with respect to the first plate. The first and ninth plates are equipped with sub-wavelength structure for antireflection (Fig. 2). The pitch is 300 μm, height 2000 μm, tip 20 μm, and bottom 20 μm.

The polarization state of an incident plane wave for this simulation is assumed to be linear. The frequencies for simulation are 119 GHz, 161 GHz, and every 5 GHz from 125 GHz to 155 GHz. This frequency range corresponds to the 140 GHz band for LiteBIRD. The incident angles to a HWP are 0 and 10 degrees, which corresponds to the maximum angle of LiteBIRD LFT optical design. Note that the normal incident case was calculated only at 140 GHz.

In order to simulate our HWP model including the sub-wavelength structure, we used DiffractMOD [14] based on Rigorous Coupled-Wave Analysis (RCWA) [15]. Since RCWA is an algorism dedicated to a periodic structure, RCWA is suitable for calculating the electromagnetic field of the complicated but periodic structure as shown in Fig. 2.

The computer resources we used for this analysis were a CPU operating at 2.6 GHz with 14 threads, RAM with about 110 GB. It took 30 hours for a Mueller matrix at a specific frequency for a normal incidence and 60 hours for an oblique incidence.

TABLE I
CATEGORIZING C_{XY}.

group	Coefficients	Effect on Polarimetry
1	$C_{QQ}, C_{QU}, C_{UQ}, C_{UU}$	modulation efficiency and polarization angle
2	$C_{II}, C_{VI}, C_{VQ}, C_{VU}$	reducing modulation efficiency
3	$C_{IV}, C_{QV}, C_{UV}, C_{VV}$	ignorable
4	$C_{IQ}, C_{QI}, C_{IU}, C_{UI}$	spurious polarization from unpolarized light

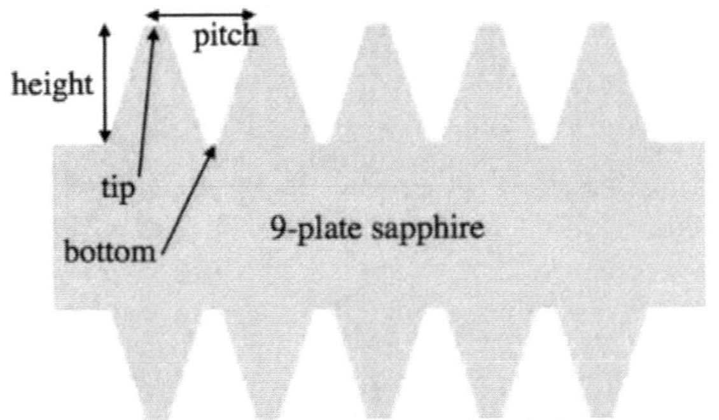

Fig. 2. HWP model with sub-wavelength structure for antireflection.

IV. Results and Discussions

This section shows the simulation results of the normal incident case at 140 GHz, the results of the oblique incident case at 140 GHz, and, in addition, the results of the oblique incident case as a function of frequency.

A. Normal incident simulation at 140 GHz

Fig. 3 shows each element of Mueller matrix as a function of the HWP rotation angle ρ for the normal incidence at 140 GHz. The 4-ρ terms can be found only in M_{QQ}, M_{QU}, M_{UQ}, and M_{UU} (cf. group 1 in Table 1). Therefore, spurious polarization does not appear in this case. M_{IQ}, M_{IU}, M_{QI}, and M_{UI} behave as a 2-ρ term with an amplitude of about 10^{-3}. This is because Fresnel reflection differ between two polarization states. The 2-ρ terms in M_{QV}, M_{IU}, M_{QI}, and M_{UI} comes from the difference of retardation between two polarization states from π radians. The 2-ρ terms in M_{IV} and M_{VI} also attribute to the retardation different from π radians as well as the difference of Fresnel reflection between two polarization states at the same time.

B. Oblique incident simulation at 140 GHz

When an incident wave enters obliquely to a HWP, the refractive index for an extraordinary ray depends on the angle between the optic axis and the normal to the incident plane, i.e. ρ - $\triangleleft$. It is given by ([16])

where n_{inc} is the refractive index in the incident space. The refractive index $n'_e(\Theta, \rho\text{-}\triangleleft)$ itself has 4-ρ dependence, which causes Fresnel reflection, refraction angle, and optical path lengths to have additional 2-ρ or 4-ρ dependence.

Fig. 4 shows the Mueller matrix elements as a function of the HWP rotation angle ρ for the 10-deg. incident case at 140 GHz. We can find the features of a 4-ρ element in the other elements as well as M_{QQ}, M_{QU}, M_{UQ}, and M_{UU}. The elements M_{IQ}, M_{IU}, M_{QI}, and M_{UI} have 4-ρ terms and higher order terms, which induces spurious polarization. They come from the combination of the difference of Fresnel reflection between two polarization states and ρ-dependence of the index $n'_e(\Theta, \rho\text{-}\triangleleft)$. Their amplitudes are about 10^{-3}. The other elements, M_{QV}, M_{UV}, M_{VQ}, M_{VU}, M_{IV}, and M_{VI}, also have 4-ρ and higher terms due to the difference of retardation from π radians and the ρ-dependence of refraction angle in addition to the ρ-dependence of Fresnel reflection and the index $n'_e(\Theta, \rho\text{-}\triangleleft)$. Their amplitudes are from 10^{-4} to 10^{-2}.

C. Coefficients A_{XY}, B_{XY}, and C_{XY}

Figs. 5-7 show the amplitudes of each harmonics for 10-deg. incident case. Since Figs. 5-7 are given as a function of frequency, they include the 140-GHz case shown in the previous subsection.

Using on A_{XY}, C_{XQ}, and $\varphi^{(4)}{}_{XQ}$, we obtain the modulation efficiency of 99.98% at 140 GHz. Fig. 8 shows the modulation efficiency η as a function of frequency. This result indicates that we can achieve an extremely high modulation efficiency of more than 99% with this multi-layer HWP design.

In terms of polarimetry, we should focus on coefficients C_{XY}, especially C_{IQ}, C_{IU}, C_{QI}, and C_{UI} which are closely related to spurious polarization. We can find their amplitudes are about 10^{-4}. Comparing C_{IQ} and C_{IU}, the phases are different by about $\pi/2$ radians. The effect of this magnitude of 10^{-4} is estimated to be comparable to that of the B-mode signal by the gravitational lensing effect. The spurious signal level is not harmful to the systematic error of measuring the CMB polarization. However, further study is needed because detectors has a sensitivity over a wide frequency range and we have to calculate a band-averaged properties such as a modulation efficiency and spurious polarization.

V. Conclusions

Many CMB polarization experiments start to employ a HWP and some experiments have already observed CMB polarization with a HWP. LiteBIRD, a next generation CMB polarization satellite, also has a plan to employ a HWP as a polarization modulator.

We have characterized the properties of a HWP by using a Mueller matrix and decomposing each element of the Mueller matrix into Fourier harmonics. In terms of polarimetry, the harmonics we should focus on is the 4 times higher frequency term than the HWP rotation frequency, 4-ρ term.

We have carried out the numerical simulation to estimate a Mueller matrix of the HWP which is based on the design for the LiteBIRD LFT. The HWP model is composed of 9 sapphire plate and sub-wavelength structure for antireflection on the first and last plate. We used DiffractMOD, commercial software, based on RCWA due to the sub-wavelength structure. The incident wave was assumed to be linear polarized plane wave. The incident angle is 0 degrees for 140 GHz and 10 degrees for 119 GHz to 161 GHz, which corresponds to the LiteBIRD 140 GHz band.

We have confirmed that there is no additional 4-ρ term for the normal incident case. On the other hand, for 10-degree incident

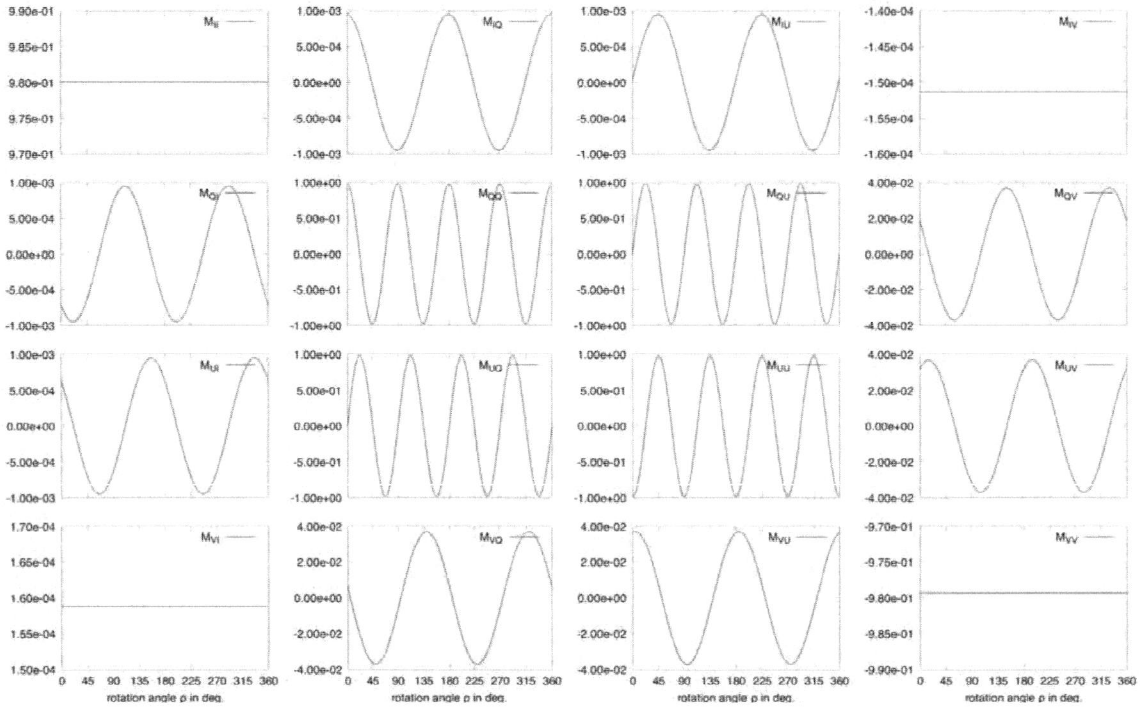

Fig. 3. Mueller matrix as a function of the rotation angle of a HWP, ρ. The results are obtained for the normal incidence at 140 GHz

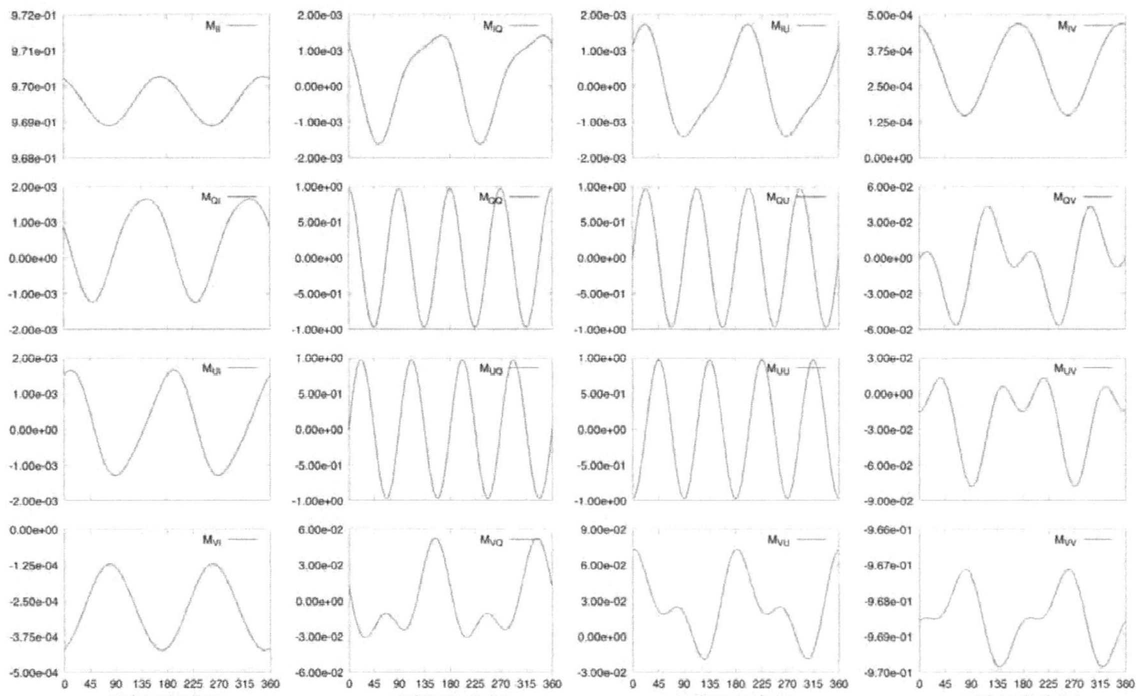

Fig. 4. Mueller matrix as a function of the rotation angle of a HWP, ρ. The results are obtained for the 10-deg. incidence at 140 GHz.

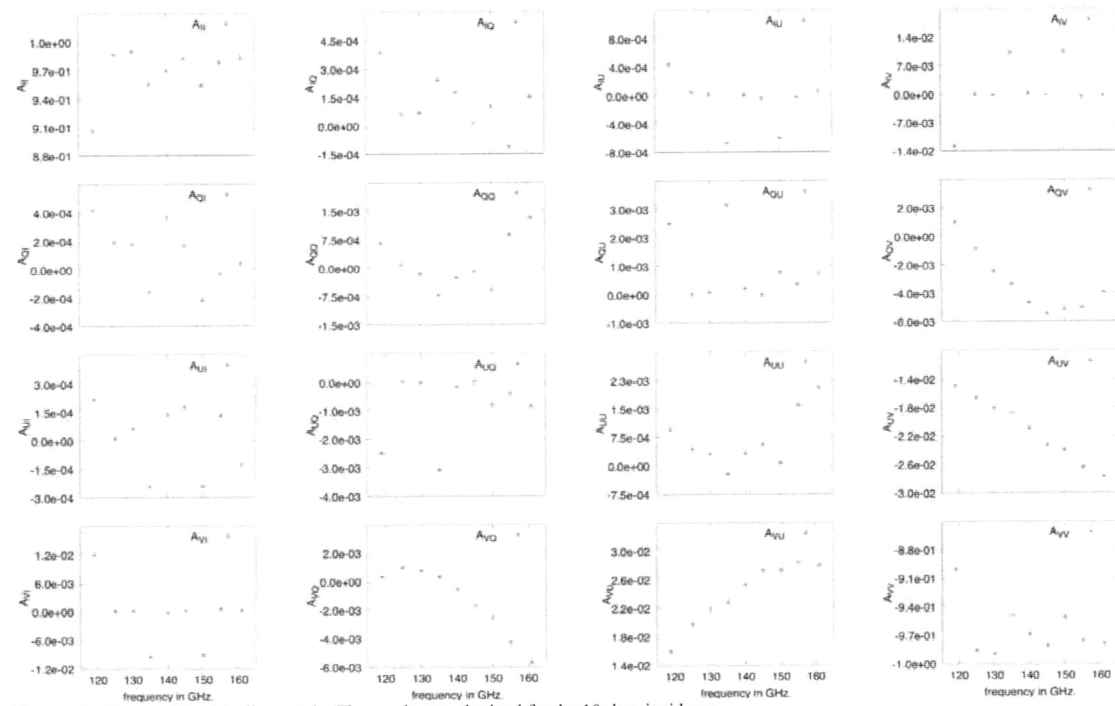

Fig. 5. The constant terms of the Mueller matrix. The results are obtained for the 10-deg. incidence.

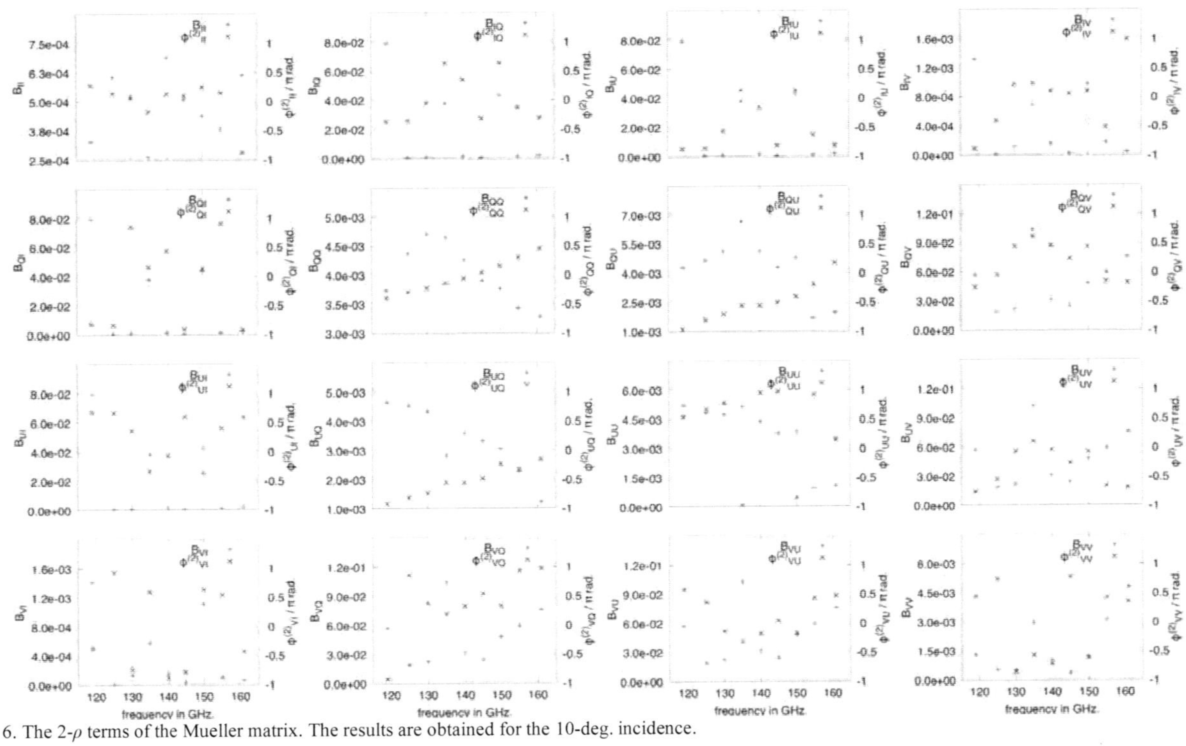

Fig. 6. The 2-ρ terms of the Mueller matrix. The results are obtained for the 10-deg. incidence.

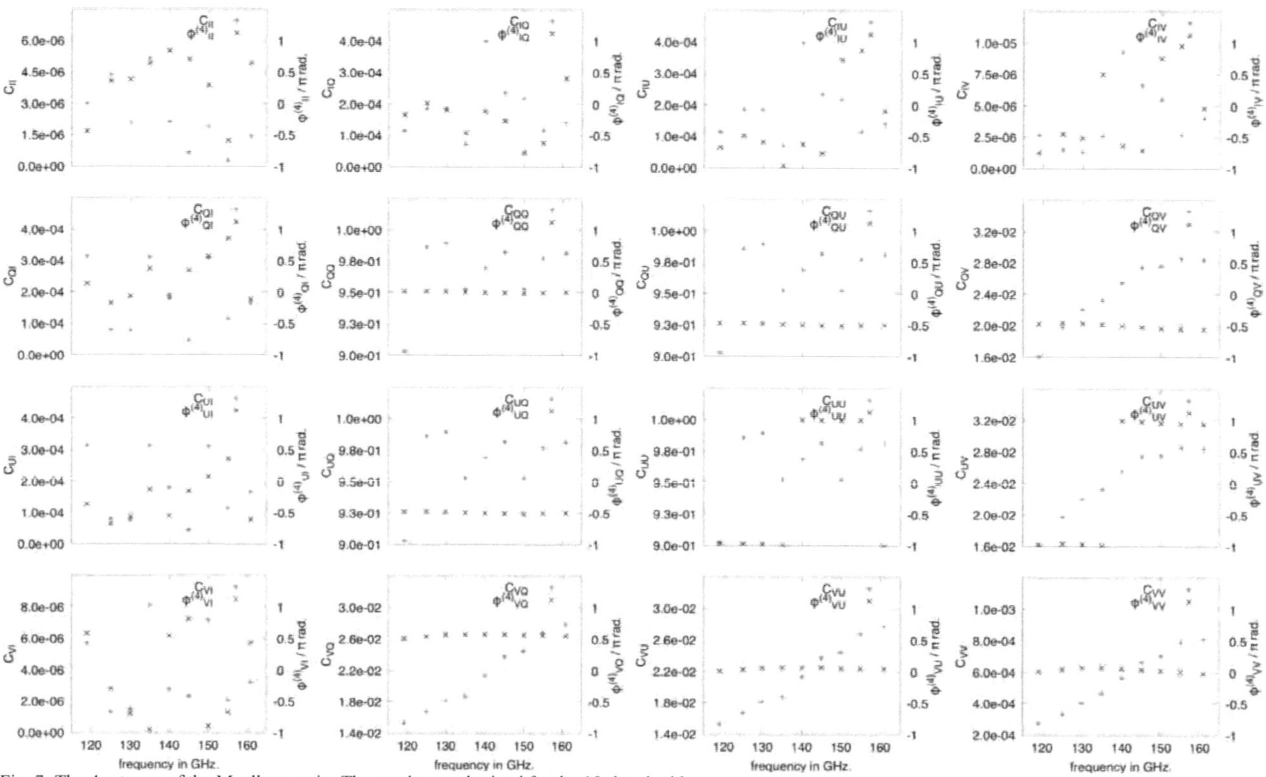

Fig. 7. The 4-ρ terms of the Mueller matrix. The results are obtained for the 10-deg. incidence.

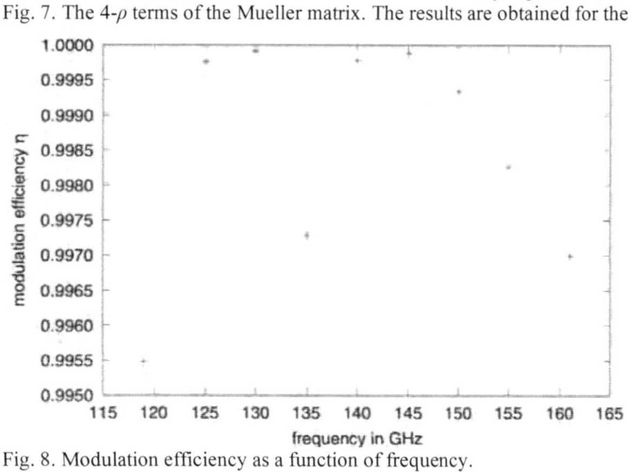

Fig. 8. Modulation efficiency as a function of frequency.

case, 4-ρ term can be observed, which causes spurious polarization. The spurious polarization attributes to the ρ-dependence of the refractive index of an extraordinary ray. The magnitude of 4-ρ terms relating to the spurious polarization is about 10^{-4}. The effect of this spurious polarization is estimated to be comparable to that of the gravitational lensing B-mode polarization. It is not harmful but further study is needed. We have also confirmed the modulation efficiency is high enough to observe CMB polarization. This design has the potential to achieve a highly efficient polarization modulator.

Acknowledgment

This work was supported by JSPS KAKENHI Grant Number 15H05891 and 15H05441.

References

[1] H. Ishino, Y. Akiba, K. Arnold, *et al.*, "LiteBIRD: Lite satellite for the study of B-mode polarization and Inflation from cosmic microwave background Radiation Detection," in *Proceedings of SPIE Astronomical Telescopes + Instrumentation*, 99040X, 2016.

[2] J. Delabrouille, et al. "Exploring Cosmic Origins with CORE: Survey requirements and mission design." *Journal of Cosmology and Astroparticle Physics*, 2018.04:014, 2018.

[3] P. A. R. Ade, Z. Ahmed, R. W. Aikin, *et al.*, "Improved constraints on cosmology and foregrounds from BICEP2 and Keck array cosmic microwave background data with inclusion of 95 GHz band," *Phys. Rev. Lett.*, 116, 031302, 2016.

[4] P. A. R. Ade, *et al.*, "A Measurement of the Cosmic Microwave Background B-mode Polarization Power Spectrum at Subdegree Scales from Two Years of POLARBEAR Data," *The Astrophysical Journal*, 848.2:121, 2017.

[5] T. Louis, *et al.*, "The Atacama Cosmology Telescope: two-season ACTPol spectra and parameters," *Journal of Cosmology and Astroparticle Physics* 2017.06:031, 2017.

[6] A. Kusaka, *et al.*, "Results from the Atacama B-mode Search (ABS) Experiment," *arXiv preprint arXiv*:1801.01218, 2018.

[7] G. Simard, *et al.,* "Constraints on Cosmological Parameters from the Angular Power Spectrum of a Combined 2500 deg^2 SPT-SZ and Planck Gravitational Lensing Map," *The Astrophysical Journal*, 860.2:137, 2018.

[8] S. Pancharatnam, "Achromatic combinations of birefringent plates Part I. An Achromatic Circular polarizer," in *Proceedings of the Indian Academy of Sciences*, vol. XLI, no. 4, sec. A, pp. 130-136, 1955.

[9] C. J. Koester, "Achromatic Combinations of Half-Wave Plates," *JOSA*, 49, 4, pp. 405-409, 1959.

[10] G. Pisano, *et al.*, "Achromatic half-wave plate for submillimeter instruments in cosmic microwave background astronomy: experimental characterization," *Applied Optics*, 45, 27, pp. 6982-6989, 2006.

[11] T. Matsumura, *et al.*, "Performance of three- and five-stack achromatic half-wave plates at millimeter wavelengths," *Applied Optics*, 48, 19, pp. 3614-3625, 2009.

[12] K. Komatsu, *et al.*, "Prototype design and evaluation of the nine-layer achromatic half-wave plate for the LiteBIRD low frequency telescope," in *Proceedings of SPIE Astronomical Telescopes + Instrumentation*, in press., 2018.

[13] E. Hecht, *Optics Fifth edition*, Edinburgh: Pearson Education Limited, 2017.

[14] *RSoft DiffractMOD v2017.03 User Guide*, Synopsys, Inc., 2017.

[15] M. G. Moharam, T. K. Gaylord, "Rigorous coupled-wave analysis of metallic surface-relief gratings," *J. Opt. Soc. Am. A* 3, 1780, 1986.

[16] S. Orfanidis, *Electromagnetic Waves and Antennas*, S. Orfanidis, 2016.

Optical Design of the Submillimeter Wave Instrument on JUICE

Mikko Kotiranta[1,*], Axel Murk[1], Karl Jacob[1], Hyunjoo Kim[1], and Paul Hartogh[2]
[1]Institute of Applied Physics, University of Bern, 3012 Bern, Switzerland
[2]Max Planck Institute for Solar System Research, 37077 Göttingen, Germany
*Contact: mikko.kotiranta@iap.unibe.ch

***Abstract*— The upcoming Submillimeter Wave Instrument on the JUICE spacecraft is a passive dual-beam heterodyne radiometer operating in the frequency bands 530 – 625 GHz and 1080 – 1275 GHz. The instrument will observe Jupiter's atmosphere as well as the atmospheres and surface properties of its moons. This work presents the optical design and analysis of the instrument that has been carried out using Gaussian beam mode analysis and physical optics simulations. The optics consists of a mechanically steerable off-axis Cassegrain telescope, relay optics and two feedhorns. Frequency independent operation of the 600-GHz channel is predicted by the simulations. The 1200-GHz channel shows some frequency dependency because of the selected feedhorn type. The scanning of the telescope affects mainly its cross-polarization level.**

I. Introduction

The JUpiter ICy moons Explorer (JUICE) is an ESA large-class mission to Jupiter and its satellites. The launch is scheduled for 2022. The spacecraft carries 10 instruments on-board, one of them being the Submillimeter Wave Instrument (SWI), which is a passive dual-beam heterodyne radiometer for the frequency bands 530 – 625 GHz and 1080 – 1275 GHz [1]. The telescope and receiver unit (TRU) of the instrument is depicted in Fig. 1. Additionally, the instrument consists of an electronic unit with different high resolution and broadband spectrometers, and a radiator unit which provides passive cooling down to a temperature of 120 K for the Schottky diode mixers and low-noise amplifiers of the receiver unit.

The scientific objectives of SWI include the investigation of the temperature distribution, chemical composition and dynamics of Jupiter's stratosphere, as well its coupling to other atmospheric regions. The atmospheres, their interaction with the Jovian magnetosphere, and the surface properties of the icy moons Callisto, Europa, and Ganymede will be characterized as well. SWI will therefore provide information that is complementary to the data provided by the Microwave Radiometer (MWR) on the Juno spacecraft [2] that is currently performing observations of the lower parts of Jupiter's atmosphere in the frequency band 0.6 – 22 GHz.

The optics of the SWI has been first discussed in [3]. Notable design changes have since taken place, the most important of them being the inclusion of a 1200-GHz receiver channel. Other changes initiated by refined structural requirements include adjustments of the optical element sizes and positions. This paper presents the current optical design and analysis that is based on Gaussian beam mode analysis [4] and physical optics (PO) [5] simulations.

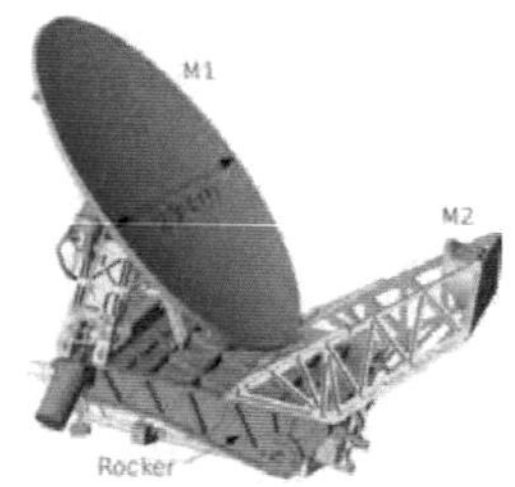

Fig. 1. CAD-model of the Submillimeter Wave Instrument.

II. Optical Design

SWI receives the signal of interest with an off-axis Cassegrain telescope. Its paraboloidal main mirror (M1) has a projected diameter of 29 cm while the hyperboloidal secondary mirror (M2) has a projected diameter of 6 cm. The telescope is mounted on a rocker that can be tilted ±4.3° to allow scanning perpendicular to the spacecraft orbital plane (cross-track). Scanning in the orbital plane (along-track) is possible by rotating the mirror M1 up to ±72°. The scanning mechanisms enable observations along Jupiter's limb, mapping of the icy moons with variable incidence angles and polarizations, as well as the use of cold sky for instrument calibration without the need for spacecraft maneuvers.

The beam from the mirror M2 is directed into the receiver unit (RU) box through a cutout in the rocker. The contents of the RU box are shown in Fig. 2. After entering the RU box, the beam meets a planar mirror M3 and an elliptical mirror M4 which reside on the rotational axis of the rocker structure. As the telescope is steered in the cross-track direction, the mirror M3 rotates together with the rocker and the telescope in such a way that the remaining beam path inside the RU box remains fixed. A free-standing wire grid (WG) splits the beam into two perpendicular polarizations. The transmitted polarization component is reflected from the elliptical mirror M5T to the smooth-walled spline-profiled feedhorn of the 1200-GHz double sideband (DSB) Schottky-mixer receiver [6]. The reflected polarization component is coupled via the elliptical

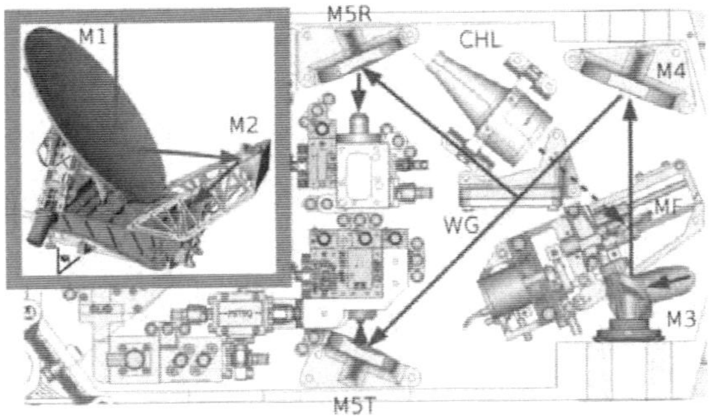

Fig. 2. Path of the optical beam in the receiver unit box (view from below, components attached to the RU box top cover).

mirror M5R to the corrugated feedhorn of the 600-GHz DSB Schottky-mixer receiver [7].

A conical blackbody calibration target (CHL) [8] acts as the hot temperature reference during the calibration. To allow the receivers to view the CHL, the beam between M3 and M4 is redirected by activating a planar flip mirror (MF).

A. Gaussian Beam Mode Analysis

Frequency independent operation has been one of the main design drivers which is achieved by imaging the aperture of the feedhorn onto the aperture of the primary mirror M1. In terms of Gaussian beam mode analysis, this means that the Gaussian beam phase shift between the feedhorn and M1 should be a multiple of π. Another constraint for the design is the edge taper requirement of at least 17 dB at M1 which together with the mirror diameter defines the maximum Gaussian beam radius.

The Gaussian beam best fitting to the simulated radiation pattern of the 600-GHz channel feedhorn has a beam radius of 0.851 mm and a radius of curvature of 8.35 mm at the feed aperture. The respective values for the 1200-GHz channel are 0.478-0.468 mm and 2.482-4.992 mm, depending on the frequency.

Table I presents the properties of the optical components and the results of the Gaussian beam mode analysis. Here, $f = (1/R_{in} + 1/R_{out})^{-1}$ is the effective focal length, d the distance to the previous component, R_{in} and R_{out} the radii of curvature of the conic mirror sections, w_i the incident beam radius at 530 GHz (row 1) and 1080 GHz (row 2), and θ_i the beam incidence angle. In case of M5R (M5T), d is the distance to the 600-GHz (1200-GHz) feedhorn aperture.

TABLE I
RESULTS OF GAUSSIAN BEAM MODE ANALYSIS

Comp.	f [mm]	d [mm]	R_{in} [mm]	R_{out} [mm]	w_i [mm]	$2\theta_i$ [deg]
M1	348.937	300.0	348.937	inf	98.848 98.872	90
M2	-64.0	241.0	209.179	-49.006	13.874 13.869	53.646
M3	inf	89.414	inf	inf	3.758 2.663	80.839
M4	59.714	72.186	105.854	136.998	9.081 8.397	45
WG	inf	92.814	inf	inf	3.108 3.108	90
M5R M5T	19.759 12.630	25.0 14.681	27.372 16.067	71.048 59.049	6.287 4.276	45 45

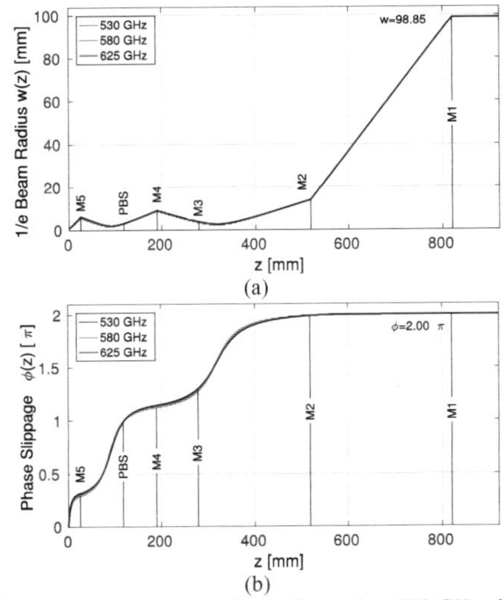

Fig. 3. Gaussian beam propagation along the 600-GHz beam path: a) radius, b) phase shift.

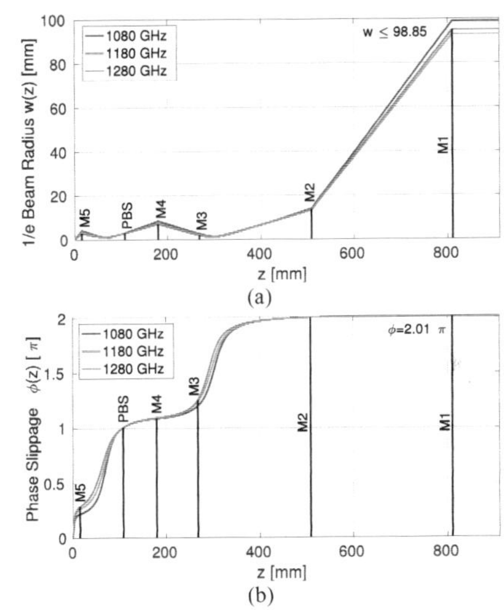

Fig. 4. Gaussian beam propagation along the 1200-GHz beam path: a) radius, b) phase shift.

Fig. 3 and Fig. 4 show the Gaussian beam radius and phase shift from the feedhorn aperture to the primary mirror M1 for the 600-GHz and 1200-GHz channels, respectively. The edge taper at M1 is approximately 18.7 dB which corresponds to a spillover loss of 1.4 %. The beam propagation in the 600-GHz channel is frequency independent, whereas the 1200-GHz channel exhibits some frequency dependency due to the frequency dependent input beam radius and radius of curvature.

B. Physical Optics Analysis

The Gaussian beam mode analysis does not include higher order modes or diffraction. The contribution of these effects has been accounted for in the PO analysis. Instead of the best-fit

Gaussian beams, the actual simulated far-field radiation patterns of the feedhorns have been used as an input.

First, the sizes of the optical components inside the RU box have been optimized for minimum spillover loss under mass and size constraints. The final projected radii are [R_{M3}, R_{M4}, R_{WG}, R_{M5R}, R_{M5T}] = [10, 19, 10, 16, 12] mm. Then the far-field radiation characteristics of the instrument have been simulated for the edge and center frequencies of each receiver channel. These are shown in Fig. 5 when the instrument points at the nadir, meaning that the telescope scan angles in the cross-track direction ϕ and in the along-track direction θ are zero. The key performance parameters obtained from the PO simulations are summarized in Table II.

The accumulated coupling efficiency at the mirror M1, η_{M1}, is > 98 %. Such a value is expected when one considers the M1 spillover loss of 1.4 % predicted by the Gaussian beam mode analysis. Losses due to the finite electrical conductivity of the optical surfaces are not included in η_{M1}. The maximum sidelobe and cross-polarization levels are below -20 dB. The latter is dominated by the 90° bend angle at M1 which is inherent to the instrument scanning concept. The absolute pointing errors of the 600-GHz and 1200-GHz beams in cross- and along-track directions are $\leq$ 0.16 arcmin. The requirement set by the science goals is less than 0.5 arcmin, but this includes the error contribution from the motors of the scanning mechanism and the alignment tolerance of optical components as well. The boresight direction of the 600-GHz and 1200-GHz beams on the sky agree to within 0.25 arcmin, the requirement being better than 1 arcmin.

The effect of telescope scanning in the cross-track direction is negligible, whereas scanning in the along-track direction causes the maximum cross-polarization level to increase: at the extremes of +72° and -72° the degradation is 1.2 dB and 2.5 dB, respectively. The pointing offsets, and consequently the coalignment of the beams, change as well, but the absolute errors remain approximately the same as for the nadir pointing.

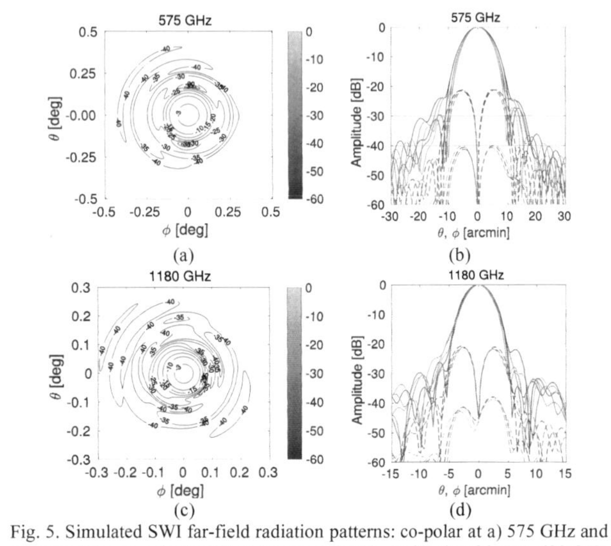

Fig. 5. Simulated SWI far-field radiation patterns: co-polar at a) 575 GHz and c) 1180 GHz; co-polar (solid) and cross-polar (dashed) at b) 530, 575, 620 GHz and d) 1080, 1180, 1280 GHz. The white lines in a) and c) indicate the polarization direction. The patterns in the planes $\theta = 0$ and $\phi = 0$ are included in b) and d).

TABLE II
RESULTS OF PHYSICAL OPTICS ANALYSIS

Freq. [GHz]	530	575	620	1080	1180	1280
Gain [dB]	62.63	63.41	64.07	68.94	69.21	69.56
η_{M1} [%]	98.14	98.34	98.61	98.02	98.03	98.07
FWHM [arcmin]	8.57	7.84	7.28	4.12	3.96	3.72
Max. cross-pol. level [dB]	-21.19	-21.11	-21.11	-20.80	-21.23	-21.30
Offset $\Delta\phi$ [arcmin]	-0.05	-0.04	-0.03	0.15	0.01	0.16
Offset $\Delta\theta$ [arcmin]	0.03	0.01	0.01	-0.11	-0.04	-0.10

CONCLUSIONS

The optical design and analysis of SWI on the JUICE spacecraft has been presented. The optical configuration is frequency independent in case of the 600-GHz receiver channel, whereas some frequency dependency is present in the 1200-GHz channel due to the frequency dependent properties of the beam launched by the 1200-GHz feedhorn. The pointing error due to the optics alone is the well below the specification of 0.5 arcmin. However, the contributions from the telescope scanning mechanism and the manufacturing and mounting tolerances of the optical components will be added to the error budget as well. Steering the telescope beam in the cross-track direction does not influence the optical performance, but along-track scanning increases the cross-polarization level up to 2.5 dB. Fortunately, the maximum degradation occurs at extreme scan angles that are only used for cold sky calibration that is not affected by the cross-polarization level.

ACKNOWLEDGEMENTS

The work at the University of Bern has been funded by the Swiss National Science Foundation under the Grant No. 200020-165744 and the ESA PRODEX program.

REFERENCES

[1] P. Hartogh *et al.*, "The submillimetre wave instrument on JUICE", in *Proc. EPSC 2013*, 2013, paper EPSC2013-710.

[2] M. A. Janssen *et al.*, "MWR: Microwave Radiometer for the Juno Mission to Jupiter," *Space Sci Rev,* vol. 213, no. 1-4, pp. 139-185, Nov. 2017.

[3] H. Kim, A. Murk, P. Hartogh, and SWI Team, "Optical Design Study for Submillimeter Wave Instrument for the Jupiter Mission," in *Proc. ISSTT 2015*, 2015, paper P-20.

[4] P. F. Goldsmith, *Quasioptical Systems*. Piscataway, NJ: IEEE Press, 1998.

[5] TICRA. GRASP 10.5.0. [Online]. Available: http://www.ticra.com/software/grasp

[6] A. Maestrini *et al.*, "1080-1280GHz Schottky Receiver for JUICE-SWI with 1600-2600K DSB Receiver Noise Temperature," in *Proc. ISSTT 2018*, 2018, paper M2.1.

[7] K. Jacob *et al.*, "Characterization of the 530 GHz to 625 GHz SWI Receiver Unit for the Jupiter Mission JUICE," in *Proc. 36th ESA Antenna Workshop,* 2015.

[8] K. Jacob, A. Schröder, M. Kotiranta, and A. Murk, "Design of the calibration target for SWI on JUICE," in *Proc. IRMMW-THz 2016*, 2016.

Session T1: **Invited II**

Future prospects for the high-resolution space far-infrared spectroscopy

Dariusz C. Lis*

LERMA, Observatoire de Paris, PSL Research University, CNRS, Sorbonne Universités, UPMC Univ. Paris 06, F-75014, Paris, France

*Contact: darek.lis@obspm.fr

Abstract—The far-infrared spectral region enables studies of the key atomic and molecular cooling lines, revealing the nature of the cool, dense universe, hidden from view at shorter wavelengths. High-resolution spectroscopy provides the velocity information needed to understand fully the 3D motions of the gas and, in solar system objects, it allows the retrieval of the vertical abundance and temperature profiles, as well as measurements of the wind patterns. I summarize the approved and planned European spectroscopic missions for astrophysics, planetary, and Earth science, focusing on the unique aspects of the FIR.

Session T2: **SMM Instruments**

A 95 GHz FMCW thermal-noise-limited radar: sensitivity and range-Doppler measurements

Raquel R. Monje*, Ken B. Cooper, Robert J. Dengler, Corey J. Cochrane, Stephen L. Durden, Adrian Tang, and Mathieu Choukroun

[1]*Jet Propulsion Laboratory, California Institute of Technology, 4800 Oak Grove Dr., Pasadena, CA, USA 91109-8099*

*Contact: rmonje@jpl.nasa.gov

***Abstract*—** We present measurement results of a compact monostatic 95 GHz frequency-modulated continuous-wave (FMCW) radar that demonstrate state-of-the-art radar sensitivity limited only by the receiver thermal noise, for a watt-level transmit power and a 540 K receiver system noise temperature. Thermal noise limited radar performance is achieved by a combination of high transmit/receive (T/R) isolation and an RF circuit architecture with a large degree of phase noise cancelation. High T/R isolation better than 85 dB is achieved combining high-directivity and low-sidelobes T/R horns, with a high efficiency quasioptical transmit-receive duplexer system.

We also present outdoor radar measurements using a variety of targets at different range and velocities, such as rain, freeway cars, clouds, irrigation sprinklers and hillsides, to test the radar range and Doppler capabilities. For a given radar bandwidth and timing operating mode the measurements demonstrate a range and velocity resolution better than 10 m and 0.1 m/s, respectively, and a maximum unambiguous range of 5 km and velocity limit of 46 m/s. The maximum unambiguous velocity limit is tested with the detection of cars in the freeway moving at velocities greater than 30 m/s. Measurements of rain, cars and hillsides verify the maximum unambiguous range, with targets detected in the 4-5 km range frame. Rain measurements at such fine velocity and range resolution allow for the dynamics and potentially the formation of different raindrop populations to be distinguished at different ranges. Detection also of a non-precipitating cloud and its velocity structure, stress the radar's sensitivity and high velocity resolution capabilities.

The radar sensitivity and range-Doppler measurement results present the first demonstration of a thermal-noise-limited FMCW radar with high (1 Watt) transmit power and long-range Doppler detection capabilities. These experimental results together with the compact design, with a single 15 cm diameter primary antenna, makes this 95 GHz FMCW radar uniquely suitable for future space missions for Earth and Planetary Science applications.

A CMOS-Based 90 – 105 GHz Pulsed-Echo Spectrometer: New Approaches for Highly-Mobile and Low-Power *in situ* Chemical Detections

Deacon J. Nemchick[*1], Brian J. Drouin[1], Adrian J. Tang[2], Yanghyo Kim[2], Gabriel Virbila[3] and Mau-Chung Frank Chang[3]

[1]*Laboratory Studies and Atmospheric Observation, Jet Propulsion Laboratory, Pasadena, CA 91109, USA*
[2]*Submillimeter Wave Advanced Technology, Jet Propulsion Laboratory, Pasadena, CA 91109, USA*
[3]*Department of Electrical Engineering, University of California at Los Angeles, Los Angeles, CA 90025 USA*
[*]Contact: Deacon.J.Nemchick@jpl.nasa.gov

Abstract—Cavity enhanced pure rotational spectroscopy has long been a potent laboratory tool for the elucidation of structure and dynamics in isolated molecular systems where sensitive pulsed-echo techniques are routinely performed up to frequencies as high ~50 GHz. Although the associated narrow linewidths (~800kHz), wide-bandwidth (often >10 GHz), and long optical path lengths have long been identified as a desirable combination for sensitive and specific gas sensing, the unaccommodating size and power requirements of traditional microwave optics/electronics are unsuitable for the stringent demands required for *in situ* deployment. Additionally, efforts to drive pulsed-echo techniques into millimeter and submillimeter wavelength regimes, where the size of optics can be reduced without suffering large diffraction losses, have failed largely due to inefficiencies of injecting radiation into the resonant optical cavity.

Recent pursuits at the Jet Propulsion Laboratory to realize compact, low-power devices capable of *in situ* chemical detections on extra-terrestrial objects have found success in calling upon novel transmitter and receiver sources built from CMOS architectures commonly employed in the high-speed communications industry. Combining custom low-power integrated circuit chipsets with novel antennae embedded optical devices allows for the realization of cavity based instruments where *all* source and detection electronics are hosted by a single 4in. x 4in. printed circuit board. The current talk will present a full system description of this miniaturized CMOS-based pulsed-echo rotational spectrometer, which has an operational bandwidth (90-105 GHz) that includes many astrophysically relevant target compounds, along with future plans to extend the bandwidth to other wavelength regions.

Differential absorption radar near the 183 GHz water absorption line for inside-cloud humidity profiling

Richard J. Roy*, Ken Cooper, Matthew Lebsock, Jose V. Siles, Luis Millán, Raquel Rodriguez Monje, Robert Dengler

Jet Propulsion Laboratory, California Institute of Technology, Pasadena, CA 91109
*Contact: richard.j.roy@jpl.nasa.gov

Abstract— Conventional passive remote sensing platforms have limited ability to perform high-resolution observations of vertical water vapor profiles in the presence of clouds, contributing to deficiencies in numerical weather and climate change prediction capabilities. The nascent Vapor/Ice Profiling Radar (VIPR) project at JPL aims to demonstrate a solution to this observational problem by performing airborne radar remote sensing measurements of humidity content with high vertical resolution inside boundary layer clouds. Based on the mature Differential Absorption Lidar (DIAL) method, the VIPR instrument utilizes a frequency-tunable G-band transmitter in a frequency-modulated continuous-wave (FMCW) radar configuration to obtain range-gated differential absorption measurements across the low-frequency (~170 GHz) flank of the 183 GHz water line. These measurements will be inverted using a novel retrieval algorithm to obtain in-cloud vertical humidity profiles. The FMCW radar design is based on previously demonstrated technology developed at JPL for science and security applications in the 95-700 GHz frequency range. The instrument design employs state-of-the-art frequency multipliers and low-noise amplifiers, as well as high transmit/receive isolation to achieve high signal-to-noise FMCW radar measurements. Preliminary ground-based differential absorption measurements which demonstrate the technique will be presented, including a discussion of the retrieved humidity error stemming from radar measurement uncertainty.

An Integrated G-Band 4-Channel Direct Detection Radiometer for the TROPICS Mission

Eric W. Bryerton* and Jeffrey L. Hesler

Virginia Diodes, Inc., Charlottesville, VA 22902, USA
*Contact: bryerton@vadiodes.com

Abstract—The TROPICS (Time Resolved Observations of Precipitation structure and storm Intensity with a Constellation of Smallsats) mission, led by Dr. William Blackwell at MIT, is a constellation of CubeSats intended to dramatically increase the temporal resolution of several key measurements required to better determine the intensity and evolution of tropical storms. One of the instruments on this platform is a G-band radiometer with three channels near the 183 GHz water vapor absorption line to measure water vapor profiles and one channel at 206 GHz for cloud ice measurements. Because the CubeSat is already a small platform with limited volume and available power, and because the same CubeSat will also contain instruments for 118 GHz and 90 GHz measurements, the G-band radiometer must be contained in a very small form factor and consume a small amount of power.

We developed a prototype G-band radiometer to meet these stringent requirement including three stages of RF low-noise amplifiers, a noise diode for calibration, 2 GHz wide waveguide bandpass filters for each of the four channels, and diode detectors for each of the four channels, along with all the necessary regulation and video amplification. The prototype integrated radiometer fits in a single 1.49" x 2.26" x 0.50" split-block housing. The prototype was designed, assembled, and measured. Design and measurements will be presented in the final proceedings and presented at the Symposium. This includes measurements of filter passbands, receiver noise, NET (noise-equivalent temperature), and noise source injection. Noise temperature and noise source injection are also characterized as a function of ambient temperature from -40 C to +40 C. At room temperature, the measured noise temperature ranges from 900-1200K for the four channels. The noise source injects between 200-300K for each of the channels through a 9dB coupler in front of the LNAs. While the noise temperature and noise source measurements meet expectations, excess 1/f noise is limiting the achievable NET at the 10ms required integration time. Follow-on studies are proceeding to identify and mitigate the sources of excess 1/f noise, including bias voltage regulators, video amplifiers, detector diodes, and the HEMT LNAs. The results of this study will also be presented.

Session T3: **Superconducting Heterodyne Detectors I**

MgB_2 Hot Electron Bolometers for Array Receivers

Daniel Cunnane*[1], Narendra Acharya[2], Wenura K. Withanage[2], Xiaoxing Xi[2], Jonathan Kawamura[1], and Boris S. Karasik[1]

[1]*Jet Propulsion Laboratory, Pasadena, CA 91109, USA*
[2]*Temple University, Philadelphia, PA 19122, USA*
*Contact: daniel.p.cunnane@jpl.nasa.gov

Abstract—MgB_2 Hot Electron Bolometers (HEBs) offer a larger IF bandwidth and higher operating temperature than NbN HEBs. Recent results from JPL and Chalmers University have shown IF bandwidths from 7-11 GHz and operation at 15-20 K without significant changes in sensitivity, which is currently within a factor of 3 compared with reports on NbN HEB mixers. Scientifically, the large bandwidth is more significant for high-resolution line spectroscopy across the galaxy as high frequencies (> 3 THz) where the radial velocity spread up to 600 km/s leads to a large Doppler broadening. The higher operating temperature of the mixer should allow for cryogen-free heterodyne instruments which will enable long duration space telescopes for high-resolution THz heterodyne spectroscopy in the future. The required local oscillator (LO) power is necessarily higher than that for the state-of-the-art NbN devices. This can be offset by making the MgB_2 device area of the order of 10^{-2} μm^2 while still maintaining the large enough normal resistance required for the impedance match with a micro-antenna. Operating at higher temperature will also reduce the LO power requirements. In our recent experiments, we operated some mixer devices using an LO power < 1 μW, with a noise temperature ≈ 2000 K from 0.6 THz to 4.3 THz. With more work on the material optimization, the sensitivity of the MgB_2 mixers is expected to improve.

A challenge for implementation of these MgB_2 devices in the future instruments is to develop a suitable array architecture in order to simplify LO distribution for multi-pixel instruments. The scientific community pushes for a 100+ pixel heterodyne receiver for efficient mapping of the sky. Here quasi-optical mixers become bulky and the small packaging enabled by a waveguide architecture is necessary.

In this work, we take the first step towards waveguide coupled MgB_2 HEBs by moving to devices fabricated on Si substrates. In the past, this was not possible because of the high pressure, high temperature process used to synthesize the high quality MgB_2 thin films, but recent advances by the group at Temple University have enabled deposition on high resistivity Si substrates. These films were used to fabricate antenna-coupled HEBs to show the devices still have the same performance as the devices made previously on SiC and Sapphire substrates. We will present the results from these devices and our plan to move forward towards waveguide-coupled devices made on SOI substrates.

Prospects with low noise and wide bandwidth MgB_2 HEB THz mixers

Sergey Cherednichenko*, Evgenii Novoselov, Usman Ul-Haq, Narendra Acharya

Terahertz and Millimetre Wave Laboratory, Department of Microtechnology and Nanoscience, Chalmers University of Technology, SE-412 96 Gothenburg, Sweden
*Contact: serguei@chalmers.se

Abstract—Research in low noise terahertz (THz) mixers is motivated by great interest from astronomical community. As THz wave sensors in heterodyne receiver, such mixers would allow for detection of faint emission of molecular and atomic lines of multiple species in interstellar medium, stars, comets, planetary atmospheres, etc. A frequency of 1THz has become so to say a border line between technologies. Below 1THz, either SIS mixers (close to quantum limit noise, wide IF bandwidth, 2-4K operation) or Schottky mixers (yet low noise, operation up to room temperature) are of prime interest for astronomical applications. Above 1THz, noise in both SIS and Schottky mixers increases rather significantly. On contrary, superconducting hot-electron bolometer (HEB) mixers show a low THz frequency variation of noise, which becomes much lower than for any other mixer technologies. Examples of successful HEB mixer applications are NbN HEB devices in Herschel Space Observatory, SOFIA, APEX, SAO Receiver Lab instruments, etc. Despite the progress, a major limitation for HEB mixers was a limited IF bandwidth (3-5GHz). This limitation originates from electron energy cooling rate. The key material properties for HEB mixers are: fast inelastic electron-phonon scattering, critical temperature at least above 7K, ability to deposit very thin films (<10nm).

Ten years ago, MgB_2 HEB mixers emerged with a promise of much wider IF bandwidth. Since then, both MBE and HPCDV deposition technologies for MgB_2 films have been explored. HPCVD allows for reliable deposition of films as thin as 5-8nm, with a Tc>28K. Such films have been obtained either by thinning down thicker films, or by direct deposition at a low deposition rate (our group). Using our in-house film and HEB fabrication processes we have obtained an unprecedented for HEBs 11GHz noise bandwidth and a noise temperature of 930K at 1.63THz LO frequency.

In this presentation we will discuss our studies MgB_2 HEB mixers of various thicknesses at different LO frequencies (from 700GHz to 4.7THz), and at temperatures from 4K to 30K. With a 5nm film, we obtain a 13GHz noise bandwidth. Optimizing HEBs for high temperature operation, we achieve low noise performance up to 30K operation. A concept of a compact (portable) cooling system for MgB_2 HEB mixers will also be presented. Finally, we will discuss our view on prospects of MgB_2 HEB mixers for astronomy and other application.

Development of THz Superconducting Mixers and Detectors for DATE5

Sheng-Cai Shi[1,2,*], Jing Li[1,2], Wen Zhang[1,2], Wei Miao[1,2], Zhen-Hui Lin[1,2,3], Jin-Ping Yang[1,2], Dong Liu[1,2], Wen-Ying Duan[1,2], Zheng Wang[1,2], Zheng Lou[1,2], Qing Shi[1,2,4], Zhi Li[1,2,4], Kang-Ming Zhou[1,2], Ming Yao[1,2,3], Jie Hu[1,2,3], Yun Ren[1,2], and Qi-Jun Yao[1,2]

[1]*Purple Mountain Observatory, Chinese Academy of Sciences, Nanjing, China*
[2]*Key Laboratory of Radio Astronomy, Chinese Academy of Sciences, Nanjing, China*
[3]*University of Chinese Academy of Sciences, Beijing 100049, China*
[4]*University of Science and Technology of China, Hefei 230026, China*
[*]Contact: scshi@pmo.ac.cn

Abstract—The terahertz (THz) and FIR band is a frequency regime to be fully explored in astronomy. However, water vapor renders the terrestrial atmosphere opaque to this band over nearly all of the Earth's surface. Dome A in Antarctic – the site for China's Antarctic Observatory, with an altitude of 4093 m and temperature below -80 Celsius degree in winter, offers the best possible access for ground-based astronomical observations in the THz and FIR band. China is planning to build a 5-m terahertz telescope (DATE5) there. We will briefly introduce the site survey results by an unmanned Fourier transform spectrometer (FTS) and the DATE5 telescope, and then present the latest development of superconducting SIS and HEB mixers for 0.85 THz and 1.4 THz, respectively, and large-format TES and MKIDs detectors for a superconducting imaging camera named TeSIA.

Preliminary design study of a 4×2 HEB array at 4.7 THz for GUSTO

J. R. Silva[1,2,*], R. Farinha[3], D. J. Hayton[1,4], W. Laauwen[1], B. Mirzaei[3], N. More[1], A. Young[5], C. Kulesa[5], C. Walker[5], J. R. Gao[1,3]

[1] SRON Netherlands Institute for Space Research, Groningen/Utrecht, the Netherlands
[2] Kapteyn Astronomical Institute, University of Groningen, 9747 AD, Groningen, The Netherlands
[3] Kavli Institute of NanoScience, Delft University of Technology, Delft, the Netherlands
[4] Jet Propulsion Laboratory, California Institute of Technology, CA 91109, USA
[5] Steward Observatory, 933 N Cherry Ave., Rm N204, University of Arizona, AZ 85721, USA
*Contact: j.r.g.d.silva@sron.nl

Abstract— Here we report on the design of the 4×2 HEB quasi-optical mixer array at 4.7 THz for GUSTO. Two studies are presented. The first is a statistical analysis of some of the key parameters of HEB devices within a single batch. In a population of 10 randomly selected devices we show a state of the art noise temperature of 720 K at 2.5 THz with only 3 % spread, while at the same time meeting LO uniformity requirement. The second study discusses the impact of different diameter elliptical lens on receiver performance. We conclude that 10 mm lens offers the best performance with the lowest risk to the project. At the same time we confirmed the value of 11.4 for the ξ_{Si} of silicon at 4.2 K. Furthermore, we describe the mixer array mechanical design resulting in a compact monolithic unit.

I. Introduction

The THz range of frequencies (0.1 to 10 THz) is known to be rich in astronomically important fine atomic/molecular lines. By using high resolution spectroscopic techniques it is possible to determine parameters such as density, temperature and velocity that can help unveil the dynamics and chemical processes that rule star forming regions. Nevertheless, only in the last decades, quantum and superconductive technology developments as well as local oscillator (LO) technology advances, have made it possible to shed some light on these regions. Currently, NbN HEBs are the most suitable mixers for high resolution spectroscopic terahertz astronomy at frequencies above 1 THz. High sensitivity and low LO power requirement make them unique at super terahertz, although the intermediate frequency (IF) bandwidth is still limited. So far, HEBs have been used in a wide range of astronomical observatories in order to observe different lines of terahertz radiation [1]–[3]. The use of multi-pixel receivers instead of a single pixel receiver improves the mapping speed of the telescope significantly. Thus, heterodyne arrays at THz frequencies are now demanded for airborne (SOFIA), balloon borne (STO-2, GUSTO) or possible future satellite (Origin space telescope, OST) THz observatories.

A. GUSTO Science

GUSTO, the Galactic/Extragalactic ULDB Spectroscopic Terahertz Observatory, follows up on the STO-2 mission's successful flight which demonstrated the feasibility of a balloon borne terahertz telescope. GUSTO is a Class D NASA balloon borne observatory mission. The University of Arizona as PI is responsible for the instrument design. The GUSTO platform is planned to be launched from McMurdo, Antarctica, in late 2021 for a flight duration of 100-170 days.

Gusto aims to:

1. Determine the constituents of life cycle of interstellar gas in the Milky Way
2. Witness the formation and destruction of star-forming clouds
3. Understand the dynamics and gas flow to and in the Galactic Center
4. Understand the interplay between star formation, stellar winds and radiation, and the structure of the interstellar medium (ISM) in the Large Magellanic Cloud (LMC)
5. Construct Milky Way and LMC templates for comparison to distant galaxies.

To achieve these goals GUSTO will survey ~124 square degrees of the Milky Way and the Large Magellanic Cloud (LMC) using three highly sensitive 4×2 HEB heterodyne array receivers to detect the 3 brightest interstellar cooling lines: [CII] at 1.9 THz, [OI] at 4.7 THz and [NII] at 1.4 THz.

B. GUSTO instrument description

The GUSTO instrument concept can be found in Fig. 1. The sky-beam from the telescope is split into two beams using a dichroic. The first beam (transmitted) will keep the information for the 1.4 and 1.9 THz channel while the second beam (reflected) will be used to detect the 4.7 THz line. The sky-beams are then reimaged onto 4x2 HEB mixer arrays, one for each channel, in the focal plane. The arrays for 1.4 and 1. 9 THz are placed side by side and therefore image slightly offset parts of the sky. The LO signals are folded into the arrays using beamsplitters, one for 1.4/1.9 THz and one for 4.7 THz. After mixing, the output signal contains the same phase and frequency information as the original sky signal. The output signals from each pixel are amplified in individual cryogenic LNA's and then further processed in the warm IF and spectrometer backend of the instrument.

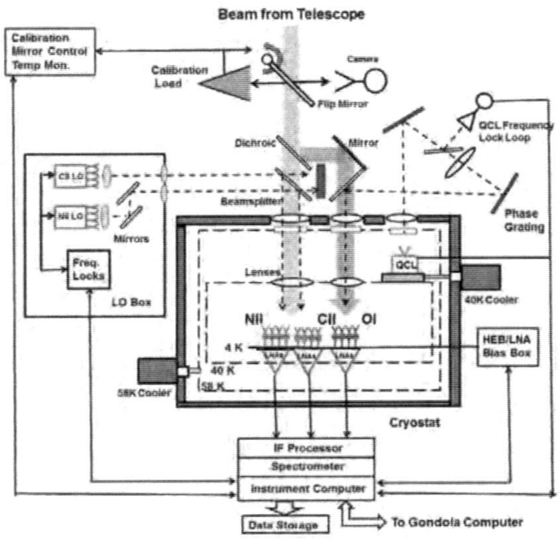

Fig. 1 GUSTO instrument block diagram, which was in the original proposal. It is still in progress with regard to, in particular, the 4.7 THz LO.

In the past, other HEB arrays have been developed [4] but these were designed with each pixel as a separate block. As GUSTO will implement the highest pixel count so far and demands a high level of integration, it requires a compact monolithic design. In addition to the need for device performance uniformity, both in sensitivity and LO requirement, it is also of utter importance to ensure the correct pointing of the individual pixels. Furthermore, because of the high integration level, challenges regarding cross-talk have to be addressed. Conceptually, each pixel will be used in a quasi-optical configuration to couple the radiation from the sky and LO into the HEB, through the use of a lens-antenna system. Each device will then be connected to a transmission line, at the end of which a connector will allow to extract the IF signal and supply dc-bias to the HEB.

For the two lower frequency channels multiplier-chain based solid state LOs will be used. At 4.7 THz it requires the use of a Quantum Cascade Laser (QCL) [5] as LO since this is the only applicable solid state source with enough power at such a high frequency.

In this paper, we will focus on the 4×2 HEB mixer array being designed for the novel array receiver the 4.7 THz channel of GUSTO.

II. Performance Evaluation

Preliminarily to the array design we realized two complementary studies where we evaluated/optimized the future performance of the array. The first study is a statistical analysis of some of the key performance parameters within a batch of devices. The second is trade study of design parameters of the Si lenses against performance.

A. Device performance statistics

In the device selection for GUSTO we aim for the highest possible performance and uniformity while having some flexibility. The selected HEBs, see Fig. 2 (a), are thin superconductive bridges of NbN with 2×0.15 μm² dimensions between the pads of tight wound spiral antennas. Such a combination has been shown to have good performance at 4.7 THz [6]. The use of a broadband spiral antenna allows for flexibility since it is possible with a single design to populate all the arrays.

For the device selection several fabrication batches were studied and one preliminary was selected. From this batch, 10 different devices were randomly picked and characterized. The measurements were performed in a vacuum setup at 2.5 THz using AR coated lenses. The measurement setup is similar to the one in [6]. The figures of merit for the evaluation were the sensitivity (noise temperature), the LO power requirement and the uniformity of these parameters among devices. The results are presented in Fig. 2 (b) and (c). The average noise temperature measured was 720 K with a 3% standard deviation. For the LO power requirement an average of 227 nW was measured with a standard deviation of 6%. The sensitivity measured reflects the state of the art, and the LO power requirement is low enough that the expected next generation QCLs will be able to pump the entire array (see next section). Moreover, the low deviation found for both parameters, 3 and 6%, indicates the potential for very uniform arrays.

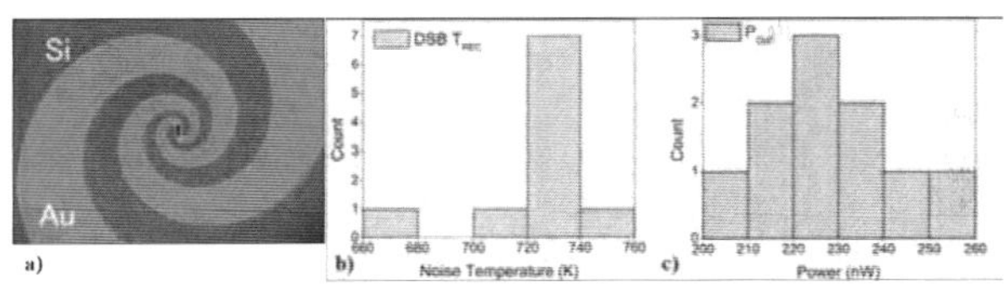

Fig. 2 a) SEM image of a tight wound spiral antenna HEB; B) Measured Noise temperature distribution; c) Measured LO power distribution

B. Lens Study

At high frequencies (>2 THz), the manufacturability of good quality waveguide structures becomes difficult due to the reduced size of the features. Therefore, the use of a quasi-optical coupling becomes the best solution to obtain high sensitivity. In our quasi-optical system, we usually combine an extended 10 mm elliptical lens made of high purity Si with the spiral-antenna present on the HEB chip. While for a single pixel receiver the lens size is not a limitation, when converging to arrays with high pixel count, the spatial footprint can become a problem. The most obvious solution for this is to simply reduce the lens diameter.

In this study two lens diameters, 5 and 3.1 mm, were compared with an existing 10 mm lens (see Fig. 3). These lens were designed as classical elliptical lenses as described in [7].

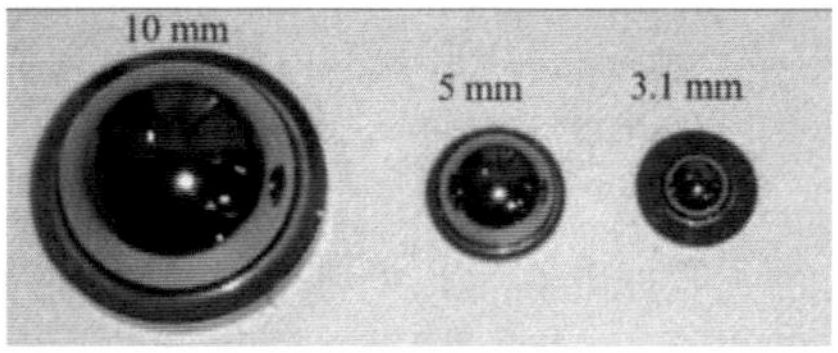

Fig. 3. Si Lens used on the experiments.

The value of ξ_{Si} is one of the parameters required for the design. In the past it has been thought this value at cryogenic temperature to be the same as room temperature, 11.7. However, more recently [8] it has been shown that the correct value should be around 11.4. Since there is still some uncertainty on this matter and we desire to optimize our system, different lenses for each set were designed. For the 5 mm lens we designed four different lenses L1, L2, L3 and L4, designed for a ξ_{Si} of 11.2, 11.3, 11.4 and 11.5 respectively. In the case of the 3.1 mm diameter we designed three lenses, L1, L2 and L3 designed for 11.2, 11.4 and 11.7 ξ_{Si} respectively. The real parameters of these lens were measured and are summarized in Table 1. In the 3.1 mm set we measured some relevant variations on the lens dimensions compared to the design, while for the 5 mm set the measurements matched the design.

Table 1. Measured lens parameters

Diameter	Lens	Ellipticity	Extension (incl. HEB)
10 mm	L1	1.0456	1.582
5 mm	L1	1.0480	0.783
	L2	1.0476	0.779
	L3	1.0468	0.775
	L4	1.0464	0.772
3.1mm	L1	1.0458	0.502
	L2	1.0471	0.498
	L3	1.0477	0.506

To evaluate the lens performance we measured the noise temperature using the same HEB-device for all the measurements. Moreover, a misalignment no bigger than 5 µm in a given direction was achieved, ensuring the performance differences were due to the lenses only. The experimental setup is similar to the one used in the device selection, only this time we used an air setup and no AR coating. As baseline we consider the noise temperature for the 10 mm lens, which best measured value was 1285 ± 25 K. The results for the noise temperature measurements for the other sets of lens can be seen in Fig. 4. For the 3.1 mm set the best noise temperature was measured for L2, with a value of 1447 ± 5 K, while for the 5 mm lens we measured 1368 ± 11 K for L4, although L2 show very similar values. Moreover L3 in the 5 mm set seems to be a clear outlier. For both these set of lenses the best noise temperature is found for designs assuming a ξ_{Si} close to 11.4 as expected. However, since we had some variations in the fabricated lenses, these could shift the real performance of the lenses. Therefore, each lens was simulated in PILRAP, a lens antenna simulation tool for the submillimeter range, using the real lens dimensions. From this tool we used as figure of merit the lens efficiency which is defined as the product of the multiple efficiencies within the lens (e.g. spillover, aperture, etc). For the simulations we assumed the real lenses dimensions and tested different dielectric values, founding that the best value that matched our experimental results was 11.4 confirming the previous conclusions. The curves for the simulated efficiency are plotted in blue in Fig. 4.

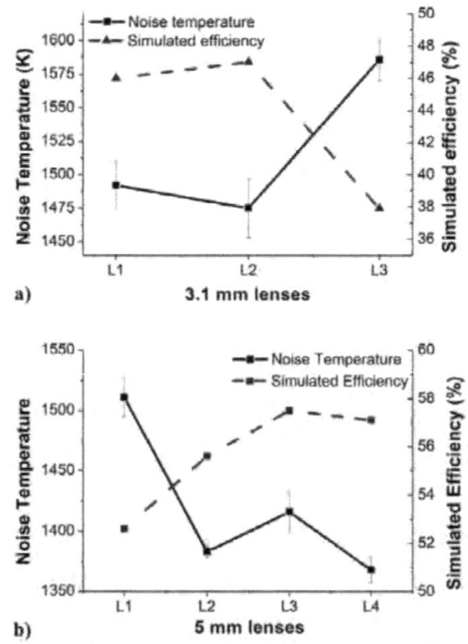

Fig. 4. Noise temperature measurements and PILRAP efficiency simulations. a)3.1 mm lenses. b) 5 mm lenses.

Besides knowing which dielectric value should be used in the design process, it is also important to know what are the tolerances acceptable in the fabrication. For the 3.1mm lenses we can see that a deviation on the extension of 5 µm can have a very big impact on the noise temperature (7%) increase while for the 5 mm lenses the variation between L2 and L4 in relationship to L3 is also 5 µm and we see a maximum absolute difference from the best (L4) of 4% (L3). Therefore drawing the conclusion that the bigger the lens the more tolerant to fabrication errors.

In Fig. 5 we compare the best lens for each set in terms of noise temperature and the predicted lens efficiency for that specific lens. To avoid confusion it should be noticed that a perfect rescaling of the same lens design has the same lens efficiency. Looking at the noise temperature measurements we see a clear trend where the noise temperature increases as the diameter is reduced. On the other hand the predicted efficiency doesn't match this behavior and seems to be decoupled from the lens diameter in study. Based on this, and because there is a clear difference on the noise temperature between sets, some other effect must be in play. When studying what is affected by the rescaling of lenses it was found the lens diameter affects the beam pattern. The smaller the lens the wider the beam pattern becomes. A hypothesis is that the noise temperature increase might have to do with the side lobes coupling improperly to the calibration loads (e.g. the side lobes are always looking at a cold or hot load). Therefore, for the smaller diameter, we should see the widest beam pattern and worse performance, matching the measurements. Further measurements should be performed minimizing the poor coupling of side lobes in order to verify this hypothesis.

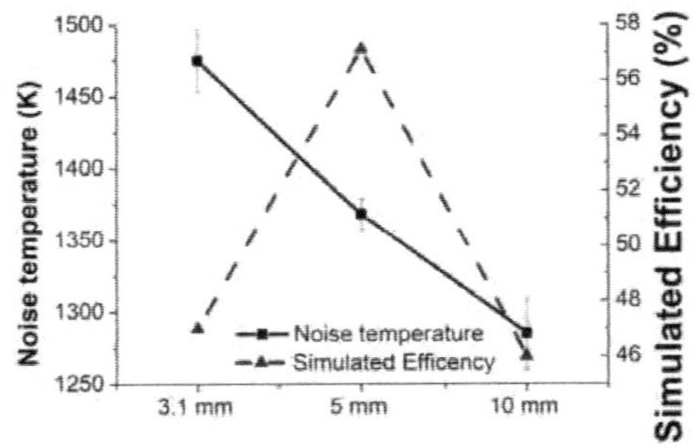

Fig. 5 Comparison of the best lens of each set and respective predicted efficiency.

Since we aim to implement eight pixel arrays, a relative low count, and the smaller lens diameter needs further study, due to the timeline it is chosen to fly 10 m lens as baseline. This choice reduces the risk for the mission since this diameter yielded the best performance from the different sets studied, and will allow to keep the relative fabrication error at the lowest possible. Furthermore this size has been flown previously in STO-2, with good flight performance. Moreover, based on our results, we will assume 11.4 ξSi for the baseline design.

III. The 4×2 HEB mixer array design

The mixer array is designed as an eight pixel array in a 4×2 configuration. The 4.7 THz HEB array is going to share the same architecture with the other two lower frequency arrays in order to reduce the development time.

The preliminary array design can be seen in Fig. 6 (a) represents the fully assembled array as it will be placed on the cold plate. The main requirements determining the design are as follow: 1) 4×2 pixel configuration; 2) 10 mm diameter Si lenses; 3) 12 mm pitch size between each individual pixel; 4) side by side placement of the two lower frequency channels while maintaining the pitch size, emulating a 4x4 array; 5) optical and IF path as similar as possible for all the pixels. 6) minimized optical and electrical crosstalk. 7) mechanical stability.

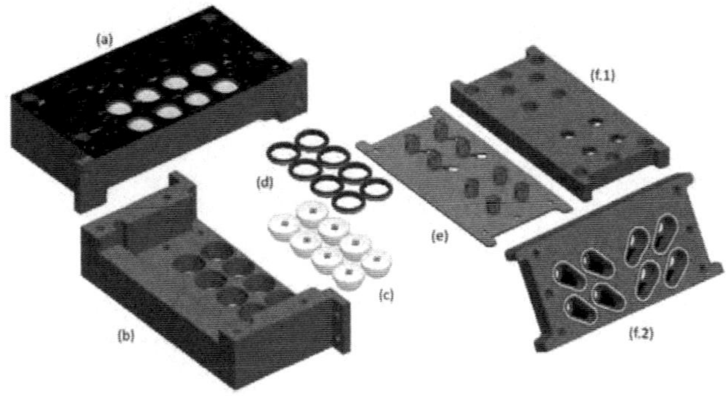

Fig. 6 GUSTO 4x2 HEB mixer array mechanical design

The copper piece identified as (a-front, b-back) in Fig. 6 is the array housing. It is the core structure of the array and defines the placement of the pixels. It will be coated with a layer of stycast mixed with SiC grains to reduce reflections. In this piece eight 10 mm Si lenses (c) will be placed with pre-aligned HEB chips and locked in place with retainer rings seen in (d). On the back of the lenses and covering them, a common PCB (e) is placed having eight independent co-planar waveguide (CPW) transmission lines and respective connectors. The devices will be connected to the CPW lines using bond wires. On top of the PCB another copper part (f.1-back, f.2-front) will be placed with the function to isolate each pixel electrically and optically. For this a spiral shield ring will be used (white in f.2) around the area of each CPW to create a faraday cage that avoids radiation to leak into adjacent pixels. In this same piece 8 stycast/SiC layers (black material in f.2) will be put to absorb any radiation that is not directly coupled into the HEB, reducing the potential for optical cross-talk in the array.

IV. Summary

In this paper we have summarized the preliminary design study for the 4×2 HEB array receiver being developed for GUSTO. We discussed the study regarding HEB device performance statistics which yielded state of the art sensitivity of 720 K at 2.5 THz, low LO power requirement of 227 nW and very good uniformity of 3 and 6% standard deviation respectively, thus matching the requirements for the array performance. Furthermore a lens optimization study was conducted from which we concluded to choose to fly 10 mm lens instead of smaller diameter lenses. These lenses showed the best performance from the set of lenses measured and are expected to the most tolerant to fabrication deviations. Moreover the smaller diameter lens needs more research before being properly understood. Furthermore, the experimental data indicates the correct ξ_{Si} at cryogenics to be 11.4.

For the 4x2 HEB mixer array we introduced the mechanical design required to support 8 pixels in a 4 by 2 configuration and with a 12 mm pitch size. It will use a quasi-optical configuration with the pre-selected 10 mm elliptical Si lenses.

Acknowledgment

We would like to thank Willem Jellema and Brian Jackson for their help regarding the PILRAP simulations. The first author would also like to thank Floris Van der Tak for his support and guidance. This publication has received funding from the European Union's Horizon 2020 research and innovation program under grant agreement No 730562 [RadioNet].

References

[1] C. Walker *et al.*, "The Stratospheric Terahertz Observatory (STO)," in *Proc. 19th International Symposium on Space Terahertz Technology*, pp. 28–32, 2008.

[2] S. Cherednichenko, V. Drakinskiy, T. Berg, P. Khosropanah, and E. Kollberg, "Hot-electron bolometer terahertz mixers for the Herschel Space Observatory.," *Rev. Sci. Instrum.*, vol. 79, no. 34501, 2008.

[3] S. Heyminck, U. U. Graf, R. Güsten, J. Stutzki, H. W. Hübers, and P. Hartogh, "GREAT: the SOFIA high-frequency heterodyne instrument," *Astron. Astrophys.*, vol. 542, no. L1, pp. 1–7, 2012.

[4] C. Risacher *et al.*, "First Supra-THz Heterodyne Array Receivers for Astronomy with the SOFIA Observatory," *IEEE Trans. Terahertz Sci. Technol.*, vol. 6, no. 2, pp. 199–211, 2016.

[5] B. S. Williams, "Terahertz quantum-cascade lasers," *Nat. Photonics*, vol. 1, pp. 517–525, 2007.

[6] D. J. Hayton *et al.*, "A 4.7 THz Heterodyne Receiver for a Balloon Borne Telescope," in *Proc. SPIE , Millimeter, Submillimeter, and Far-Infrared Detectors and Instrumentation of Astronomy VII*, vol. 91531R, 2014.

[7] D. F. Filipovic, S. S. Gearhart, and G. M. Rebeiz, "Double-Slot Antennas on Extended Hemispherical and Elliptical Silicon Dielectric Lenses," *IEEE Trans. Microw. Theory Tech.*, vol. 41, no. 10, pp. 1738–1749, 1993.

[8] W. Zhang *et al.*, "Optimized sensitivity and beam pattern of a twin-slot antenna coupled NbN HEB mixer at 1 . 6THz," in *Proc. of SPIE*, vol. 8452, pp. 1–7, 2012,

Session T4: **Invited III**

The Renaissance of Submillimeter Astronomical Spectroscopy

Paul F. Goldsmith*

Jet Propulsion Laboratory, California Institute of Technology Pasadena CA 91109, USA
*Contact: Paul.F.Goldsmith@jpl.nasa.gov

Abstract— High resolution spectroscopy is recognized as a powerful astrophysical tool. For understanding many critical aspects of the formation and evolution of interstellar clouds and how new stars are formed within them, the most important spectral lines are at submillimeter wavelengths, corresponding to the terahertz frequency range between 300 and 6000 GHz, (0.3 to 6 THz). In many astronomical situations, line widths resulting from Doppler shifts arising from gas motions can be $\leq$ 1 km/s. Thus, extracting the full information available about the kinematics of gas in the source being observed requires resolution $R = f/\Delta f \geq 3x10^5$. This requires a heterodyne system, which can readily achieve the required frequency resolution. Exploiting the wealth of ionic, atomic, and molecular transitions has been hampered by the nearly total absorption by the Earth's atmosphere and the relatively low sensitivity of available detectors.

The situation has improved dramatically over the last decade. The development of the Heterodyne Instrument for Far Infrared (HIFI), on board the Herschel Space Observatory launched in 2009 encouraged a continuing technological transformation that includes vastly improved mixers, local oscillators, optics, and the development of focal plane arrays. The data from Herschel have inspired improved laboratory measurements generating spectral line catalogs, quantum calculations, and measurements of collision rates and chemical reaction rates for key astrophysical species. New observing platforms including the SOFIA airborne observatory and long-duration balloons flying at 40 km altitude have resulted in a stream of data of ever-increasing frequency coverage and range and extent of sources observed. Anticipated ultra-long-duration balloons and the possibility of space missions ranging from small satellites to a large-aperture major (Flagship) mission are generating great excitement about possible future observations.

In this talk I will review how the overall picture has evolved in recent years. I will highlight recent astronomical results, including velocity-resolved large-area studies of Giant Molecular Clouds and nearby galaxies in the 158 micron fine structure transition ([CII]) of C^+, high-sensitivity observation of water in collapsing cloud cores, and multi-transition observations of the CO ladder through the entire submillimeter range. I will indicate some of the most exciting areas for further exploitation in the years ahead, in these and other transitions including HD at 112 microns and the fine structure transitions ([NII]) of N^+ at 205 microns and 122 microns wavelength. The formation of new stars, particularly massive stars, is a key aspect of the evolution of galaxies, and understanding how new stars form requires observation of many different spectral lines that probe the physical conditions and chemical composition of interstellar clouds, including those regions that are actually collapsing to form a single protostar or a cluster of protostars.

The Submillimeter Wave Instrument on JUICE

Paul Hartogh* and the SWI team

[1]Max Planck Institute for Solar System Research, Göttingen, 37077, Germany
*Contact: hartogh@mps.mpg.de

Abstract—The Submillimeter Wave Instrument (SWI) is part of the payload of ESA's L1 mission JUICE (JUpiter ICy moons Explorer). It sounds the middle atmosphere of Jupiter and the exospheres and surfaces of the Galilean satellites with a 290 mm telescope, movable in azimuth and elevation using two heterodyne spectrometers covering the frequency ranges of 530 to 625 GHz and 1080 to 1275 GHz. The heterodyne receivers with tunable solid-state local oscillators and subharmonically pumped mixers are passively cooled. The performance of the receivers has been steadily improved during the last 2-3 years. Presently the receiver temperatures are about 1100 and 2000 K DSB (600 and 1200 GHz respectively). The intermediate frequency output of 3.5 to 8.5 GHz is analyzed by two real-time spectrometer backends consisting of broadband autocorrelators, high resolution multi-channel Chirp Transform Spectrometers and continuum channels. In order to fulfill the very stringent mass limitations of the JUICE payload, the structure of the Telescope and Receiver Unit (TRU) is made of AlBeMet 162. The instrument is now in the phase between PDR and CDR and the design has mostly been completed. This presentation will address the science objectives, the technology challenges and recent break-throughs, the instrument hardware development status and future plans until the launch of the JUICE mission. Finally potential applications of SWI-like instruments for other explorations of the solar system will be addressed.

Session T5: **LNAs and Backends**

Traveling-wave Superconducting Parametric Amplifiers

Peter K. Day[1*], Byeong Ho Eom[1], Henry G. Leduc[1]

[1]Jet Propulsion Laboratory, Pasadena, CA 91109, USA
[*]Contact: Peter.K.Day@jpl.nasa.gov

Abstract— Traveling-wave parametric amplifiers based on superconducting NbTiN films are being developed that provide gain over nearly an octave of bandwidth at microwave frequencies. The amplifiers are non-linear transmission lines, where the nonlinearity comes from the current dependence of the kinetic inductance. Amplification results from three and four-wave mixing processes, and phase matching over a wide frequency range is achieved by engineering the dispersion characteristics of the transmission line. Measurements of the noise of these amplifiers in the GHz frequency range demonstrate that they may approach the quantum limit for a phase preserving amplifier. We will discuss the possibility of extending the operating band to of this type of amplifier to the millimeter-wave band.

A 1.5 W 3 GHz Back-End Processor in 65-nm CMOS for Sub-millimeter-wave Heterodyne Receiver Arrays

Yanghyo Kim[1,2*], Yan Zhang[2], Adrian Tang[1,2], Theodore Reck[1], and Mau-Chung Frank Chang[2]

[1]*Jet Propulsion Laboratory, Pasadena, 91109, USA*
[2]*University of California, Los Angeles, Los Angeles, 90095, USA*
[*]Contact: rod.kim@jpl.nasa.gov

Abstract—As compact sub-millimeter wave heterodyne receiver technology continues mature rapidly, astronomers have begun to implement array-based instruments with full receiver arrays up to 64 pixels already demonstrated. While this impressive technology solves many of the key problems related to heterodyne arrays (LO generation and distribution between pixels, antenna array fabrication with well controlled beam patterns, and IF chains with low inter-pixel crosstalk), one challenge that remains in these and even larger arrays is that of back-end processing implementation. While several FPGA-based spectrometer processors exist, these systems consume power on the order of 30-50 W making their accommodation a challenge when a large number of pixels is required, as the back-end power is inflated by the number of pixels.

CMOS system-on-chip (SoC) technology is the advanced electronics technology used in mobile electronics such as smartphones/laptop, and provides an avenue to address this back-end power challenge head on. Several studies describe how a dedicated CMOS SoC offers 10-100X less power consumption compared to an FPGA implementation performing the same digital operation. This is primarily as a CMOS SoC is not required to carry the additional circuitry required to support re-configurability, greatly reducing the capacitance, and in turn, power consumption per logic cell. Additionally, CMOS SoCs allow for not only digital, but also analog and mixed-signal circuitry to be co-integrated within a single chip solution, enabling a drastic reduction of power consumption related to interfaces as the large capacitances of off-chip PCB traces and device packages no longer have to be driven with high-speed digital signals. This because an SoC-based spectrometer allows the digitizer can be integrated directly with the digital processing core into a single chip.

This work presents a fully integrated 3 GHz bandwidth (6 GS/s) back-end processor, which is implemented in advanced 65-nm CMOS process, for sub-millimeter heterodyne spectroscopic instrumentation. The spectrometer processor contains a 3-bit/interleaved-by-2 flash analog-to-digital converter (ADC), a 4196-point fast Fourier transform (FFT) processor, and a 4-billion-count data accumulator (ACC). For accurate digital timing, the chip contains an internal programmable frequency synthesizer providing a timing-skew controlled clock to each system block. In order to overcome process/temperature variations, a replica ADC is designed, offering key-components' bias-conditions. Additionally, integrated resistor-based 8-bit digital-to-analog converter (DAC) provides programmable reference levels to the ADC, reducing constant DC power and offering ways to calibrate mismatch issues between interleaved ADC channels. The chip consumes a total of 1.5 W and is embedded within a compact PCB module and a USB connection for data readout and configuration of chip settings.

Poster Session

HYper Spectral Microwave Sounder *(HYMS)*

O. Auriacombe*[1], M. Henry[1], B. N. Ellison[1], M. Jarret[2], J. Charlton[2], S. Parkes[3], I. Rule[4]

[1]*Millimetre Wave Technology Group, Chilton, UK*
[2]*JCR Systems Ltd., South Gloucestershire, UK*
[3]*STAR Dundee Ltd., Dundee, UK*
[4]*Met Office., Exeter, Devon, UK*
*Contact: olivier.auriacombe@stfc.ac.uk

Abstract— Space-borne heterodyne microwave 'Hyper-Spectral Imaging' (HPI) of Earth's atmosphere offers significant potential to greatly improve operational meteorological weather forecasting accuracy. By performing high-spectral resolution measurements of key atmospheric species (O_2 and H_2O) from space, a more accurate evaluation of global atmospheric temperature and water content distribution can be achieved. This information is used, in turn, to substantially increase the precision of advanced Numerical Weather Prediction (aNWP) models that are crucial in defining and predicting rapidly evolving weather phenomena [1].

Unfortunately, current and planned space-borne operational meteorological microwave remote sensing systems lack the necessary precision to support aNWP and, as a consequence, a serious gap exists in observational capability related to high altitudes (>40kM) [2]. To rectify this situation, future deployment of a fleet of microwave HPI spectrometers with spectral resolution order of better than 10^5 will be required. However, the appropriateness and benefits of the HPI technique must first be demonstrated. In order to achieve this key step, we are currently developing an airborne HPI microwave sounder operating in the (50 to 68) GHz frequency range and targeting the fine spectral structure of O_2 with a maximum spectral imaging capability of 1 MHz.

Our HYper-spectral Microwave Sounder (HYMS) will use a high-gain radio frequency (RF) low noise amplifier (LNA) followed by a frequency down-conversion stage comprising a pair of sub-harmonic mixers operating in a double sideband configuration. This will provide two RF spectral 'windows' that encompass (50.3 to 57.3) GHz and (63.3 to 67.9) GHz and the O_2 emission signature. High-spectral resolution is achieved through the use of a new ultra-wideband (8 GHz) digital Fast Fourier Transform (FFT) spectrometer. The required radiometric sensitivity is <0.4 K and this places significant demands on the system noise and requires the use of state-of-the-art component technology. In addition describing the application more detail, we will describe the radiometer system concept, design and development progress, and will present preliminary measurement results.

[1] Aires, F. ; Prigent, C. ; Orlandi, E. et al. (2015) "Microwave hyperspectral measurements for temperature and humidity atmospheric profiling from satellite: The clear-sky case". Journal of Geophysical Research: Atmospheres, vol. 120(21), pp. 11334-11351. http://dx.doi.org/10.1002/2015JD023331

[2] Aires, F., and Pringent, C., "Use of spectral information at Microwave Region for Numerical Weather Prediction", ESA Contract 4000105721/12/NL/AF

29th IEEE International Symposium on Space THz Technology (ISSTT2018), Pasadena, CA, USA, March 26-28, 2018
29th IEEE International Symposium on Space Terahertz Technology, Pasadena, California, USA, March 26-28, 2018

A dual-polarization sideband separating Schottky based receiver for ALMA Band 2+3 Warm Cartridge Assembly

B. Thomas[1*], J. Ceru[1], F. Villa[2], P. Yagoubov[3]
[1]Radiometer Physics GmbH, Werner-von-Siemens Strasse 4, 53340 Meckenheim, Germany.
[2]INAF / OAS Bologna via Gobetti, 101, 40129, Bologna, Italy.
[3]European Southern Observatory, Karl-Schwarzschild-Str. 2, 85748 Garching, Germany.
*Contact: bertrand.thomas@radiometer-physics.de

***Abstract*— We present in this paper the development of a dual-polarized sideband separating receiver using semiconductor Schottky and MMIC technology operating at room temperature from 67 GHz to 116 GHz to meet the extended frequency range of ALMA Band 2+3. Preliminary measurements on a single polarization receiver prototype have previously been reported, exhibiting a SSB receiver noise temperature ranging between 3000K and 8000K over the entire 67-116 GHz, and a side band separation between 13dB and 17dB [3]. An upgrade of the side-band separating receiver for dual-polarization is described, featuring a single powerful LO source to pump all four fundamental balanced mixers operating in quadrature mode. Test results including noise figure, gain flatness, sideband ratio, as well as spectral purity are described hereafter.**

I. Introduction

ALMA Band 2+3 receivers development covering the extended frequency range 67-116 GHz is currently on-going in international collaboration coordinated by ESO [1]. This combination of bands is made possible by the continuous technological developments in cryogenic MMIC HEMT low noise amplifiers in order to reach ultra-broadband and ultra-low noise performance, allowing for the relaxation of the constraints in terms of noise temperature on the following stages of the receiver chain. In turns, it enables GaAs semi-conductor technologies such as Schottky diode based mixers and ferrite based waveguide isolator to be used as room temperature operated receiver inside the Warm Cartridge Assembly.

II. Design & Architecture

The Schottky based 2SB receiver presented here is located in the Warm Cartridge Assembly. Each polarization channel of the side-band separating receiver consist of a broadband WR-10 waveguide Y-junction followed by two extended-W band ferrite isolators covering 67-116 GHz, two planar Schottky based fundamental balanced mixers operating in the range 67-116 GHz pumped by a common 75-110 GHz active sextupler and booster W-band amplifier source split and 90° phase shifted via a full-band 3dB waveguide hybrid coupler. IF signals are recombined with an hybrid coupler and amplified.

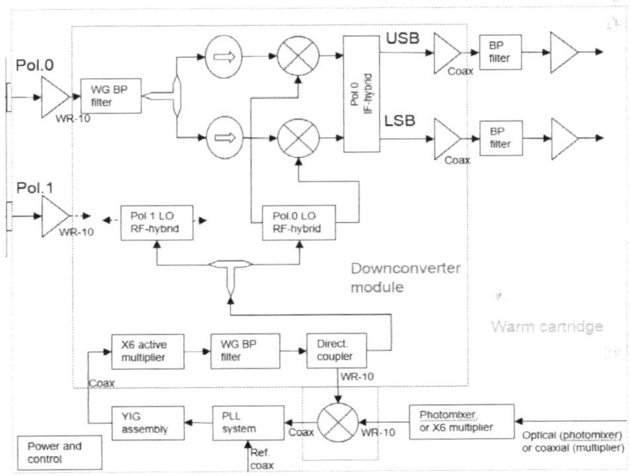

Fig. 1 Schematic diagram of the Warm Cartridge Assembly architecture (orange box), with the 67-116 GHz dual-polarization 2SB receiver (green box) developed by RPG.

III. Component Level Characterization

A fundamental balanced mixer based on cross-bar topology has been designed to cover the extended WR10 waveguide band, from 67 GHz up to 116 GHz. In order to achieve such a broadband RF matching, the use of a ridged waveguide in the RF input waveguide to couple the RF signal to the diodes, as well as the natural mode decoupling of the RF and LO signal is necessary. Measured performance of the 67-116 GHz fundamental balanced mixer are reported in [3], with SSB conversion losses of 7 dB in average over the 67-116 GHz RF range.

Figure 2 shows the measured S-parameters of the RPG 67-116 GHz isolator. The insertion losses are below 1dB up to 100 GHz, and below 2 dB up to 116 GHz. Return losses are <-17dB and the isolation >20dB over the complete 67-116 GHz range.

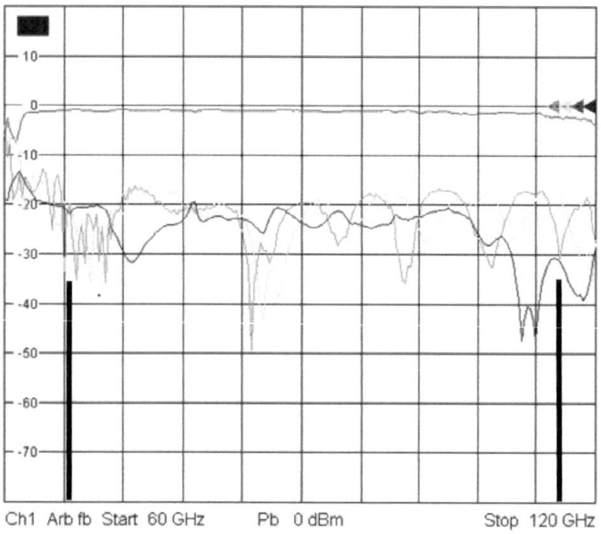

Fig. 2 Measured S-parameters of the extended WR-10 waveguide isolator used in the 2SB receiver. Red: insertion losses, blue: reverse isolation. Orange and yellow: input/output return losses.

A thorough characterization of the LO sextupler source for spurious content has been performed and the results are presented in Figure 3. Following the sextupler, a booster amplifier and a specially designed 79-104 GHz bandpass filter are used to amplify the desired 6th harmonic of the input signal and reject further the out-of-band 4th, 5th, 7th and 8th harmonics of the sextupler. The 5th and 7th harmonics of the fundamental cannot be filtered out as they appear inside the LO band, but exhibit output power levels of -40dBc and -30dBc respectively, in compliance with the ALMA requirements.

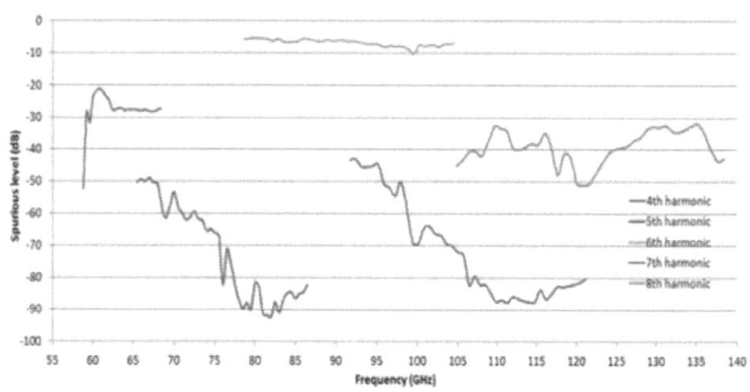

Fig. 3 Measured harmonic content of the sextupler. Top green curve shows the desired 6th harmonic signal. An external filter ensures additional rejection of the 4th and 8th harmonics.

IV. 67-116 GHz 2SB RECEIVER TESTING

A view of the complete 67-116 GHz dual-polarization 2SB receiver is shown in the Fig. 4.Extensive testing has been performed on the 2SB receiver in order to verify the sensitivity of the receiver at room temperature and the gain response. To do so, a traditional Y-factor measurement has been performed using an ambient and LN2 cooled calibration target. The fundamental LO signal is provided externally.

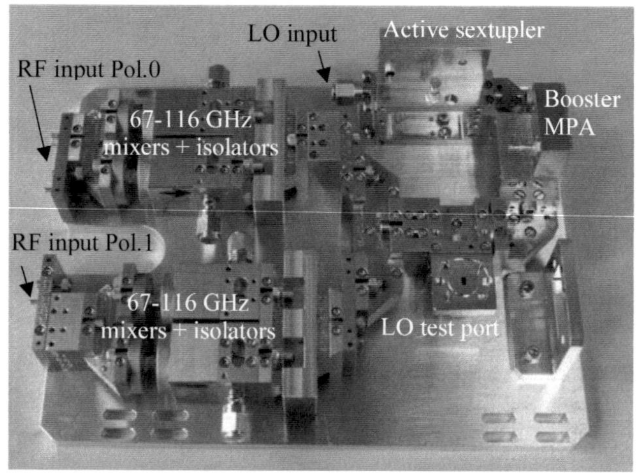

Fig. 4 Photo of the complete dual-polarization 2SB receiver featuring the power LO source, 4x fundamental balanced mixers and isolators, and LO/RF filters.

A. Gain and sideband ratio response

The gain slope and sideband ratio of the complete 2SB receiver has been characterized every 3 GHz over the entire 79-104 GHz LO range, and corresponding RF range 63 GHz to 120 GHz, as the IF hybrid coupler and amplifiers extend up to 16 GHz. Example of test results showing the gain & USB ratio for the V-polarization (top graph) and H-polarization (bottom graph) are presented below. The gain slope (in blue on the graph) is approx. 5dB over 15 GHz bandwidth. The side-band ratio (in green on the graph) is >15dB for the V-polarization, and >10dB for the H-polarization

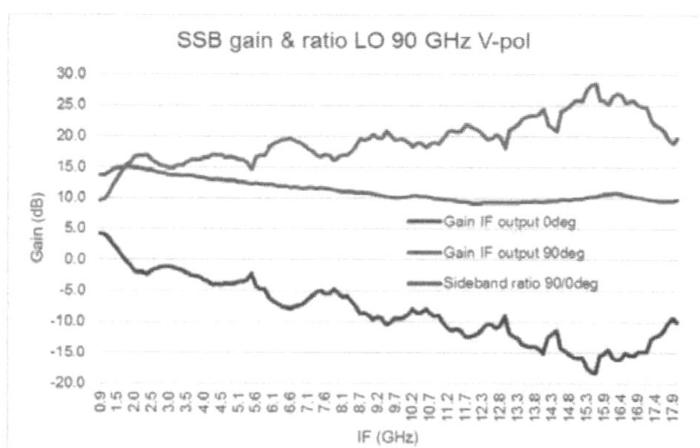

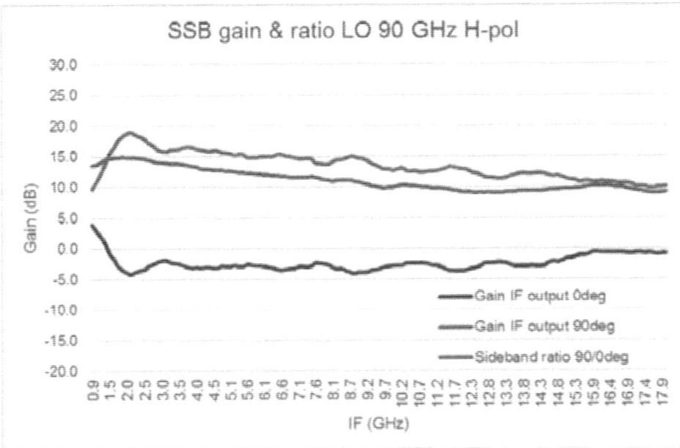

Fig. 5 Measured gain and sideband ratio of the 2SB receiver for a fixed LO frequency of 84 GHz, at the IF outputs of the V-polarization channel (top graph), and the H-polarization channel (bottom graph).

B. *Noise figure response*

Noise figure of the complete 2SB receiver for each polarisation has been characterised over the full 63-120 GHz range for each polarization. An example of test results showing the receiver noise temperature for both IF outputs (LSB & USB) of the H-polarization channel are presented in Fig. 6. The measured noise temperature is between 3000K and 8000K in the 4-12 GHz IF band.

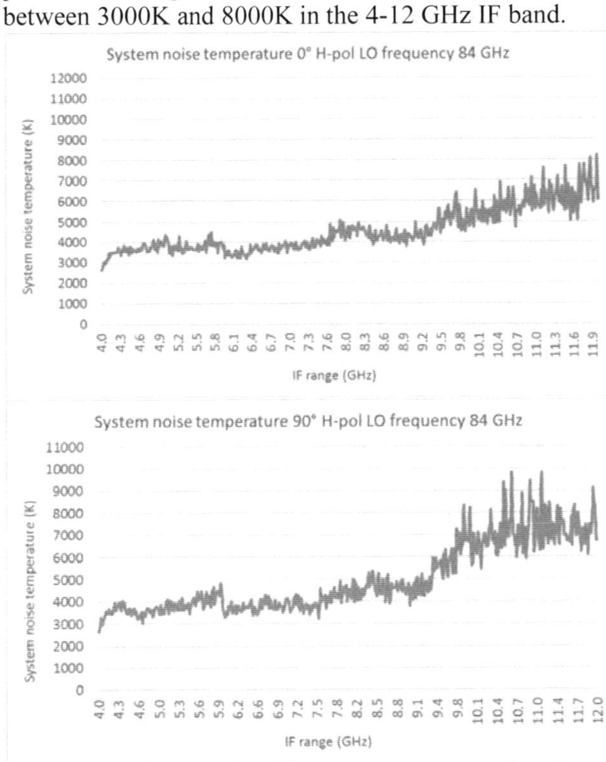

Fig. 6 Measured receiver noise temperature of the 2SB receiver for a fixed LO frequency of 84 GHz, at both IF outputs (LSB & USB) of the H-polarization channel.

C. *Spurious emission at RF input*

The input spurious emission at the RF waveguide input for both polarization in the WR10 and WR05 range using R&S spectrum analyser extender mixers. Results are presented in Table I. At the 6th harmonic of the fundamental LO signal, the maximum spurious level measured is -35 dBm on H-polarisation channel at 90 GHz. At the 12th harmonic of the fundamental LO signal, the maximum spurious level measured is -52 dBm on H-polarisation channel at 192 GHz. The performance of the out-of-band filter introduced at the RF input of the 2SB receiver has proven to reject the 2nd harmonic of the LO signal better than 22 dB.

TABLE I

LO input frequency	H-spurious 6*LO	V-spurious 6*LO	H-spurious 12*LO	V-spurious 12*LO
13.1 GHz	-41.5 dBm	-53 dBm	-68.3 dBm	-68.8 dBm
13.5 GHz	-41.4 dBm	-52.5 dBm	<-79 dBm	<-80 dBm
14 GHz	-41.8 dBm	-51 dBm	<-77 dBm	<-80 dBm
14.5 GHz	-31.5 dBm	-45.3 dBm	<-80 dBm	<-80 dBm
15 GHz	-35 dBm	-46.3 dBm	<-80 dBm	-70.8 dBm
15.5 GHz	-36.4 dBm	-46.4 dBm	<-80 dBm	<-80 dBm
16 GHz	-39 dBm	-61.2 dBm	-52 dBm	-64.9 dBm
16.5 GHz	-54.9 dBm	-60.2 dBm	-69.7 dBm	-65.6 dBm
17 GHz	-48.5 dBm	-61.1 dBm	-66.8 dBm	-64.6 dBm
17.3 GHz	-43.9 dBm	-63 dBm	-60.7 dBm	-66.1 dBm

CONCLUSION

An ultra-wideband 67-116 GHz dual-polarization sideband separating Schottky based receiver has been defined, prototyped and tested. The performance have been measured over the entire 67-116 GHz range, confirming the suitability of this technology for the ALMA Band 2+3 Warm Cartridge Assembly requirements. Further testing is on-going to characterize the short term gain stability (1/f noise).

ACKNOWLEDGEMENTS

The authors would like to thank specially Mr. Gibson from GMD for the bandpass filters design and the RPG mechanical department for the quality of the blocks manufacturing. This work has been funded by ESO under contract number 023566.

REFERENCES

[1] F. Villa, "The Status of the Band 2+3 receiver for ALMA". *IV Workshop sull'Astronomia Millimetrica in Italia,* Instituto di Radioastronomia Bologna, Italy. 7-10 November, 2017

[2] M. Kotiranta, et al."Cryogenic Low Noise Amplifier Development for 67-116 GHz", prcoeedings of the 39th International Conference on Infrared, Millimeter, and Terahertz waves (IRMMW-THz), Tucson, USA, 14-19 September 2014.

[3] B. Thomas, H. Gibson, J.H Freerksema, A. Walber, P. Yagoubov "67-116 GHz Single Sideband Schottky-based Receiver operating at room temperature", The 7th ESA Workshop on Millimetre-Wave Technology and Applications, Aalto University, Espoo, Finland, 6-8 June, 2016.

A Deployable 600–720 GHz ALMA-Type Sideband-Separating Receiver Cartridge

Ronald Hesper[1*], Andrey Khudchenko[1], Michel Lindemulder[2], Mariëlle Bekema[1], Rob de Haan-Stijkel[1], Jan Barkhof[1], Joost Adema[1] and Andrey Baryshev[1]
[1]*Kapteyn Astronomical Institute, University of Groningen, The Netherlands*
[2]*KVI — Center for Advanced Radiation Technology, University of Groningen, The Netherlands*
*Contact: R.Hesper@astro.rug.nl

***Abstract*—We designed, built and tested a new ALMA-type dual polarization, sideband-separating (2SB) receiver cartridge for the 600–720 GHz band (ALMA Band 9). It meets all the major technical specifications, most of them with ample margin. Key performance results: SSB noise temperature typically 200 K or better, with a slight upturn to about 250 K at the upper band edge; image rejection typically better than 15 dB with a few in-band excursions to about 13 dB; the IF pass-band ripple over any 2 GHz subband is about 3 dB. The four 4–12 GHz IF outputs provide a total of 32 GHz bandwidth.**

Two copies are produced initially, one for deployment in the SEPIA front-end at APEX (Chile), the second for the future LLAMA telescope situated in Argentina. Apart from being deployable instruments in their own right, the two receiver cartridges act as prototypes and demonstrators for a future upgrade of the ALMA Band 9 array. In addition, the new cartridges are designed in such a way (including heritage from the ALMA Band 5 receivers built in collaboration with GARD, Chalmers University of technology, Sweden) that other high-frequency bands above, say, 200 GHz could be implemented with relative ease in the existing structure. This would enable, for instance, the production of 230 GHz receivers for various projected observatories that aim to participate in the Event-Horizon Telescope consortium.

I. Introduction

In atmosphere-dominated sub-millimeter wavelength ranges like ALMA Band 9 (600–720 GHz, $\approx$500μm) and higher, using sideband-separating (2SB) receivers can have major advantages for the observation efficiency when compared to doule sideband (DSB) ones. We estimated before, using typical “Band 9” atmospheric conditions at sites like ALMA and the typical noise performance of the currently installed receivers there, that integration times for spectral line observations can roughly be halved when going from the DSB case to 2SB [1]. This improvement applies both to single-dish telescopes and interferometers. For continuum observations (total power), there may be little gain for interferometers (everything else being equal) and possibly a slight loss for single dish observatories.

Of the ALMA-style receiver cartridge described here, two are being built for single-dish observatories: APEX in Chile, where it will be part of the triple-band SEPIA facility instrument [2] and LLAMA in Argentina, which is currently under construction. The first cartridge, for APEX, is undergoing final qualification at the time of writing, and is scheduled to be delivered mid-July 2018.

Apart from their purely scientific role in these two observatories, the receivers also act as demonstrators and test cases for the use of 2SB receivers at these high frequencies in general, as there are no other deployed receivers in this region to the best of our knowledge. From a technical point of view, there are several innovations in the cartridge design intended to facilitate future extensions to other wavelengths, and finally, they are also a test case (already successful, judging from the test results) of techniques that should help in making arrays of sideband-separating mixers.

II. Construction

The base structural approach for the 650 GHz Cold Cartridge Assembly (CCA, the part residing in the vacuum of the cryo-

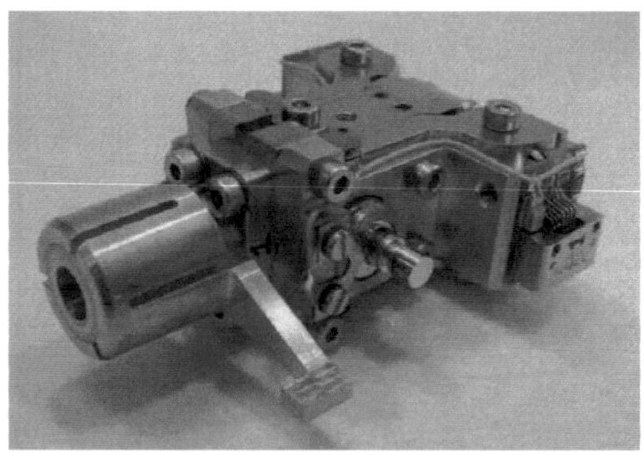

Fig. 1. The Band 9 sideband-separating mixer (excluding the IF hybrid) that was used in the new cartridge. An interface “collet” (left), which comes in place of the original mixer holder, interfaces the horn mechanically and thermally to the optics assembly. On the opposite end another horn (not visible) couples in the LO signal. In the center, one of the two backpieces containing the SIS junctions is visible, here with a protective short on the GPO-type IF connector. Magnetic field for Josephson suppression is supplied to the junctions by magnet coils housed in the two bulges left and right on the far end of the block, transported by magnet conductors visible on the top surface (and, symmetrically, on the bottom).

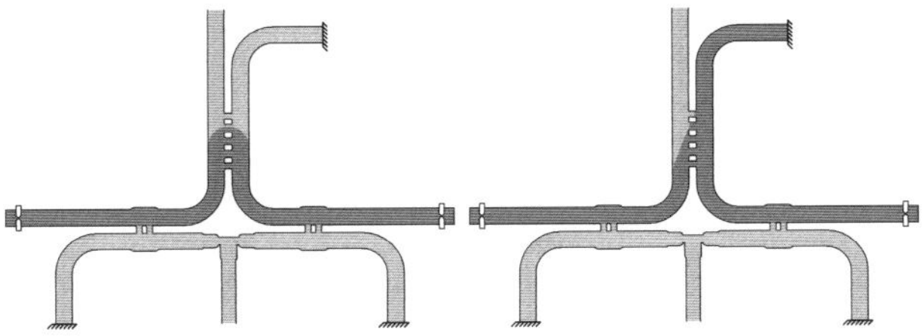

Fig. 2. The two reflection paths in the RF waveguide structure, "U-turn" (left) and "Y-branch" (right), that cause most of the deterioration of the image rejection once the contributions of the amplitude and phase balance are minimized. Similar paths exist in the IF system, as discussed in the main text, where the same labels are used for the analog cases.

genic front-end) is very similar to the Band 9 ones produced in series for the ALMA observatory [3]. Apart from the mixers themselves, there are modifications to the optics block, optics support structure and LO injection optics. Another important change is the doubling of the IF transport channels, from two to four, across all temperature levels. The most important design features are discussed in the following sections.

One of the new developments in this receiver's design, compared to the currently operational ALMA Band 9 receivers, is the direct coupling of the SIS mixer outputs to the cryogenic low-noise amplifiers, without intermediate circulators to suppress standing waves, but at the same time without tightly integrating the mixers and amplifiers. This leads to a certain deterioration of the image rejection ratio (IRR), as explained below, but because of the mixer's good performance and the greatly improved performance of some IF components, we still obtain a quite acceptable IRR (better than 15 dB for most frequencies). The fact that the IF components are not tightly integrated allows for convenient independent testing, while the ability to place the IF amplifiers a good distance away from the mixers will be beneficial in the development of 2SB mixer arrays.

A. Mixers

The receiver is based on the 2SB SIS mixers developed in our group over the last few years [4], shown in Fig. 1. The RF quadrature hybrid is optimized for high isolation (a.k.a. directivity) and input return loss, as these parameters are the main limiting factors in obtaining good image rejection once the phase and amplitude are adequately balanced. The other waveguide components (LO couplers, LO splitter and matched loads) received similar optimizations. Apart from the waveguide design, it was also found that the right choice of material and first-class machining are instrumental for obtaining the theoretically expected performance[1]. With this design and an IF system incorporating circulators to isolate the mixers from the LNA input reflections, image rejection ratios in excess of 20 dB over most of the band were demonstrated [4].

[1]The operational RF hybrid blocks were machined by GARD, (Group for Advanced Receiver Development), Chalmers University of Technology, Gothenburg, Sweden. The material used was a copper-tellurium alloy, *not* gold plated.

B. IF System

After optimization of the RF part of the 2SB mixers as mentioned in the previous section, the next limitation of the image rejection ratio (IRR) is mainly due to the performance and matching of the components in the IF system. Although the frequencies are much lower, very similar arguments apply here, namely that once the phase and amplitude balances are adequate, the IRR is dominated by the return losses of the components at the end-points of the chain, and the isolation of the, in this case, IF hybrid.

To recapitulate, in the RF part of the 2SB system, imperfections in these properties cause image band leakage mainly by two parasitic signal paths from mixer to mixer (illustrated in Fig. 2): one by way of the hybrid and the RF load that terminates its idle port ("Y-branch"), and the other ("U-turn") directly mixer-to-mixer because of the finite isolation of the hybrid. Both of these paths lead to the presence of an error signal at the mixers that is always in counter-phase with the direct signal, and therefore maximally destructive to the image rejection. To mitigate this, we focussed on optimizing the RF load and the isolation of the hybrid, keeping an eye on the phase and amplitude balance al well, of course.

As mentioned before, similar mechanisms play a role in the IF system, and if we want to get rid of the circulators (which are bulky, lossy and expensive, as well as limiting in bandwith), while keeping high IRR values, these should be addressed. The main qualitative difference in the IF case is that U-turn and Y-branch paths are present on *both* the mixer-side and amplifier-side of the hybrid, yielding two U-turns and four Y-branches, because both the reflections of the mixer outputs and the amplifier inputs are typically non-negligible. In contrast, in the RF part, when the feedhorn is well-matched to the hybrid, all but one of each kind of reflection path are eliminated.

The IF hybrids were developed and produced by Centro Astrónomico de Yebes (CAY) [5]. The hybrid's isolation has not been measured cryogenically, but at room temperature it is reported to be very similar to the input reflections (S_{nn}) [6], something we also found generally the case when designing the RF hybrid. There is no reason to believe that this changes at low temperature. The S_{nn} parameters *were* measured cryogenically, and shown to be below -24 dB almost everywhere, with a few points touching -22 dB. Because the U-turn reflection paths also depend on high reflection coefficients of both the mixers and amplifiers, we conclude that with this isolation performance of the hybrid their contribution to the IRR is minor.

On the mixer side, new electromagnetic simulations of the SIS devices [7], which come from the same batches as the DSB production mixers, show them to be much better matched on the IF side than previously estimated. Typical values around -8 dB were found. We plan to verify these calculations experimentally in the near future, but until new Band 9 junctions can be produced there is not much we can do to improve this, and the obtained performance shows it to be adequate.

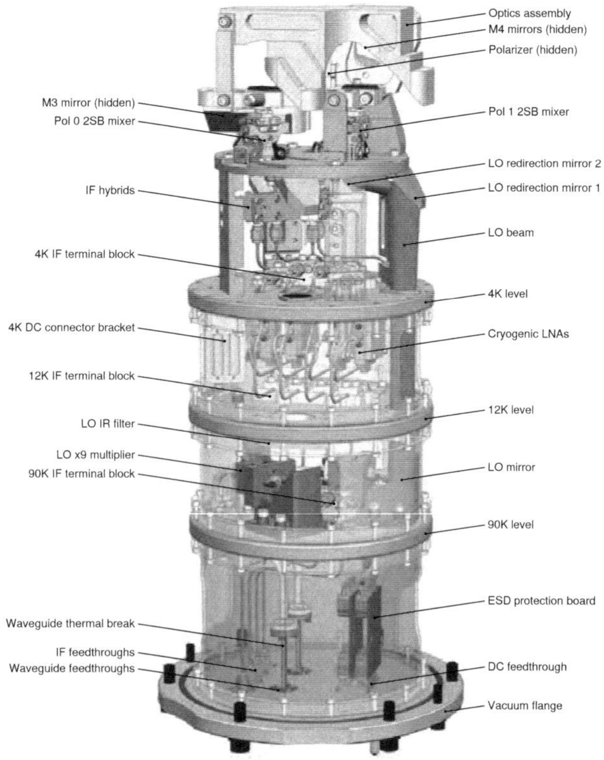

Fig. 3. Annotated 3D model of the Band 9 sideband-separating cartridge (with the fiberglass thermal insulation rings rendered transparant). The basic structure is derived from the Band 9 DSB production cartridge. The optics assembly is retained (with slight reworking to make space for the larger mixers), although it now contains two superfluous LO mirrors that originally projected the LO beams by way of beamsplitters into the mixers. Instead, each LO beam is now re-focussed into the bottom of the mixers by a combination of ellipsoidal and hyperboloidal mirrors. To facilitate later re-use of the cartridge structure for other frequencies, the LO and IF regions are spatially separated as much as possible.

Fig. 4. Left: the completed Band 9 2SB cartridge from roughly the same angle as Fig. 3. Top right: detail of the 4 K level with the two 2SB mixers just below the optics assembly and one of the IF hybrids clearly visible in the center. On top of the 4 K plate is the thermally anchored IF terminal block, that allows easy mating and de-mating of the IF connection between the 4 K and 12 K levels. Bottom right: detail of the 90 K level with the two LO multipliers and a similar IF thermal anchor block. On all levels the interconnects to the underlying level are accessible from above.

This leaves the input matching of the cryogenic LNAs as the main target for optimization. The LNAs used here, also developed and produced by CAY, are a further evolution of the three-stage discrete HEMT amplifiers used in the ALMA DSB Band 9 receivers [8]. The main improvement is in the input reflection, which was decreased from about -4 dB to better than -10 dB over most of the band.

These three factors together make that the total image rejection is only slightly reduced with respect to the former configuration with circulators, to about 15 dB in the lowest points over most of the band (see section III-B).

In the DSB Band 9 cartridge, the SIS mixers were biased through the load ports of the circulators (which were DC-transparent) by resistive bias networks housed in small boxes mounted on the 4K level. In order to improve the integration of IF components, the new LNAs feature not only a bias-T but also the SIS bias network. The (high level) bias voltages are supplied to the same connector as the bias for the LNA HEMTs, greatly simplifying the DC cable harness as well.

C. Cartridge

Because of the doubling of the number of IF channels, the IF system was mechanically completely redesigned compared to the DSB case. Some design features were kept, especially the interstage couplings, which are always above the temperature plates and double as thermal anchors, facilitating easy disassembly of the cartridge for modification or repair.

The two produced 2SB cartridges are partially constructed out of left-over components of the DSB production cartridges, so not all original features are functional anymore. For instance, the optics block could be simplified considerably, as the LO mirrors and beamsplitters are superfluous now. As an aside, while re-use of the old optics would save a certain amount of cost if a full ALMA Band 9 upgrade is undertaken, there is also an opportunity to redesign the optics completely with a much improved polarisation purity compared to the original Band 9 optics.

In the new design,the IF components are clustered in the core cylinder of the cartridge, while reserving the outer space to the LO components. This should make the cartridge more or less universal from the 4 K level downwards, allowing us to reuse large parts of the design for other frequency bands, for instance 230 or 345 GHz for use in the Event Horizon Telescope effort or any other observatories using ALMA-style front ends.

An annotated 3D model is shown in Fig. 3, and a few photographs of the completed cartridge in Fig. 4.

D. SIS Device Selection

As important as the phase and amplitude balance within the RF waveguide structure is the matching between the SIS mixer devices. The phase behavior of the mixers is likely to be quite uniform within one design and wafer run (or batch). The gain, however shows more variation, even within one batch. The gain of the mixers as function of RF frequency was determined during their DSB testing with a standard hot-cold load measurement, and candidate pairs selected by eye.

Besides the mixer's gain, another important parameter is their LO power requirement. With the LO splitter integrated in the hybrid block, there is no way to level the LO power to the mixers individually. Since the required LO power depends, within one design, largely on the normal-state resistance R_N and the RF coupling coefficient (and hence the noise temperature T_N), both were additional criteria for mixer pairing. Finally, there are considerations of tunability, especially important for a facility instrument. Since the mixers were harvested from left-over production junctions, it may be obvious that the ones easiest to operate were already gone. Nevertheless, a sufficient number of candidate pairs were identified.

After this pre-selection, no sure-fire way for mixer pairing based on DSB data has been found yet, as described in [4]. More than a dozen candidate combinations were tried, and finally four sets with both good noise temperature and image rejection, and decent tunability, were set apart for the deliverable receivers.

III. Measurement Results

A. Noise Temperature

The SSB noise temperature of the first Band 9 2SB receiver cartridge, measured above the front-end window and heat filters, is shown in Fig. 5. For the deliverable to the SEPIA instrument, a request was made to investigate the performance in an extended band, 6 GHz beyond both edges of the original ALMA specification (602–720 GHz). Especially the low-frequency extension, which is interesting for high-redshift observations of several spectral lines, shows quite a usable performance. In fact, the noise temperature over an even larger range (say, 580–730 GHz) stays fully within the ALMA specifications. Since the measurements were performed in the lab at about sea level, a significant part of the upturn beyond the ALMA band edges is likely to be due to the deep and broad water vapor absorption lines at 570 and 750 GHz. At the observation site, the extended band performance is expected to be better still.

B. Image Rejection

Figure 6 shows the image rejection ratio as function of input (on-sky) frequency for both polarizations. In most places,

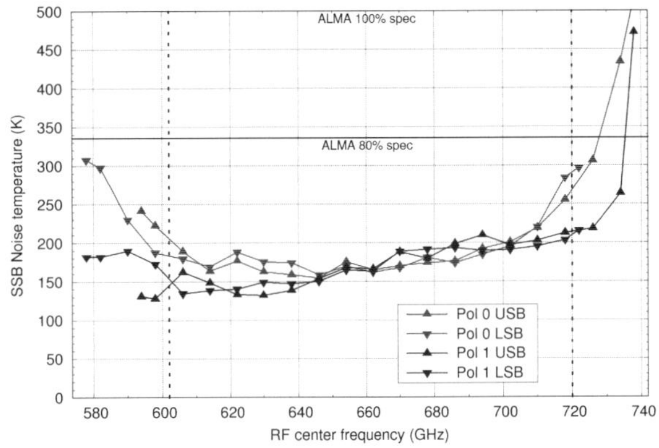

Fig. 5. The SSB noise temperature as function of input frequency of the first Band 9 2SB receiver cartridge. The regions outside the 602–720 GHz range (vertical dashes) form the extended band. The two levels are indicated below which, respectively, 80% and 100% of the points have to be below in order to meet the original ALMA Band 9 DSB specification (doubled to represent corresponding SSB values).

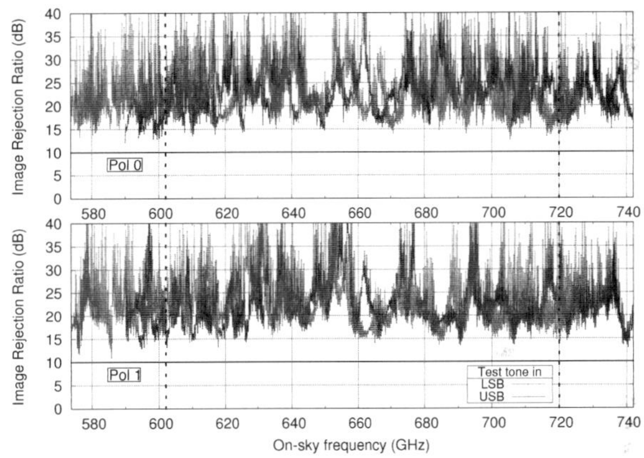

Fig. 6. The image rejection ratio as function of input frequency for both polarizations, and over the extended frequency band (vertical dashed lines). The horizontal axis refers to frequencies within the sideband that does *not* contain the test tone, *i.e.*, the sideband used for observation. The ALMA specification is 10 dB (horizontal line).

the rejection is above 15 dB. Also here, the band is extended beyond the original ALMA limits with quite usable results. Although the *average* IRR is definitely lower than the one obtained with IF circulators [4], the baseline values are actually very comparable. It is mainly the variations upwards that are much faster now, being dominated by the long IF runs.

C. IF Spectrum

Because of the absence of IF circulators, the IF ripple is of special concern. A typical IF output spectrum is shown in Fig. 7. Despite the absence of circulators, the IF ripple due to standing waves (the 0.5–1 GHz periodicity visible in the figure) is not much larger than, say, 1.5 dB.

Since most current back-ends digitize the IF signal in 2 GHz subbands, the maximum ripple over *any* 2 GHz interval within

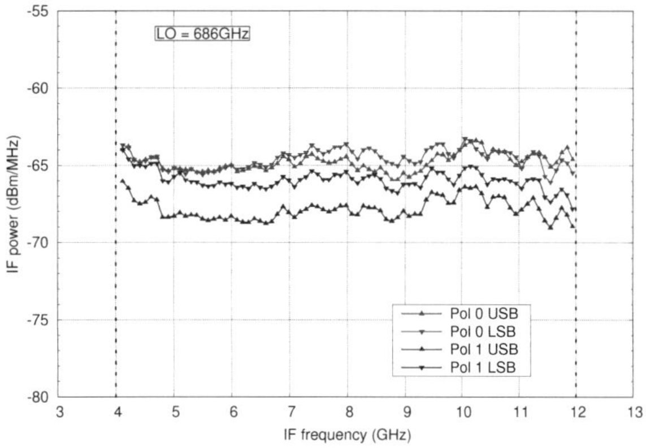

Fig. 7. IF output spectrum with the mixer looking at a 300 K load. The LO frequency is 686 GHz, which puts the CO_{6-5} line (691 GHz) in the USB. Despite the absence of circulators between the mixers and the LNAs, the IF ripple due to standing waves is not much larger than about 1.5 dB.

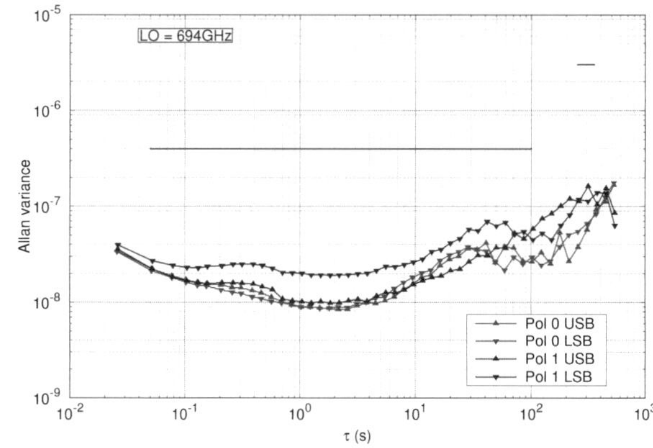

Fig. 9. Allan variance of the IF output power as function of integration time, looking at a 300 K black-body load. The LO frequency is 694 GHz. The horizontal bars indicate maximum allowed values in the ALMA specifications.

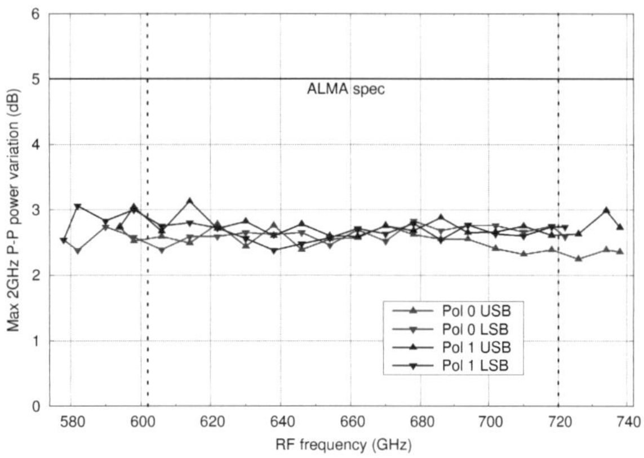

Fig. 8. The maximum peak-to-peak IF ripple in any 2 GHz subband of the IF band, for all tested (extended range) LO settings. The ALMA specification allows 5 dB of ripple.

the IF band is an important specification, as it determines the loss of effective number of bits (ENOBs), and therefore dynamic range, due to IF baseline ripple. In Fig. 8 this is plotted for all tested LO frequencies. As can be seen, the ripple is of the order of 3 dB, resulting in about 1/2 bit loss, and quite constant over the band. Of course, when quantization happens with a sufficient number of bits, a larger part of the IF band (or even the entire band) can be digitized in one go without the need of equalization of the subbands. It is likely that upcoming back-ends will have this capability.

D. Stability

Amplitude stability is an important parameter to determine the optimum on-source integration time, especially in a single-dish telescope where random gain fluctuations are not correlated away. The Allan variance as function of integration time is shown in Fig. 9, and turns out to be roughly an order of magnitude better than the ALMA specification. This is also a good indication of the quality of all active and passive parts of the signal chain (mixers, hybrids, LO, amplifiers), as all of them have overriding influence on the Allan variance.

E. Performance Overview

The 2SB Band 9 cartridge went through the full set of qualification measurements, in this case to be delivered to APEX for the SEPIA front-end[2]. Not all results are detailed here, but Table I contains the most important specifications, along with the obtained results. Where several resulting values or ranges are given, these pertain to different polarizations or IF channels. To enable direct comparison with the ALMA specifications, the results do not include the extended RF frequency range. However, *all* specifications are also met in the extended range. One small exception is the beam squint, which is marginally out of spec, but has been accepted by the SEPIA project nonetheless.

IV. Conclusion

We produced a fully deployable ALMA Band 9 (602–720 GHz) sideband-separating receiver cartridge with a total IF bandwidth of 32 GHz. Apart from some minor optical issues due to the use of a reject optics block, it meets all ALMA specifications, and in case of the crucial ones (sensitivity, image rejection, IF passband ripple and stability), with large margins, and even over an extended frequency range of about 580–730 GHz. It proves that removing the IF circulators, even when the mixers and amplifiers are far apart, can still meet state-of-the-art specifications when careful matching of the IF components is observed, which is important for future array applications. The cartridge design is modular, easy to service and designed with the adaptation to other bands in mind.

[2]These results were obtained during the in-house commissioning phase. Since then, some modifications have been made to the cartridge configuration to ensure long-term stability, and the tabled values may deviate slightly from those at final delivery, which were still being measured at the time of writing.

After commissioning in August 2018 on APEX, it should start proving its worth on the sky.

TABLE I
KEY SPECIFICATIONS AND RESULTS FOR THE 2SB BAND 9 RECEIVER.

Requirement	Specification	Result
RF frequency range	602–720 GHz	586–730 GHz
IF frequency range	4–12 GHz	
Number of IF channels	4	
Polarizations	2 (linear)	
Noise temperature		
80% of the RF band	$\leq$ 335 K	$\leq$ 200 K
100% of the RF band	$\leq$ 500 K	$\leq$ 284 K
Image rejection		
100% of the RF band	$\geq$ 7 dB	$\geq$ 12.6, 13.3 dB
$\geq$10 dB	90% of the RF band	100%
$\geq$15 dB	goal	96%, 93%
IF ripple		
4–12 GHz	$\leq$ 7 dB p-p	2.8 – 3.8 dB p-p
any 2 GHz subband	$\leq$ 5 dB p-p	2.8 – 2.9 dB p-p
Amplitude stability σ^2_{Allan}		
$0.05 \leq \tau \leq 100$ s	$\leq 4 \times 10^{-7}$	$5.5 - 7.1 \times 10^{-8}$
$\tau = 300$ s	$\leq 3 \times 10^{-6}$	$1.1 - 1.5 \times 10^{-7}$
Gain compression		
373 K vs. 77 K	$\leq$ 3%	$\leq$ 0.6%
Beam performance		
Aperture efficiency	$\geq$ 80%	$\geq$ 81.2%
Polarization efficiency	$\geq$ 97.5%	$\geq$ 97.5%
Beam squint	$\leq$ 10% FWHM	$\leq$ 10.2% FWHM

ACKNOWLEDGEMENT

This work was supported by the Netherlands Research School for Astronomy (NOVA) under a NOVA-IV ALMA R&D grant.

REFERENCES

[1] A. Khudchenko, R. Hesper, A. M. Baryshev, J. Barkhof, and F. P. Mena, "Modular 2SB SIS receiver for 600–720 GHz: Performance and characterization methods," *IEEE Transactions on Terahertz Science and Technology*, vol. 7, no. 1, pp. 2–9, Jan 2017.

[2] Belitsky, V., Lapkin, I., Fredrixon, M., Meledin, D., Sundin, E., Billade, B., Ferm, S.-E., Pavolotsky, A., Rashid, H., Strandberg, M., Desmaris, V., Ermakov, A., Krause, S., Olberg, M., Aghdam, P., Shafiee, S., Bergman, P., Beck, E. De, Olofsson, H., Conway, J., Breuck, C. De, Immer, K., Yagoubov, P., Montenegro-Montes, F. M., Torstensson, K., Pérez-Beaupuits, J.-P., Klein, T., Boland, W., Baryshev, A. M., Hesper, R., Barkhof, J., Adema, J., Bekema, M. E., and Koops, A., "SEPIA — a new single pixel receiver at the APEX telescope," *A&A*, vol. 612, p. A23, 2018. [Online]. Available: https://doi.org/10.1051/0004-6361/201731458

[3] A. M. Baryshev, R. Hesper, F. P. Mena, T. M. Klapwijk, T. A. van Kempen, M. R. Hogerheijde, B. D. Jackson, J. Adema, G. J. Gerlofsma, M. E. Bekema, J. Barkhof, L. H. R. de Haan-Stijkel, M. van den Bemt, A. Koops, K. Keizer, C. Pieters, J. Koops van het Jagt, H. H. A. Schaeffer, T. Zijlstra, M. Kroug, C. F. J. Lodewijk, K. Wielinga, W. Boland, M. W. M. de Graauw, E. F. van Dishoeck, H. Jager, and W. Wild, "The ALMA band 9 receiver - design, construction, characterization, and first light," *A&A*, vol. 577, p. A129, may 2015.

[4] R. Hesper, A. Khudchenko, A. M. Baryshev, J. Barkhof, and F. P. Mena, "A high-performance 650-GHz sideband-separating mixer — design and results," *IEEE Transactions on Terahertz Science and Technology*, vol. 7, no. 6, pp. 686–693, Nov 2017.

[5] I. Malo-Gómez, J. D. Gallego-Puyol, C. Diez-González, I. López-Fernández, and C. Briso-Rodríguez, "Cryogenic hybrid coupler for ultra-low-noise radio astronomy balanced amplifiers," *IEEE Transactions on Microwave Theory and Techniques*, vol. 57, no. 12, pp. 3239–3245, Dec 2009.

[6] I. Malo-Gómez, *private communication.*

[7] K. Rudakov, *private communication.*

[8] R. I. Amils, J. D. Gallego, C. Diez, I. López-Fernández, A. Barcia, S. Muñoz, J. L. Sebastián, and I. Malo, "Low input reflection cryogenic low noise amplifier for radio astronomy multipixel receivers," *Journal of Instrumentation*, vol. 11, no. 10, p. P10018, oct 2016. [Online]. Available: http://stacks.iop.org/1748-0221/11/i=10/a=P10018

A Low-Power Compact Millimeter-wave Radiometer for a Weather Cubesat

Sean Bryan[*], Philip Mauskopf[1], and Christopher Groppi[1]

[1]Arizona State University, Tempe, AZ, 85281, USA
[*]Contact: sean.a.bryan@asu.edu

Abstract—Millimeter-wave atmospheric sounding satellites measure the temperature, humidity, and oxygen content across the globe and in profiles resolved by altitude. Considered separately from calibration or modulation subsystems, current heterodyne radiometer technology has significant power and size requirements. This either requires a large satellite such as ATMS, or limits the channel count and frequency coverage of small satellites or cubesats such as MicroMAS-2. In this talk, we present a significantly improved radiometer system based on patent-pending technology developed at ASU, and sensitivity forecasts used to optimize the design for collecting weather data. Using a millimeter wave LNA as the first element in the radiometer system, followed by a millimeter-wave filter bank and power detectors, has several advantages. Since the system does not use a local oscillator, power requirements are reduced. Since the passband of each channel is defined by its filter bank cavity, the system will naturally have high frequency stability, unlike an LO-based system that could require temperature stabilization. Also, since the channelization is accomplished with physically small millimeter-wave filters instead of potentially larger IF lumped element filters, more channels with finer resolution are possible in a small space. We have already successfully built prototype filter banks in the context of developing the WSPEC instrument for astronomy, and are now working towards an integrated prototype of a full LNA/filter-bank/detector system in the context of future cubesat development.

*** Formatting Issue - Best Available Paper/Graphic ***

29th IEEE International Symposium on Space THz Technology (ISSTT2018), Pasadena, CA, USA, March 26-28, 2018

The 220GHz Terahertz Cloud Radar System for Atmosphere Observation

Weidong Hu[1*], Shi Chen[1], Wenlong Zhang[1], Fen Yue[1], and Leo P. Ligthart[2]

*[9]E&)F)6"#%0) &"*G"3&2!*6-<E&F%6H&-1(J*' (*'-"*GB)++)& &"K(L&"(%'3&(! & 5'3&2!%M1&E&)F)6% 9888:9 <? !) %("*

*[Ā] N&G'A %L&0) -"*G'&2!%*6-<"O 21+-"*G'+&2)2(+"$%6)%&&')6%N&G<ĀC88"@@P&!&'+(%0'*

[*]Contact: isstt2018info@gmail.com

Abstract—Clouds play an important role in the atmospheric influence on the earth's radiation budget, which is reflected both in some long-term effects of climate change and some short-term effects of sudden meteorological disasters, therefore, Cloud detection becomes more and more important nowadays.

The terahertz cloud profiling radar is an effective tool in weather observations for cloud vertical structure, cloud liquid water and ice water content, etc. In this paper, a 220GHz terahertz cloud profiling radar system for atmosphere remote sensing is designed, according to the detection requirements, system parameters are calculated as well as the system diagram is proposed. Then the prototype of terahertz cloud profiling radar is developed, and atmosphere observation experiments have been carried out outdoors.

The experimental results show that it is available to get atmosphere particles echo using terahertz wave, and the radar system is stable and sensitive capable of providing accurate information for clouds.

TECHNOLOGY DEVELOPMENT FOR LONG WAVELENGTH ASTRONOMICAL SATELLITES

Jonathan R. Hoh [2], Paul F. Goldsmith (PI)[1], Jose V. Siles [1], Adrian J. Tang[1], Christopher E. Groppi [2], and Jeremy Whitton[2]

[1]*NASA Jet Propulsion Laboratory, 91109, Pasadena CA, USA*
[2]*Arizona State University School of Earth and Space Exploration, 85287, Tempe AZ, USA*
*Contact: jrhoh@asu.edu

Abstract— Here we present the characterization of the performance of a novel design for a digital spectrometer that could be used for high resolution cm/mm/submm spectroscopy. The CMOS ASIC spectrometer design, developed at JPL and UCLA, has dramatically lower power consumption than current approaches that generally employ Field Programmable Gate Arrays (FPGAs). Particularly for space missions and for small satellites, power consumption is a major issue. The order of magnitude lower power consumption of the ASIC approach is thus critical for future missions employing large-format focal plane arrays. Our task was to evaluate this 1024 channel, 1.3-GHz bandwidth CMOS spectrometer in terms of ability to integrate and its filter shape. The chip was to be tested largely at half-maximum speed to allow for use of the polyphase filter bank.

The results of this SURP show that the ASIC spectrometer can be made to perform largely as expected based on its design parameters. However, they suggest that more testing of the spectrometer chip is required. Based on the Allan Variance plots obtained, it is possible that there is an issue with the long-term integration of the chip; however, the reduced turnoff seems to indicate that while the output may have been drifting, the relative outputs between bins was drifting to a lesser extent. The filter shape testing confirms that the spectrometer has the resolution expected, and that its polyphase filter bank works properly. The higher noise floor for certain bins is something which could warrant investigation, but a -27dB noise floor is likely sufficiently low for the applications it will be used in. Finally, the anomalous filter shapes exhibited for a clock power level of -13dBm could be an artifact of the test system used, but warrant further investigation.

Nb/Al-AlN/Nb superconducting tunnel junctions: fabrication process and characterization results

Alexey Pavolotsky[1,*], Vincent Desmaris[1], and Victor Belitsky[1]
[1] Chalmers University of Technology, Group for Advanced Receiver Development, Department of Space, Earth and Environment, Göteborg, 412 96, Sweden
*Contact: Alexey.Pavolotsky@chalmers.se

***Abstract*—Modern radio astronomy demands for broadband receiver systems. For SIS mixers, this translates into objective to employsuperconducting tunnel junctions with very low R_nA and low specific capacitance. The traditionally used Nb/AlOx/Nb junctions have largely approached their physical limit of minimizing those parameters. It is commonly recognized that it is AlN-barrier junctions, which are needed for further progressing of the broadband instrumentation for radio astronomy. In manuscript we present results of developing a fabrication process for high quality Nb/Al-AlN/Nb junctions and characterization the junctions' DC electric properties and their aging and annealing stability, as well as the junctions' specific capacitance.**

I. Introduction

Modern radio astronomy instrumentation projects, e.g. ALMA 2030 Roadmap [1], call for twofold o threefold enhancement of the RF and IF bandwidths of the SIS mixers. Wider RF band requires lower Q-factor and consequently, lower R_nA-product of the SIS junctions. Al-oxide, traditionally used as tunnel barrier material of SIS junctions approaches its physical limit: by using SIS junctions with higher current density and thus thinner tunnel layer, the tunnel barrier quality becomes hardly predictable. It is widely recognized by the SIS community that in general, the quality of the Al-oxide SIS junctions degrades and becomes unreliable, once R_nA decreases below probably ~15 Ohm·µm² for AlO_x tunnel barrier junctions. Simultaneously, a wider IF band of SIS mixer needs the junction capacitance be lower, in turn forcing shrinking the junction size, which may cause production yield problems and may lead to spreading of junction parameters [2].

An alternative SIS tunnel barrier material, aluminum nitride, AlN, present a reduced tunnel barrier height as compared to the Al-oxide tunnel barrier. Consequently, for the same current density (or R_nA), the physical thickness of the tunnel barrier is substantially larger for AlN than for Al-oxide material. This opens up two possibilities: **i**. for the given current density, the junctions with AlN barrier will have lower capacitance giving the immediate advantage of operating the junctions over much wider RF and IF bands as compared to today's junctions using Al-oxide barrier; **ii**. the junctions with AlN barrier can have much higher current density without degrading of their quality.

We have developed a process for high-quality junction fabrication based on microwave plasma nitridation. The Nb/Al-AlN/Nb junctions with $R_nA \sim 15$ Ohm·µm² demonstrate excellent quality, $R_j/R_n > 25$.

During the fabricating process and in the course of mounting/packaging, the mixer chip is exposed to the elevated temperatures. The upper temperature limit that can be accepted during junction fabricating (baking of resists or heating during deposition of the layers) and packaging (curing of glues or epoxies, or heating for wire bonding) is defined by the stability of the junction properties at the elevated temperatures. For this reason, we have carried out the study of aging and annealing behavior of Nb/AlN/Nb junctions by exposing junction samples to the aging/annealing temperature profile between room temperature and 200°C.

Finally, we measured specific capacitance of the Nb/AlN/Nb junctions using approach similar to one reported in the paper [3] and compared those with the data for $Nb/AlO_x/Nb$ junctions.

II. Nb/Al-AlN/Nb junctions fabrication process

The developed process for fabricating of Nb/Al-AlN/Nb junctions is based on the standard Nb/Al-AlOx/Nb process supported by GARD [2], [4]–[7] but with plasma nitridation of Al instead of thermal oxidation.

The earlier developed Nb/Al-AlN/Nb junctions fabrication processes first relied on either glow discharge nitridation [8]–[10]. Later, it was understood that nitrogen ion bombardment causes structure damage in the formed ultra-thin AlN layer and by that, limits the low end of accessible R_nA range. The solutions to the problem were found later [11]–[14] either in spacial separation of the sample and the plasma cathode volume (often referred as anodic plasma or remote plasma nitridation) optionally combined with active controlling of the substrate potential [15] to avoid ion bombardment in unwanted extent. Among the remote plasma sources, there were various kinds used: RF biased cathode [11] or magnetron [14], either inductively coupled plasma source [12], [13].

In the present work, a remote plasma source based on electron cyclotron resonance (ECR) was used. ECR sources are characterized by relatively high plasma density

($\sim10^{-11}\ldots10^{-12}$ cm^{-3}), low plasma potential (ca. 15...30V) and consequently, low ion energy; high density of neutrals (about 2 orders of magnitude higher than that of ions) and their high temperature (~0.2 eV) [16]. On our view, these features of ECR plasma makes is especially suitable for nitridation of AlN tunnel layer. A high concentration of hot neutrals suggests an effective nitridation of the Al, once the low energy of ionized components ensures low radiation damage to the formed AlN ultra-thin layer. Additionally, the magnetic field in the plasma source is strongly diverging, and because of that, it is only over a very limited spacial region, where the frequency of the electron cyclotron oscillations appeared in resonance with the microwave excitation frequency. That additionally guarantees a spacial separation between plasma region and the sample surface being nitridized. Also, importantly, the ECR source is quite compact, thus facilitating its integration into the deposition system.

A schematic view of the plasma nitridation arrangement is shown on the Fig. 1. The plasma source is placed relatively far from the substrate, ensuring the spacial separation between the plasma region and the sample under nitridation. There is a shutter between the plasma source and the sample, which is kept closed during transient upon plasma ignition and the parameters settling process. The sample holder is grounded through the ammeter. The current to ground during nitridation was always within +1...+3 mA confirming very low ion bombardment current density.

Facilitating the initial phase of the Nb/Al-AlN/Al multilayer process development, we used anodization profiling of the tunnel structure similar to the approach suggested in [17]. We targeted obtaining the sharp profile at the Nb/Al interface. The anodization profile of Nb/Al interface was considered "sharp enough" if it was about the same shape and sharpness as that of the known high quality Nb/Al-AO_x/Nb junctions, e.g. those used for ALMA Band 5 mixer production [5]–[7], shown on the Fig. 2 (green curve). Additionally, the height of the peak at the profiles corresponding to the AlN tunnel layer was taken as a rough indication of the layer thickness, which allowed quick converging to the nitridation process parameters providing the junction's R_nA numbers within the range of interest (the blue and red curves on the Fig. 2).

III. Electrical characterization of Nb/Al-AlN/Nb junctions

A. Current-Voltage characteristics

The fabricated junctions demonstrated current-voltage characteristics typical for high quality junctions: low subgap leakage current quantified by R_j/R_n ratio larger than 25 and high and vertical superconducting gap current onset. The junctions with lowest R_nA demonstrated the signs of overheating while recording the current-voltage characteristic. Plots at the Fig. 3 demonstrate typical characteristics for the junctions with high (ca. 120 Ohm·µm^2) and low (ca. 15 Ohm·µm^2).

B. Aging and annealing stability of Nb/Al-AlN/Nb junctions

The upper temperature limit that can be accepted during junction fabricating (baking of resists or heating during deposition of the layers) and packaging (curing of glues or epoxies, or heating for wire bonding) is defined by the stability of the junction properties at the elevated temperatures. For this reason, we have carried out the study of aging and annealing behavior of Nb/AlN/Nb junctions.

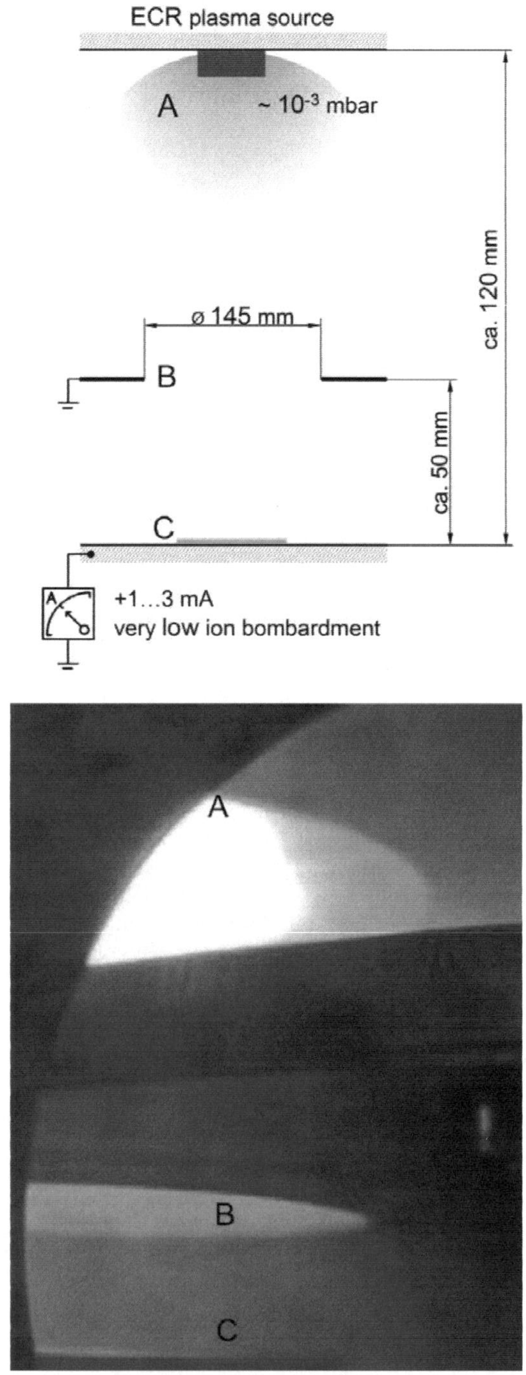

Fig. 1. Plasma nitridation arrangement: A – plasma source, B – opening in the shutter plate, C – sample holder with sample.

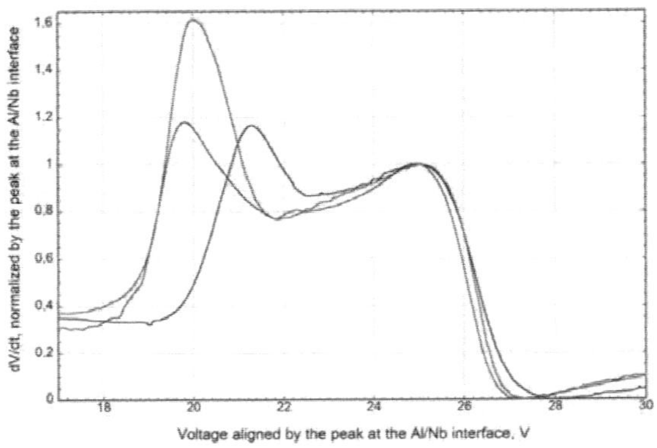

Fig. 2. Anodization profiles of the tunnel structures: green line: - Nb/AlOx/Nb, (R_nA ~ 30 Ohm·μm^2), same as those used for ALMA Band5 mixers [5]–[7]; red line - Nb/AlN/Nb, R_nA ~ 120 Ohm·μm^2; blue line - R_nA ~ 15 Ohm·μm^2.

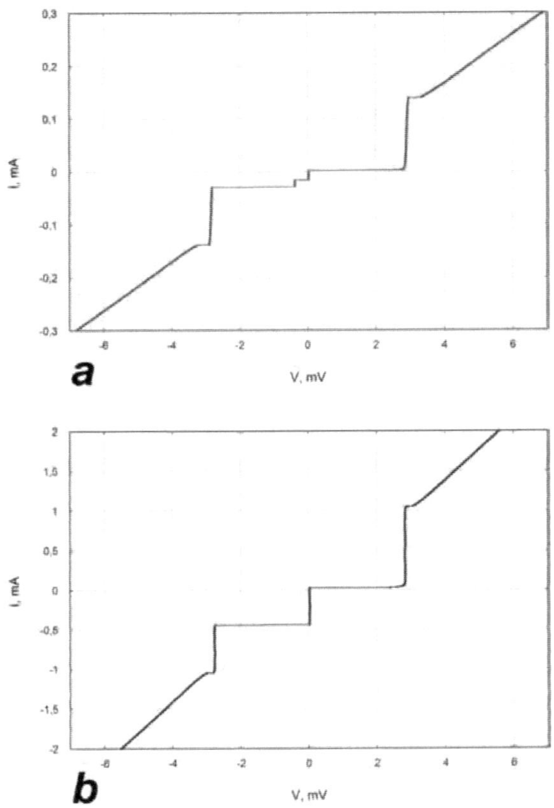

Fig. 3. Typical current-voltage characteristics of Nb/AlN/Nb junctions with ~6 mm^2 area: (a) R_nA ~ 120 Ohm·mm^2, R_j/R_n > 35; (b) R_nA ~ 15 Ohm·mm^2, R_j/R_n > 25

The wafers of junction with R_nA ~ 15 and 120 Ohm·mm^2 respectively had been exposed to the aging/annealing temperature profile between room temperature and 200°C, as shown at the Fig. 4a. Current-voltage characteristics at 4K temperature were recorded directly after the junctions fabrication and after each step of the aging/annealing temperature point. Values of normal resistance (R_n) and quality factor (R_j/R_n ratio) were extracted from all recorded current-voltage characteristics Fig. 4b and Fig. 4c present the evolution of the normal resistance and quality factor values in course of the aging/annealing.

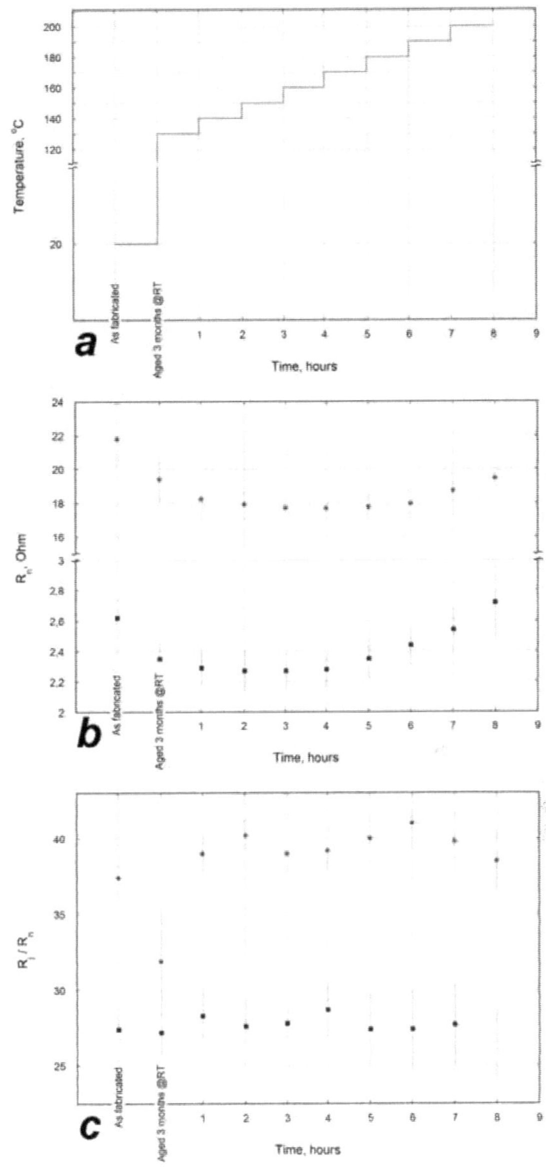

Fig. 4. Temperature profile (a) of aging/annealing of AlN-barrier junctions and evolution of R_n (b) and R_j/R_n ratio (c) for the junctions with R_nA ~ 120 Ohm·mm^2 (red) R_nA ~ 15 Ohm·mm^2 (blue).

From the plots, one can conclude that compared with the high-quality Nb/Al-AlOx/Nb junctions [4], the Nb/Al-AlN/Nb junctions are probably somewhat more temperature stable. It is safe and rather advisable to anneal junctions up to the modest 130 - 150°C and probably safe to expose to temperatures up to 180°C for the shorter time intervals. In the other words, any resist, glue or epoxy can safely be used in the fabricating and packaging processes involving Nb/Al-AlN/Nb junctions.

Moreover, heat treatment at the modest 130 - 150°C could be advisable for adjusting of the junctions' normal resistance values, as well as for stabilizing of their properties. No degradation of the junction quality was observed after that treatment. Above 180°C however, junction quality starts degrading, which also could be seen also by reducing of the superconducting gap voltage and smearing of the gap feature at the current-voltage.

C. Junction capacitance characterization

By the moment, the first measurements of the Nb/Al-AlN/Nb junctions specific capacitance have been made following the approach similar to that communicated in the paper [3], using cryogenic S-parameter measurements [18].

Specific capacitance numbers of the junctions with $R_nA \sim 6$ and 50 Ohm·mm^2 and area of 4 – 10 mm^2 were attempted to measure. So far, we have found extracting of reliable value of specific capacitance of the junctions with $R_nA \sim 6$ Ohm·mm^2 to be problematic. That is probably due to the very high conductance of the junctions. As for the junctions with $R_nA \sim 50$ Ohm·mm^2, we estimate specific capacitance to be about 45 fF/mm^2.

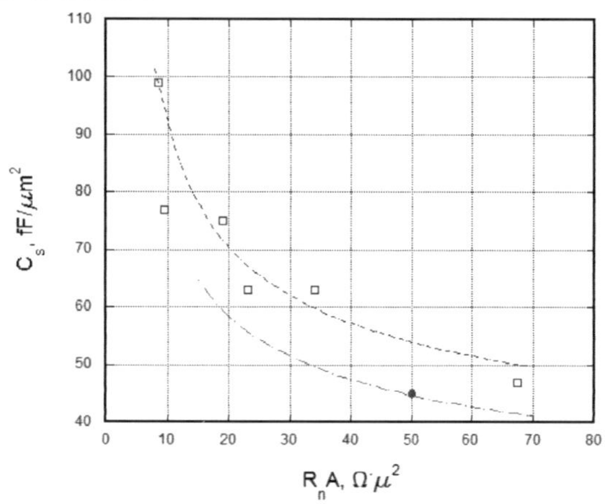

Fig. 5. Specific capacitance of Nb/Al-AlN/Nb junction (*red*) as compared with that of Nb/Al-AlOx/Nb junctions [19] (*blue*). The capacitance data for the junctions are approximated with empirical relation $C_s = a/ln(R_nA)$ [20], where a is equal to 211 [19] for the Nb/Al-AlOx/Nb junctions and to 175 for the measured Nb/Al-AlN/Nb junctions.

On the Fig. 5, the comparison between measured specific capacitance numbers of superconducting tunnel junctions with AlN and AlO_x tunnel barriers is presented. For the junctions with $R_nA \sim 50$ Ohm·mm^2, the specific capacitance of the junctions with AlN barrier is about 17% lower than that of the junctions with AlO_x barrier, which is consistent with usually referred values of dielectric permittivity of alumina and bulk aluminum nitride materials, i.e. 9.5 – 10 vs 7.9 – 8.3, e.g. [21].

CONCLUSIONS

In this work, we developed a fabrication process for the high quality Nb/Al-AlN/Nb junctions. We characterized the junctions' dc electric properties and their aging and annealing stability. We showed that Nb/Al-AlN/Nb junctions are probably somewhat more stable against aging and annealing than traditional Nb/Al-AlO_x/Nb junctions are. We present the early results of the Nb/Al-AlN/Nb junctions' specific capacitance measurements. The obtained data shows the specific capacitance of the Nb/Al-AlN/Nb junctions about 17% lower than that of the Nb/Al-AlO_x/Nb junctions with the same R_nA product value. That is in the agreement with the difference between dielectric constants of bulk aluminum nitride and alumina.

ACKNOWLEDGMENT

This work was partially supported by the European Organisation for Astronomical for Research in the Southern Hemisphere (ESO) in the frame of the Collaboration Agreement No. 73301/16/78225/OSZ.

REFERENCES

[1] A. D. Bolatto (chair) *et al.*, "A Road Map for Developing ALMA. ASAC recommendations for ALMA 2030," 2016. [Online]. Available: https://science.nrao.edu/facilities/alma/science_sustainability/RoadmapforDevelopingALMA.pdf.

[2] P. Yadranjee Aghdam, H. Marouf Rashid, A. Pavolotsky, V. Desmaris, and V. Belitsky, "Dependence of the scatter of the electrical properties on local non-uniformities of the tunnel barrier in Nb/Al-AlOx/Nb junctions," *J. Appl. Phys.*, vol. 119, no. 5, 2016.

[3] P. Y. Aghdam, H. Rashid, A. Pavolotsky, V. Desmaris, D. Meledin, and V. Belitsky, "Direct Measurement of Superconducting Tunnel Junction Capacitance," *IEEE Trans. Terahertz Sci. Technol.*, 2015.

[4] A. B. Pavolotsky, D. Dochev, and V. Belitsky, "Aging- and annealing-induced variations in Nb/Al- AlOx/Nb tunnel junction properties," *J. Appl. Phys.*, vol. 109, no. 2, p. 024502, 2011.

[5] B. Billade *et al.*, "Performance of the first ALMA band 5 production cartridge," *IEEE Trans. Terahertz Sci. Technol.*, vol. 2, no. 2, pp. 208–214, 2012.

[6] V. Belitsky *et al.*, "ALMA Band 5 receiver cartridge," *Astron. Astrophys.*, vol. 611, p. A98, Mar. 2018.

[7] V. Belitsky *et al.*, "SEPIA – a new single pixel receiver at the APEX telescope," *Astron. Astrophys.*, vol. 612, no. 31458, p. A23, Apr. 2018.

[8] T. Shiota, T. Imamura, and S. Hasuo, "Nb Josephson junction with an AlNx barrier made by plasma nitridation," *Appl. Phys. Lett.*, vol. 61, no. 10, pp. 1228–1230, Sep. 1992.

[9] R. Dolata, M. Neuhaus, and W. Jutzi, "Tunnel barrier growth dynamics of Nb/AlOx-Al/Nb and Nb/A1Nx-Al/Nb Josephson junctions," *Phys. C Supercond.*, vol. 241, no. 1–2, pp. 25–29, Jan. 1995.

[10] A. W. Kleinsasser, W. H. Mallison, and R. E. Miller, "Nb/AlN/Nb Josephson junctions with high critical current density," *IEEE Trans. Appiled Supercond.*, vol. 5, no. 2, pp. 2318–2321, Jun. 1995.

[11] B. Bumble, H. G. LeDuc, J. A. Stern, and K. G. Megerian, "Fabrication of Nb/AlNx//NbTiN junctions for SIS mixer applications," *IEEE Trans. Appiled Supercond.*, vol. 11, no. 1, pp. 76–79, Mar. 2001.

[12] T. Zijlstra, C. F. J. Lodewijk, N. Vercruyssen, F. D. Tichelaar, D. N. Loudkov, and T. M. Klapwijk, "Epitaxial aluminum nitride tunnel barriers grown by nitridation with a plasma source," *Appl. Phys. Lett.*, vol. 91, no. 23, p. 233102, Dec. 2007.

[13] T. W. Cecil, M. E. Cyberey, R. E. Matthews, J. Z. Zhang, and A. W. Lichtenberger, "Development of Nb/Al-AlN/NbTiN SIS Junctions With ICP Nitridation," *IEEE Trans. Appl. Supercond.*, vol. 19, no. 3, pp. 409–412, Jun. 2009.

[14] P. N. Dmitriev *et al.*, "High quality Nb-based tunnel junctions for high frequency and digital applications," *IEEE Trans. Appiled Supercond.*, vol. 13, no. 2, pp. 107–110, Jun. 2003.

[15] N. N. Iosad, A. B. Ermakov, F. E. Meijer, B. D. Jackson, and T. M. Klapwijk, "Characterization of the fabrication process of Nb/Al-AlNx/Nb tunnel junctions with low RnA values up to 1 W·mm^2," *Supercond. Sci. Technol.*, vol. 15, no. 6, p. 319, Jun. 2002.

[16] J. E. Stevens, "Electron cyclotron resonance plasma sources," in *High Density Plasma Sources: design, physics and performance*, O. Popov, Ed. Noyes Publications, 1995, pp. 312–379.

[17] H. Kohlstedt, S. Kuriki, T. Lehnert, D. Billon-Pierron, and K.-H. Gundlach, "Relation Between Tunnel Junction Quality, Anodization Profiles, and Low-Frequency Noise," *Supercond. Devices Their Appl. Springer Berlin Heidelb.*, pp. 218–223, 1992.

[18] H. Rashid, D. Meledin, V. Desmaris, A. Pavolotsky, and V. Belitsky, "Superconducting 4-8-GHz hybrid assembly for 2SB cryogenic THz receivers," *IEEE Trans. Terahertz Sci. Technol.*, vol. 4, no. 2, pp. 193–200, 2014.

[19] P. Y. Aghdam, H. Rashid, A. Pavolotsky, V. Desmaris, D. Meledin, and V. Belitsky, "Specific Capacitance Dependence on the Specific Resistance in Nb/Al-AlOx/Nb Tunnel Junctions," *IEEE Trans. Terahertz Sci. Technol.*, 2017.

[20] V. Y. Belitsky, S. W. Jacobsson, S. A. Kovtonjuk, E. L. Kollberg, and A. B. Ermakov, "100 GHz mixer vertically integrated (stacked) SIS junction array," *Int. J. Infrared Millimeter Waves*, vol. 14, no. 5, pp. 949–957, May 1993.

[21] J. W. Lamb, "Miscellaneous data on materials for millimetre and submillimetre optics," *Int. J. Infrared Millimeter Waves*, vol. 17, no. 12, pp. 1997–2034, Dec. 1996.

Development of the Terahertz Superconducting Imaging Array (TeSIA)

Jing Li[1,2,*], Zhen-Hui Lin[1,2,3], Jin-Ping Yang[1,2], Dong Liu[1,2], Wen-Ying Duan[1,2], Zheng Lou[1,2], Qing Shi[1,2,4], Zhi Li[1,2,4], Wen Zhang[1,2], Wei Miao[1,2], Qi-Jun Yao[1,2], and S.C. Shi[1,2]

[1]*Purple Mountain Observatory, Chinese Academy of Sciences, Nanjing, China*
[2]*Key Laboratory of Radio Astronomy, Chinese Academy of Sciences, Nanjing, China*
[3]*University of Chinese Academy of Sciences, Beijing 100049, China*
[4]*University of Science and Technology of China, Hefei 230026, China*
[*]Contact: lijing@pmo.ac.cn

Abstract—Dome A, the highest point of the cold and dry Antarctic ice sheet, offers the best access to atmospheric windows at THz/FIR wavelengths on Earth. The astronomical facilities China is planning to build there include a 5-m THz telescope named DATE5. An instrument proposed for the DATE5 telescope is the THz superconducting imaging array (TeSIA) operating at the 350-μm window, with a pixel number of 32×32 and a background-limited sensitivity (NEP) of $1\text{x}10^{-16}$ W/Hz$^{0.5}$. For the development of TeSIA, microwave kinetic inductance detectors (MKIDs) based on Al superconducting films are chosen as the detectors of the system operating at 0.3 K. In this paper, we will firstly introduce the design and performance of the system, and then present some results of video-rate imaging and testing observations on a submillimeter-wave telescope.

About effect of the temperature operating conditions on the Noise Temperature and Noise Bandwidth of the Terahertz Range NbN Hot-Electron Bolometers.

Ivan Tretyakov[1], N. Kaurova[1], B. M. Voronov[1] and G. N. Goltsman[1]
[1]Moscow State University of Education, Moscow, 119991 Russia
*Contact: ivantretykov@mail.ru

Abstract—Results of an experimental study of the noise temperature (Tn) and noise bandwidth (NBW) of the superconductor NbN hot-electron bolometer (HEB) mixer as a function of its temperature (Tb) and NbN bridge length are presented. It was determined that the NBW of the mixer is significantly wider at temperatures close to the critical ones (Tc) than are values measured at 4.2 K. The NBW of the mixer measured at the heterodyne frequency of 2.5 THz at temperature Tb close to Tc was ~13 GHz, as compared with 6 GHz at Tb = 4.2 K. This experiment clearly demonstrates the limitation of the thermal flow from the NbN bridge at Tb ≪Tc for mixers manufactured by the in situ technique. This limitation is close in its nature to the Andreev reflection on the superconductor/metal boundary. In this case, the noise temperature of the studied mixer increased from 1100 to 3800 K.

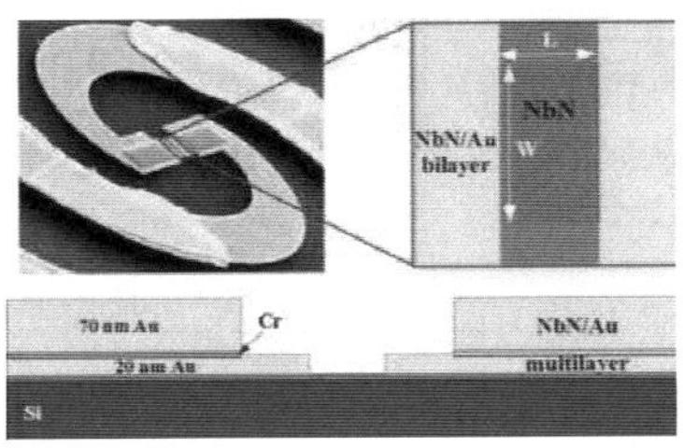

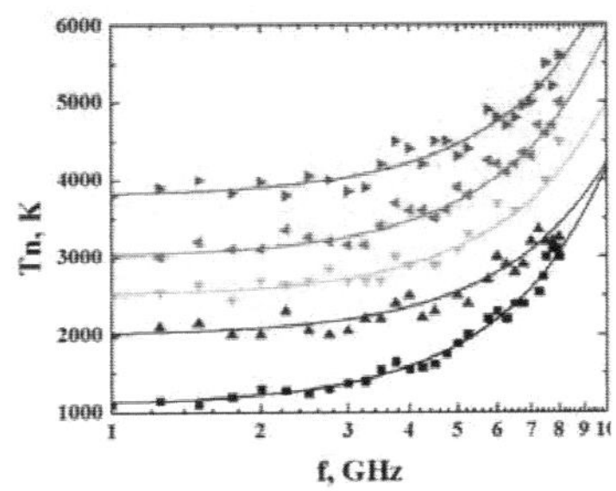

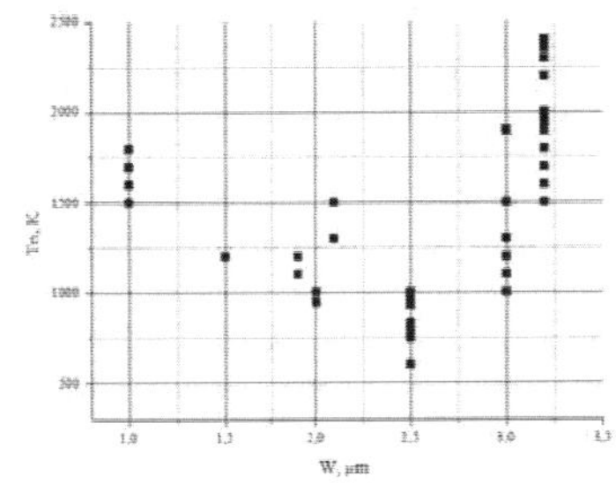

Bandwidth Measurements of NbN HEB Devices with GaN Buffer Layers

Nicholas J. Rommelfanger[1,*], Bruce Bumble[1], and Boris S. Karasik[1]
[1]Jet Propulsion Laboratory, California Institute of Technology, Pasadena, CA 91109, USA
*Contact: nrommelfanger@umail.ucsb.edu

***Abstract*—Hot electron bolometers (HEBs) made of thin films of superconducting NbN are commonly used as far-infrared (THz) heterodyne detectors. The intermediate frequency (IF) bandwidth of a detector is limited by the rate of heat transfer away from the device. Detectors were fabricated from NbN on a GaN buffer layer, instead of a MgO buffer, in hopes of forming a closer acoustic match with the NbN and decreasing the phonon escape time. Acoustic matching has previously shown promising results with MgB_2 films on SiC. Devices on both GaN/Si and MgO/Si were tested with frequencies of 15-25 GHz near T_c = 13.3 K. Results yielded an IF bandwidth for NbN on GaN/Si of approximately 2.8 GHz and of approximately 1.8 GHz for NbN on MgO/Si. These results contrast with a recent publication that measured IF bandwidth of 7.5-8 GHz for NbN HEBs on GaN [1]. It is believed that smaller devices used in the aforementioned publication might result in electron diffusion through the contacts, decreasing the overall electron energy relaxation time. The larger devices used in this experiment show no significant bandwidth improvement of GaN over Si for application in THz HEB detectors.**

I. Introduction

NbN HEBs are used to perform THz spectroscopy in the far-infrared spectrum. This region contains spectral lines that describe the composition of star-forming regions in our and nearby galaxies. HEBs are the only mixer type which can work above 1.3 THz with low noise. NbN HEBs fabricated at JPL are typically deposited on a Si substrate with a MgO buffer. Such devices usually have an IF bandwidth under 3 GHz [2]. Detection of high-frequency lines around 5 THz (e.g., [OI] line @ 4.7 THz) requires an IF bandwidth of about 8 GHz, so measurements of such lines in one shot are not currently feasible [3]. Thus, HEB technology must be improved to increase the IF bandwidth.

To increase their IF bandwidth, devices must transfer heat more rapidly. When infrared radiation is incident on a device, electrons in the NbN film are heated. These electrons pass their heat to phonons in the NbN, which subsequently pass heat to phonons in the substrate. The time it takes for heat to leave the device, the phonon escape time, contributes significantly to the overall cooling time of an NbN HEB. The phonon escape time is proportional to the film's thickness. However, since any further reduction of the thickness in NbN HEB leads to device degradation, the only way to decrease the phonon escape time (= increase the IF bandwidth) is to improve the acoustic impedance matching on the interface between the device and substrate.

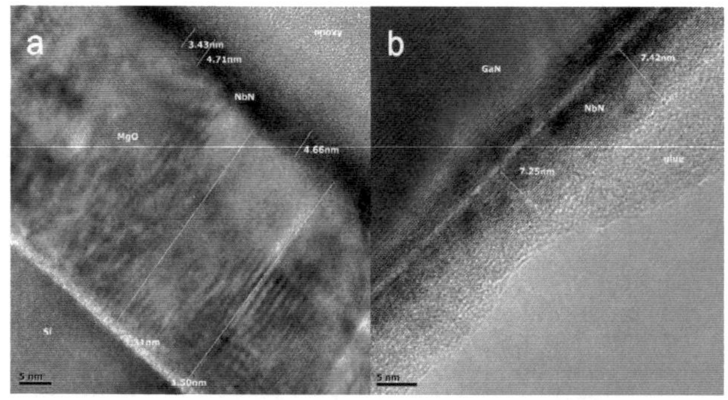

Fig. 1. TEM images of the two types of devices. (a) A device on MgO buffer. The Si substrate visible in the lower left is organized into a lattice, but the MgO buffer appears generally amorphous. The NbN film above the MgO is especially disorganized, with no clear lattice visible. Additionally, the film thickness varies by over a nanometer across the length of the image. (b) A device on GaN. Unlike the MgO buffer, the GaN buffer is neatly organized, and even the NbN film appears to be epitaxial. The NbN film thickness is also relatively constant across the length of the image.

A group from Chalmers University and Moscow State Pedagogical University tested NbN devices made with a GaN buffer, instead of the typical MgO buffer [1]. The group measured an IF bandwidth of 7.5-8 GHz with a GaN buffer. They postulated that the increase in bandwidth was the result of improved acoustic impedance matching on the interface between the GaN and NbN [1]. However, the devices the group used were very small: 3 µm x 0.3 µm x 5 nm. At this device size, electron diffusion is a non-negligible cooling method that could increase the IF bandwidth. If the contacts in a device are too close together and do not have a significant superconducting gap impeding the diffusion, hot electrons diffuse out of the device through the contacts instead of passing their heat to phonons in the device. While this results in faster devices, past experience at JPL with Nb devices has shown that diffusion-cooled devices have higher noise temperatures and less sensitivity than phonon-cooled devices [4].

In this experiment, the bandwidth of larger GaN devices was measured to determine if electron diffusion was the cause of the increased bandwidth observed with small GaN devices.

II. Experimental Methods

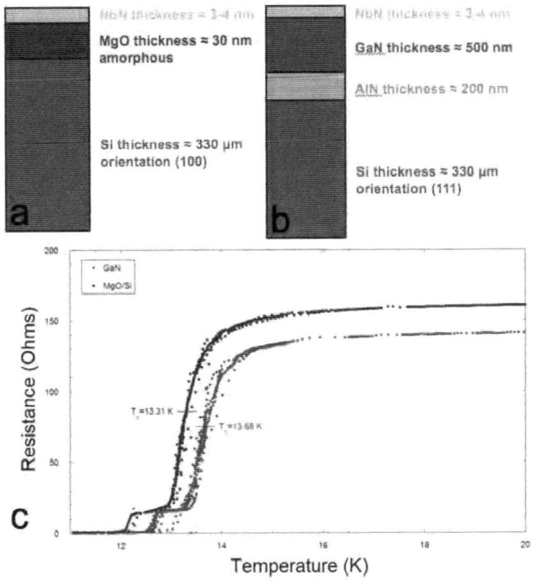

Fig. 2. (a) Diagram of the cross section of a device on MgO buffer. The NbN film is deposited on the MgO buffer, which rests on a Si substrate. (b) Diagram of a device on GaN buffer. NbN is deposited on a GaN buffer, and an AlN intermediate buffer between the GaN and Si substrate promotes epitaxial GaN growth. (c) Resistance vs. temperature curves for the two devices. The shape of the curve is similar for both devices, but the device on MgO has a higher resistance, and the device on GaN has a higher T_C.

Two types of NbN devices were fabricated: one with a MgO buffer and one with a GaN buffer. Both types of devices were 8 µm x 2 µm x 5 nm, almost an order of magnitude larger than the devices from another group mentioned above. The GaN wafer uses an intermediate buffer of AlN to promote epitaxial growth of GaN. The Transmission Electron Microscopy (TEM) images shown in Fig. 1 illustrate the difference in NbN structure caused by the two buffers. The MgO buffer and NbN film shown in Fig. 1a appear relatively amorphous, while the GaN buffer and NbN film in Fig. 1b look much better ordered into lattices.

Cross sections of the wafers used, including dimensions, are shown in Figs. 2a and 2b. R(T) curves of the HEB devices are displayed in Fig. 2c. The MgO device has a slightly higher resistance than the GaN device. The MgO device has T_C = 13.3 K, and the GaN device has T_C = 13.7 K. Otherwise, the two R(T) curves have similar structure. IV curves for the device on GaN are shown in Fig. 3. The curves were measured around T_C without pumping the device. Figure 4 is a photograph of a test device on a GaN buffer. The device itself is circled on the inset. The rest of the image shows the gold-plated circuit contact and the substrate.

Fig. 3. IV curves near T_C for the device on GaN. The device was not pumped for these measurements.

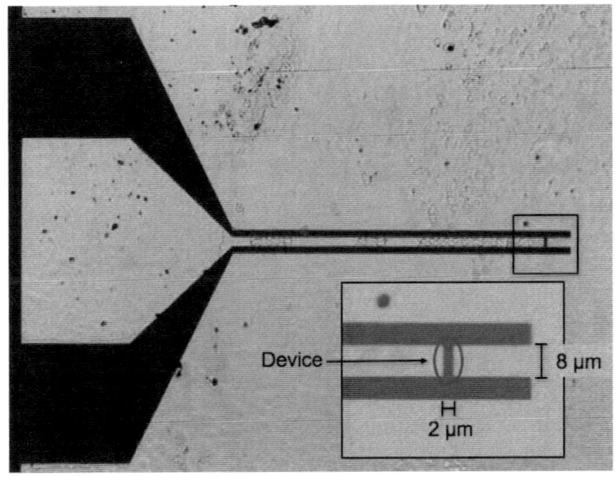

Fig. 4. Image of an HEB chip on GaN used for testing. The gold pentagon on the left is a dc/IF contact, and the HEB device is enlarged in the inset. All devices tested were 8 µm x 2 µm x 5 nm. Current flows to the right along the gold path in the center of the image and across the device into the ground plane.

The IF bandwidth of the devices was tested using microwave techniques. A block diagram of the experimental setup is shown in Fig. 5. The device was cooled below T_C, dc biased, and pumped with microwave radiation. One microwave source was fixed at 15 GHz, and another source was varied between 15-22 GHz. A 7 GHz low-pass filter set the upper limit on the measurable IF signals. The fixed and variable signals were combined and sent to the device at the end of a cryogenic dipstick in a LHe dewar. The device was positioned above the surface of the LHe at a temperature below T_C.

The device mixes the two microwave signals, and the resulting IF signal travels up the dipstick and through the low-pass filter. The IF signal passes through two amplifiers before being read by the spectrum analyzer.

III. Results and Analysis

Bandwidth measurement data of the device on MgO are shown in Fig. 6. Here, IF bandwidth is defined as the frequency at which the signal strength rolls off by 3 dB, signifying a 50% decrease in power. All data in this figure was collected at 12.60 K. The three curves were collected at 3 different bias points, marked on the IV curve in the lower left. The average IF bandwidth among the 3 curves is around 1.8 GHz. This value is expected; NbN devices with MgO buffers have been tested extensively and bandwidth is rarely above 3 GHz. Thus, the experimental setup is validated.

Figure 7 plots bandwidth measurements for the device on GaN. To account for the difference in T_C between the two buffers, data was collected at 12.90 K for GaN buffer, producing an IV curve like that for MgO buffer at 12.60 K. Again, data was collected at the 3 bias points marked on the IV

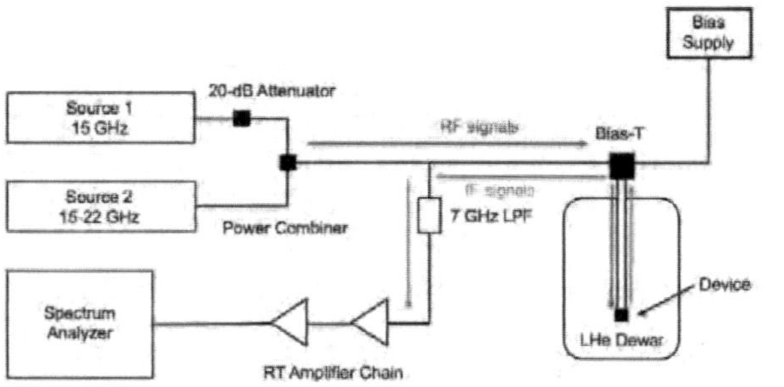

Fig. 5. The experimental setup. Signals originate from two generators: one is fixed at 15 GHz and attenuated, and the other is variable from 15-22 GHz. The signals are combined and travel to the device on the end of a dipstick in a LHe dewar. The device produces an IF signal which passes back up the dipstick, through a 7 GHz low-pass filter, and through a series of amplifiers. The signal is then read by the spectrum analyzer.

curve in the lower left. For these trials with GaN, the average bandwidth is about 2.8 GHz.

While the GaN device offers a 1 GHz improvement in bandwidth over the MgO device, it is far from the 7.5-8 GHz bandwidth measured by the group from Chalmers and Moscow State University. Thus, it appears that the large bandwidth observed in [1] may be due to electron diffusion, and not improved acoustic impedance matching. Although this group's diffusion-cooled devices exhibit a large bandwidth, previous results at JPL predict the devices' sensitivity is poor. GaN may not drastically increase IF bandwidth, but it may still be a viable buffer because it produces higher quality NbN films with well-organized lattice.

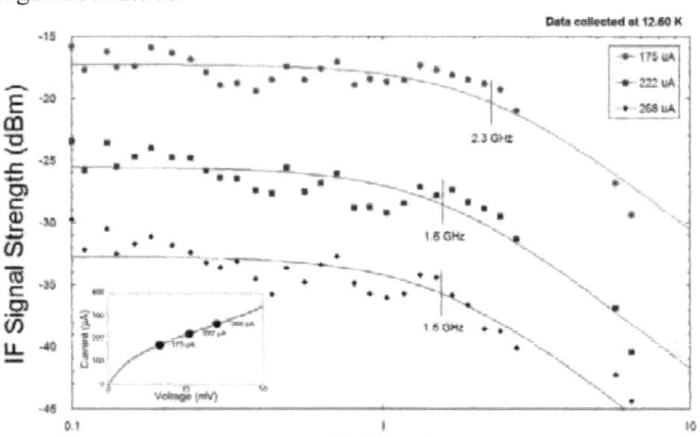

Fig. 6. Bandwidth measurements for the HEB device on MgO. All data was collected at 12.60 K. Each curve was collected at a different bias point, and the bias points are marked on the IV curve in the lower left inset.

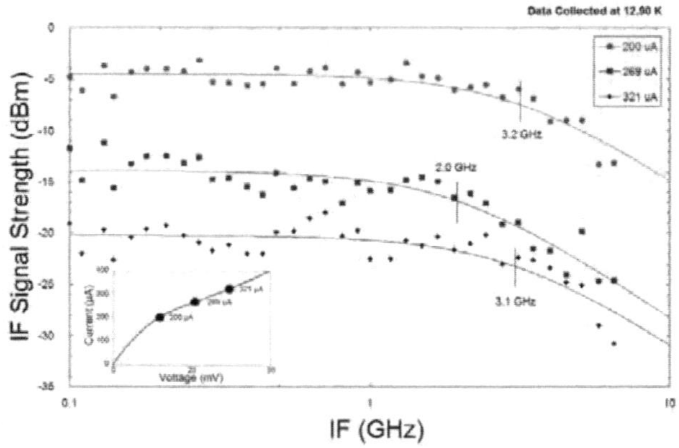

Fig. 7. Bandwidth measurements for the HEB device on GaN. This data was collected at 12.90 K, 0.3 K higher than the measurements for MgO buffer to account for the difference in T_C. Measurements were made at the three bias points marked on the IV curve in the lower left inset.

Conclusions

The IF bandwidth of NbN HEBs with GaN buffer layers was measured as almost 3 GHz, while the NbN device with MgO buffer has an IF bandwidth near 2 GHz. Thus, the large increase in bandwidth observed by another group when using GaN buffers cannot be confirmed. GaN produced well-ordered NbN films that may be useful for fabricating better quality NbN HEBs, but the increased acoustic impedance matching played little role in increasing IF bandwidth.

Acknowledgments

NR thanks the JPL SIP for organizing his internship and Dan Cunnane for experimental help. The authors also thank Carol Garland (Caltech) for TEM analysis of the film/substrate interface. This research was carried out at the Jet Propulsion Laboratory, California Institute of Technology, under a contract with the National Aeronautics and Space Administration.

References

[1] Krause, Sascha, et al., "Reduction of Phonon Escape Time for NbN Hot Electron Bolometers by Using GaN Buffer Layers," *IEEE Transactions on Terahertz Science and Technology*, vol. 7(1), pp. 53-59, Jan. 2017.

[2] A. Shurakov et al., "Superconducting hot-electron bolometer: from the discovery of hot-electron phenomena to practical applications," *Superconductor Science and Technology,* vol. 29(2), 023001, Feb. 2016.

[3] Lis, D.C., et al., "Herschel/HIFI measurements of the ortho/para ratio in water towards Sagittarius B2(M) and W31C," *Astronomy & Astrophysics*, vol. 521, L26, Oct. 2010.

[4] J. Kawamura, private communication.

All-NbN Technology on Sapphire Substrates for SIS-based THz Receivers

M. Merker[1,*], C. Bohn[1], A.Schmid[1] , K. Ilin[1] and M.Siegel[1]
[1]Institute of Micro-and Nanoelectronic Systems, Karlsruhe Institute of Technology, 76187 Germany
*Contact: Michael.merker@kit.edu

***Abstract*—Heterodyne SIS receivers for THz radiation have been investigated and successfully employed in the past. In order to push the operation frequency beyond the 700 GHz limit, superconductors other than Niobium (Nb) have to be used. We have developed a NbN/AlN/NbN technology for the fabrication of high-quality SIS junctions on sapphire substrates. The main optimization criteria are the energy gap of the superconductor along with the leakage current and the critical current density. The development of our technology resulted in gap voltages as high as 5.1 mV with critical current densities and leakage currents that are suitable for the realization of SIS mixers. Based on the material parameters that have been extracted from our technology, we have developed an RF design of a prototype for an integrated receiver using the all-NbN fabrication process. We have demonstrated transmission of 600 GHz radiation generated by a flux-flow oscillator to an SIS mixer integrated on the same chip.**

I. Introduction

Heterodyne detection of terahertz signals using Superconductor Insulator Superconductor (SIS) mixers has shown to be capable of quantum limited detection while simultaneously enabling high resolution spectroscopy [1-4]. This quantum limited operation is possible up to an upper frequency limit which is given by the energy gap of the superconducting materials used for the electrodes of the SIS devices and the RF circuits. The most popular material in the past decades has been pure Niobium (Nb) which can be operated up to 700 GHz. In order to push this limit to higher frequencies, other superconductors have to be used, such as Niobium nitride (NbN) or Niobium titanium nitride (NbTiN). These materials can exhibit twice the energy gap of pure Nb thus enabling the low loss operation up to 1.4 THz.

However, realizing these materials with the required high quality is much more challenging as compared to pure Nb. NbN based technology has been demonstrated already, achieving the highest energy gap of up to 6 mV on Magnesium oxide (MgO) substrates with high critical current densities up to 70 kA/cm^2 [5]. Other groups have demonstrated a high-quality NbN technology on sapphire or silicon substrates where, either a buffer layer had to be used or an elaborate pre-treatment of the substrate is necessary [6,7]. In order to decrease the complexity of the fabrication process, we aim at the realization of all-NbN technology directly on R-plane sapphire substrates.

Apart from the energy gap of the superconductor, there are several other requirements to the trilayer properties in order to be suitable for a low noise receiver operation, such as the leakage current through the barrier and the critical current density j_c. These two parameters, however, are not uncorrelated, but defined by the barrier itself and the interface between the barrier and the electrodes. Thus, the surface of the NbN base electrode has to be ultimately smooth in order to realize a proper barrier.

Not only the leakage is affected by the critical current density, but also the specific capacitance C' of the trilayer increases with higher j_c values. Therefore the fabrication process needs to allow for the realization of sub-µm sized mixers.

We have developed a trilayer deposition process for high gap voltages along with adequate leakage and critical current densities. In order to achieve a process allowing for the fabrication of high-quality SIS devices as well as proper RF circuits, the fabrication has been optimized considering the various requirements. The quality of the SIS trilayers is evaluated using the Josephson junction (JJ) quality parameters at an operation temperature of 4.2 K.

II. Technology Development

The development of the deposition process for high-quality SIS devices requires a careful optimization of each layer involved, as well as a close process control. We have developed a 3-chamber DC magnetron sputtering system, with a movable heater stage, enabling the in-situ deposition of the trilayers at 775 °C. The main optimization criteria are the energy gap of the superconductor ($\Delta=V_{gap}/e$), the leakages through the barrier and the critical current density j_c. As already reported in [8], the gap voltage was optimized to values up to 5.1 mV, allowing for operation frequencies above 1 THz. A measure for the leakage currents, the subgap ratio, is derived from the resistance in the retrapping branch at 3.5 mV (R_{sg}) normalized to the normal state conductance R_N, which is obtained by a linear fit in the normal state regime. Figure 1 shows the measured values over a wide range of critical-current densities. The subgap ratio clearly decreases towards higher j_c values due to the decreasing thickness of the barrier. As expected, the gap voltage (≈ 5 mV) in turn does not depend

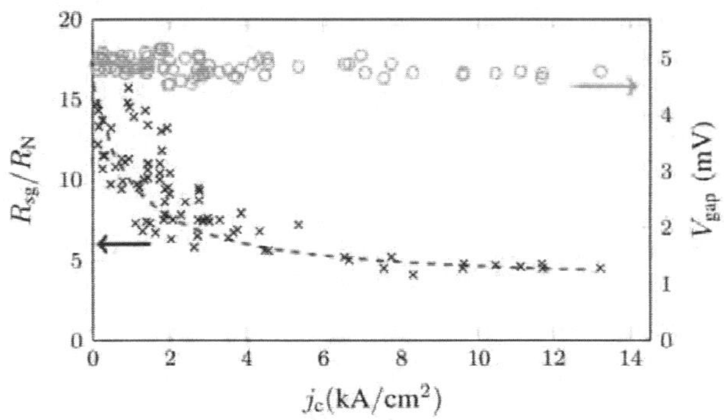

Figure 1: Dependence of the subgap ratio (R_{sg}/R_N,X) and the gap voltage (V_{gap},O) on the critical-current density j_c. The dashed red line is to guide the eye.

on j_c, but shows some fluctuations about 10 %. In our opinion, this is due to minor flucutiations of the deposition parameters which, in the current setup, cannot be avoided without some modifications to the deposition system. Nevertheless, the achieved quality of the trilayers seems suitable for the realization of an SIS receiver.

In order to fabricate high frequency devices, not only the trilayer quality, but also the peripheric materials and the patterning of each layer is of utmost importance. The fabrication process as described in detail in [9] and [10] is used, but it has required modifications mainly concerning the thickness of the different layers. The thickness of the electrodes is selected according to the London penetration depth of NbN ($\lambda_L \approx 200$ nm). The minimum realizable mixer size with this fabrication process is about $A_{mix} = 500 \times 500$ nm^2. The fabrication process must also accommodate the realization of a microstrip waveguide with an appropriate characteristic impedance range. The waveguide is formed by the base electrode of the trilayer as ground plane, the dielectric formed by the insulating layer (SiO) and the wiring layer creating the microstrip itself. Therefore, the thickness of the isolator in conjunction with the feature sizes of the wiring layer defines the characteristic impedance of the microstrip. The thickness of both, the insulator, as well as the wiring layer was selected to $d_{SiO} = d_{wiring} = 400$ nm. Optimization of the patterning resulted in minimum feature sizes in the wiring layer of 700 nm. The achievable characteristic frequency thus ranges from $\approx 70\ \Omega$ to less than 10 Ω, which is suitable for the impedance matching that is required on chip.

Apart from the geometrical properties of the wiring layer, also the superconducting properties are important. Since the wiring layer defines the microstrip line, it has to allow the THz radiation to travel from e.g. the antenna to the mixer. Therefore, the energy gap needs to be large enough, analogue to the mixer electrodes. The optimization of superconducting properties of the wiring layer led to a transition temperature of $T_c = 14$ K which is more than 1 K lower than the transition temperature of the electrode material. This results from the fact that the wiring layer is made in the very last fabrication step, and therefore, has to be deposited at room temperature as opposed to the electrode material. In addition to that, the wiring layer does not have the bare, clean substrate

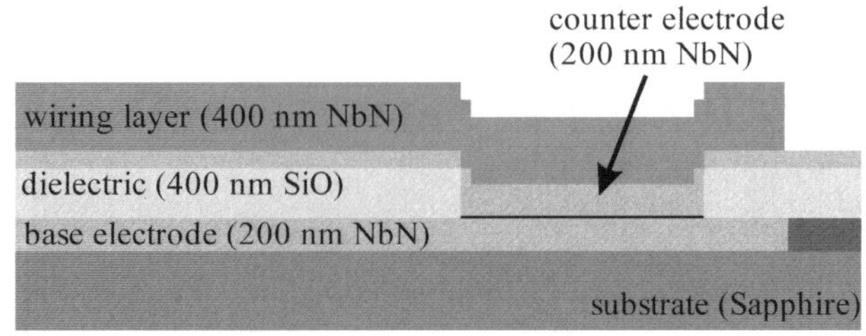

Figure 2: Schematic cross-section of the modified fabrication structure.

underneath, but it has to be deposited on top of various materials, namely the NbN of the counter electrode as well as amorphous SiO. This is why the wiring layer cannot reach the quality level of the electrode material. In the future, the wiring layer material might be replaced by NbTiN which can be deposited with high quality even on amorphous substrates [11], [12], thus allowing for low loss transmission lines on chip for ultimately high operation frequencies.

III. RF Prototyping

Considering the obtained material parameters of each single layer, RF simulations can be performed. The simulations were conducted using CST – Microwave Studio. The actual material parameters were included in the simulation. The superconductor parasitics were derived from the Mattis-Bardeen theory from DC measurements. The intrinsic mixer capacitance was obtained using the designed mixer area and the specific capacitance C' of that trilayer, which in turn was calculated by the relation between C' and the critical current density which in this case was 6 kA/cm^2.

The first prototype was designed comprising a flux-flow oscillator (FFO) and an SIS mixer along with an appropriate RF circuit that connects both of these devices. This RF circuit has to fulfill a number of requirements including the impedance matching of the FFO to the mixer, thus providing a good RF transmission. This impedance matching does not only include the transformation of the real parts of the impedances into each other, but also the imaginary part of the mixer input impedance needs to vanish for the design frequency. This imaginary part arises from the intrinsic capacitance of the SIS mixer. A tuning structure is used to tune out this capacitive imaginary part of the impedance. Figure 3 depicts a scanning electron microscope (SEM) image of a realized prototype. The tuning structure and the SIS mixer are in the left part of the image as well as in the close up. In this case, this tuning structure is realized by a butterfly stub which is connected to the microstrip waveguide in a certain distance to the mixer l_{tune}.

While providing a good RF connection between FFO and mixer, the connection needs to be interrupted for DC, since both devices have different DC operation points. This is achieved by a DC break for each electrode which can be seen in the center of figure 3. The DC break design is similar to what can be found in [13], since these designs proved to exhibit excellent performance of both, transmission level and bandwidth. Yet, the designs had to be adapted to our technology.

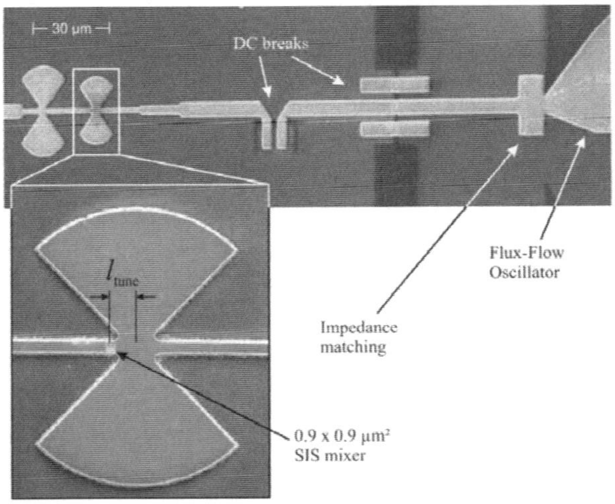

Figure 3: Scanning electron microscope image of an RF prototype comprising the FFO, the RF-circuit and the SIS mixer.

The fabricated prototypes were measured at 4.2 K immersed in a liquid helium transport dewar. Figure 4 depicts a current-voltage (IV) measurement of an, in this case, 0.9x0.9 μm² sized SIS mixer with (—) and without (—) FFO radiation. The gap voltage is at 4.5 mV with a critical-current density of 6 kA/cm² and a suitable subgap ratio of 8.

The RF design also incorporated a control line which is supposed to suppress the critical current of the mixer by magnetic field. However, the critical current of the control line was not high enough to fully suppress the critical current of the mixer. That is why the IV curve is shown here without magnetic field.

Applying FFO radiation induces Shapiro steps and photon assisted tunneling (PAT) steps in the IV curve (see —). The Shapiro steps are clearly visible up to the third order in the central part of the image. The photon assisted tunneling steps appear below as well as above the gap. Below the gap, Shapiro steps and PAT steps interfere, however, above the gap, this PAT step is clearly visible. This measurement is done at an FFO frequency of 540 GHz. Using the current in the retrapping branch at a certain voltage (e.g. 3 mV) as a measure of the RF power accepted by the mixer and conducting this measurement for many frequencies, allows to create a spectrum of transmitted power. This can be compared to the RF simulations done for that particular design as it is shown in the inset of figure 4. Except for the bandwidth, the simulation closely matches the measurement, not only the peak transmission frequency of 600 GHz, but also the shape of the frequency response is almost identical. We have successfully realized a number of designs with variable tuning length and different tuning structures on the prototype chip. The results thereof will be shown elsewhere.

Conclusions

The thorough fabrication-process development for SIS receivers based on NbN/AlN/NbN trilayer allows the realization of high quality sub-μm SIS mixers as well as microstrip waveguides with a large characteristic impedance range. Due to the optimization of the particular material

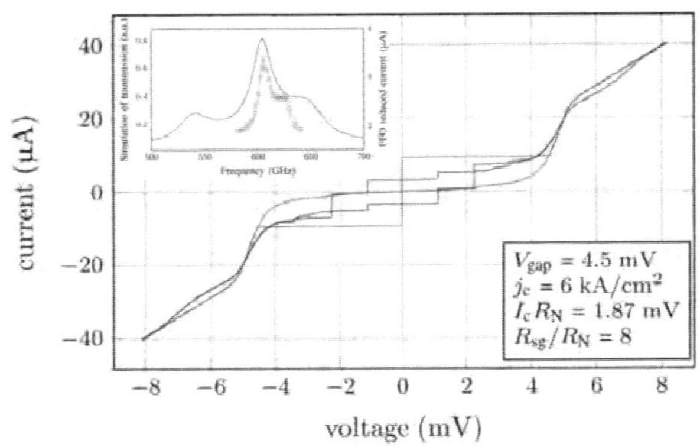

Figure 4: Current-voltage characteristic of a mixer with (—) and without (—) FFO-radiation. The inset shows the FFO induced current (-○-) versus frequency along with the simulated transmission (—).

properties, each layer should be able to carry frequencies higher than 1 THz. Furthermore, all material parasitics were studied, which is very important for the simulations of the RF circuit, in order to create a design for an SIS receiver.

The achieved gap voltages of 5.1 mV are suitable for operation frequencies above 1 THz. By extracting all necessary material parameters, RF simulations were conducted, resulting in a design of a prototype device, consisting of flux-flow oscillators integrated with SIS mixers on the same chip. The study of the frequency response of the realized structures proved a good agreement with the simulations even at frequencies as high as 600 GHz. In the future, we plan to incorporate an antenna in the design allowing the study of a fully integrated receiver.

References

[1] C. Walker, J. Kooi, M. Chant, H. G. Leduc, P. Schaffer, J. Carlstrom, and T. Phillips, International journal of infrared and millimeter waves, vol. 13, no. 6, pp. 785–798, 1992.

[2] J. W. Kooi, C. Walker, H. LeDuc, P. Schaffer, T. Hunter, D. Benford, and T. Phillips, International journal of infrared and millimeter waves, vol. 15, no. 3,pp. 477–492, 1994.

[3] K. Gundlach and M. Schicke Superconductor Science and Technology, vol. 13, no. 12, p. R171, 2000.

[4] J. Zmuidzinas, N. Ugras, D. Miller, M. Gaidis, H. LeDuc, and J. Stern, IEEE Transactions on Applied Superconductivity, vol. 5, no. 2, pp. 3053–3056, 1995.

[5] A. Kawakami, Z. Wang, and S. Miki, Journal of Applied Physics, vol. 90, no. 9, pp. 4796–4799, 2001.

[6] K. Makise, H. Terai, and Y. Uzawa, IEEE Transactions on Applied Superconductivity, vol. 26, no. 3, pp. 1–3, 2016.

[7] J.-C. Villegirr, N. Hadacek, S. Monso, B. Delnet, A. Roussy, P. Febvre, G. Lamura, and J.-Y. Laval, IEEE transactions on applied superconductivity, vol. 11, no. 1, pp. 68–71, 2001.

[8] M. Merker, C. Bohn, M. Völlinger, K. Ilin, and M. Siegel, IEEE Transactions on Applied Superconductivity, vol. 27, no. 4, pp. 1–5, 2017.

[9] J. Meckbach, M. Merker, S. Buehler, K. Ilin, B. Neumeier, U. Kienzle, E. Goldobin, R. Kleiner, D. Koelle, and M. Siegel, IEEE Transactions on Applied Superconductivity, vol. 23, no. 3, pp. 1100504–1100504, 2013.

[10] J. M. Meckbach, vol. 13. KIT Scientific Publishing, 2013.

[11] T. Matsunaga, H. Maezawa, and T. Noguchi, IEEE transactions on applied superconductivity, vol. 13, no. 2, pp. 3284–3287, 2003.

[12] K. Makise, H. Terai, M. Takeda, Y. Uzawa, and Z. Wang, IEEE transactions on applied superconductivity, vol. 21, no. 3, pp. 139–142, 2011.

[13] S. Shitov, A. Ustinov, N. Iosad, and H. Kohlstedt, Journal of applied physics, vol. 80, no. 12, pp. 7134–7137, 1996

Optical Response of LEKID Arrays

Shibo Shu[1*], Martino Calvo[2], Johannes Goupy[2], Andrea Catalano[23], Aurelien Bideaud[2], Alessandro Monfardini[2], Samuel Leclercq[1] and Eduard F.C. Driessen[1]

[1]*Institut de Radioastronomie Millimétrique, 300 rue de la Piscine, Domaine Universitaire, 38406 Saint Martin d'Hères, France*

[2]*Institut Néel, CNRS and Université Grenoble Alpes, 25 av. des Martyrs, 38042 Grenoble, France*

[3] *Laboratoire de Physique Subatomique et de Cosmologie, Université Grenoble Alpes, CNRS, 53 av. des Martyrs, Grenoble, France*

*Contact: shu@iram.fr

***Abstract*—We present an analysis of the optical response of lumped-element kinetic-inductance arrays, using the NIKA2 1mm array as an example. This array design has a dual-polarization sensitive Hilbert inductor for directly absorbing incident photons. We calculated the optical response from a transmission line model and compared it with simulation using HFSS. From the simulation, we noticed that a non-negligible part of energy is reflected as higher modes. We show the difficulty of achieving high absorption rate using aluminum. This analysis could be extended to other kinetic inductance detector array designs in millimeter, sub-millimeter and terahertz frequency bands.**

I. Introduction

To detect weak astronomical signals, superconducting devices such as kinetic inductance detectors (KIDs) [1] and transition edge detectors (TESs) [2], are widely developed for many astrophysics applications in millimeter, sub-millimeter, and far-infrared wavelength bands [3]-[7]. For total-power detection, usually a large number of pixels is required to increase the observation efficiency [8]. For example, the SCUBA2 [9] has 10^4 pixels and the OST satellite [10] requires 10^5 pixels for its far-infrared imager. In the astronomical detection field, KIDs have attracted great interest and have been developed rapidly in the last decade. Compared with TESs, the main advantages of KIDs are their simple structure and the intrinsic frequency domain multiplexing property. KIDs are based on superconducting microresonators [11], coupled capacitively or inductively to a feedline for frequency multiplexing. When the energy of incident photon is larger than the superconducting gap ($h\nu > 2\Delta$), the Cooper pairs are broken and quasi-particles are generated. This increases the kinetic inductance and the loss in the superconductor. The increased kinetic inductance shifts the resonance frequency and the increased loss decreases the resonance dip. Both can be read out from the resonance curve. Using a single feedline, hundreds or even thousands of KIDs can be read out simultaneously.

For lumped-element KIDs (LEKIDs) [12], Roesch et al. [13] have investigated the optical response of a LEKIDs array in the 2 mm atmospheric band. In their model, the array consists of a set of periodic pixels only, and the readout line is not considered. The absorption rate is calculated from the return loss measured using a vector network analyzer. In their analysis the simulation and measurement matched well, however, the simulation is done assuming a waveguide boundary and the higher modes are not considered.

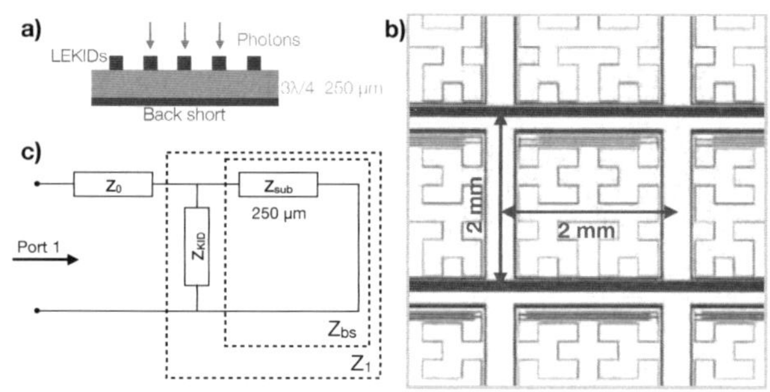

Fig. 1. a) Cut view of the LEKIDs array in the photon incident direction. b) Schematic drawing of the NIKA2 1mm array. The pitch size is 2 mm. c) Transmission line model of the optical coupling in (a). The LEKID, mainly the inductor, is represented by a sheet impedance Z_{KID}.

II. Array Layout

The NIKA2 1mm LEKIDs array [5] (Fig. 1) is simulated in this article. This array consists of 1140 pixels read out by 8 microstrip feedlines. The size of the array is a circle with radius of 40 mm corresponding to a 6.5' field-of-view on the IRAM 30-m telescope. The pixel pitch size and the inductor size are 2*2 mm^2 and 1.6*1.5 mm^2, respectively. The pixels are placed repeatedly with the same inductor design but different capacitor finger lengths for different designed resonance frequencies. The back of the wafer is covered with 200 nm thick aluminum, and acts at the same time as the ground plane for the MS readout feedline and as the backshort to optimize the optical coupling.

This array uses the bare LEKID design, which has no antenna structure on the focal plane. The incident light illuminates the array directly. The detection band of the

LEKIDs is determined by the backshort distance, which is the thickness of the silicon substrate. The band central frequency (wavelength) is designed to be 260 GHz (1.15 mm), the central frequency of the 260 GHz atmospheric window. The backshort is designed to be $3\lambda_{Si}/4$=250 μm, where the wavelength in silicon $\lambda_{Si} = \lambda/\sqrt{\varepsilon_r}$=333 μm, the wavelength in free space λ=1.15 mm and the relative permittivity of silicon ε_r=11.9.

The inductor is designed with a 3rd-order Hilbert curve [14] for dual-polarization sensitivity. The inductor width and the interval distance are designed as s=4 μm and w=240 μm, respectively. Since the wavelength is much larger than the structure, the inductor can be simplified as a sheet impedance Z_{KID}. The optical coupling can be modeled using the transmission line model [15], shown in Fig. 1 (c). The impedance of the backshort can be expressed as

$$Z_{bs} = jZ_{sub}\tan(\beta l),$$

where $\beta = 2\pi/\lambda_{Si}$ and l is the backshort distance (250 μm). The effective impedance of the KIDs together with the backshort is

$$Z_1 = \frac{1}{\frac{1}{Z_{KID}}+\frac{1}{Z_{bs}}}.$$

Assuming the reactance is zero, the impedance of KID is

$$Z_{KID} = R_{sq}/(s/w).$$

Then the absorption rate is calculated as

$$\text{absorption rate} = 1-|S11|^2 = 1-\left|\frac{Z_0-Z_1}{Z_0+Z_1}\right|^2,$$

where the vacuum impedance $Z_0 = 377\ \Omega$. Given the sheet resistance $R_{sq} = 1.6\ \Omega/\text{sq}$ of the used 20 nm aluminum film, $Z_{KID} = 96\ \Omega$ and the maximum absorption is 64.7% at the band center. This is only a simple estimation for the absorption rate. Actually, away from the band center, the absorption rate largely depends on the reactance of Z_{KID}, which is usually not zero and cannot be evaluated analytically. Therefore, we use the electromagnetic simulation as a useful tool for focal plane array design.

III. Simulation

The simulations are done using HFSS [16]. A Floquet port is assigned to Port 1 for exciting plane waves (Fig. 2). Considering the simulating frequency from 150 to 350 GHz, 18 modes are included in the analysis. Two TEM modes, TE00 and TM00 with *y* and *x*-axis polarizations, respectively, are simulated as the incident signal.

The surfaces of all components of pixel are assigned with an impedance boundary in the simulation. The simulated S-parameters are shown in Fig. 2 (b) and (c) for incident polarization 1 and 2, respectively. For the clarity of the discussion, only the modes with maximum values larger than -20 dB are considered and plotted.

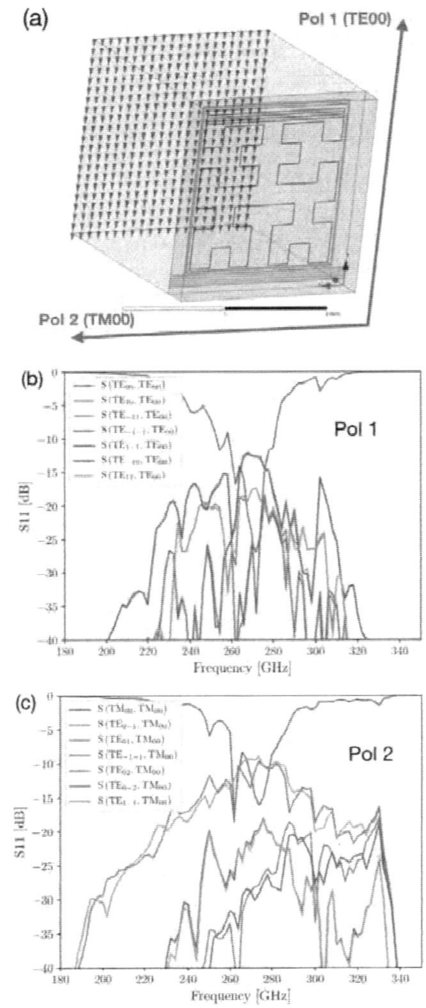

Fig. 2. a) The simulation model of a single pixel in HFSS. Two polarizations (TE00 and TM00) are stimulated as the incident signal. The arrows show the direction of the electric field in the two dominant modes. b) S-parameter results for a TE00 excitation (Pol 1). Only the modes with a maximum return loss higher than -20 dB are plotted for clarity (same for (c)). The TE10 and TE-10 modes have more reflect energy than other higher modes. c) S-parameter results for a TM00 excitation (Pol 2). TE0-1 and TE01 modes reflect more energy than other higher modes.

From the S11 of the dominant mode, the band center is -36 dB at 272 GHz and -17.5 dB at 274 GHz for Pol 1 and Pol 2, respectively. Compared with the design central frequency 260 GHz, there is a 12 GHz difference. The total absorption rate has a maximum at 262 GHz for both polarizations, which is consistent with the design. This analysis shows that actually a non-negligible part of the energy is reflected into higher modes, which are not considered in earlier studies [14].

The two dominant reflected higher modes are TE10 and TE-10 for Pol 1 and TE0-1 and TE01 for Pol 2. The maximum reflection of all higher modes together is 15.1% and 27.2% for Pol 1 and Pol 2, respectively. The average reflection in band from 230 to 290 GHz of all higher modes is 6.3% and 13.0% for Pol 1 and Pol 2, respectively. A large portion of the reflected higher modes is due to the reflected energy by impedance mismatch, 35.3% at peak calculated using the transmission line model. We also find that the electric field distributions of these higher modes for both polarizations are consistent with the geometry of the Hilbert curve. This suggests that part of the higher modes is caused by the non-uniformity of the inductor geometry. Compared to the wavelength 1.15 mm, the segment length of 240 μm is quite large ($\lambda/5$). Ideally the typical geometry should be much smaller than the wavelength to be treated as a uniform sheet ($<\lambda/10$) [17]. However, decreasing the segment length will also decrease the inductor width, resulting in a decrease of array yield and uniformity. For example, if we increase the third-order Hilbert inductor to a fourth order, the inductor width will be around 1 μm, and the total length of the inductor will be 27 cm. To fabricate such long and narrow line is a challenge for the yield of fabrication.

CONCLUSIONS

We have presented a transmission line model for our test LEKID array. Simulation results show that there is a non-negligible part of energy reflected to higher modes. Using low resistivity material, like aluminum, there may be a fabrication problem for achieving the maximum absorption rate. We are currently working on the extraction of the energy absorbed by the individual components, like the inductor, and optical response measurements. Further results are under investigation.

ACKNOWLEDGMENT

We thank Alvaro Gonzalez, Tom Nitta and Wenlei Shan for useful discussion and Yutaro Sekimoto for encouragement. The HFSS simulations were performed at National Astronomical Observatory of Japan.

REFERENCES

[1] P. K. Day, H. G. LeDuc, B. A. Mazin, A. Vayonakis, and J. Zmuidzinas, "A broadband superconducting detector suitable for use in large arrays," Nature, vol. 425, no. 6960, p. 817, 2003.

[2] K. D. Irwin and G. C. Hilton, "Transition-edge sensors," in Cryogenic particle detection. Springer, 2005, pp. 63–150.

[3] A. Monfardini, L. J. Swenson, A. Bideaud, F. X. Desert, S. J. C. Yates, A. Benoit, A. M. Baryshev, J. J. A. Baselmans, S. Doyle, B. Klein, M. Roesch, C. Tucker, P. Ade, M. Calvo, P. Camus, C. Giordano, R. Guesten, C. Hoffmann, S. Leclercq, P. Mauskopf, and K. F. Schuster, "NIKA: A millimeter-wave kinetic inductance camera," Astronomy and Astrophysics, vol. 521, p. A29, Oct. 2010.

[4] J. Sayers, C. Bockstiegel, S. Brugger, N. G. Czakon, P. K. Day, T. P. Downes, R. P. Duan, J. Gao, A. K. Gill, J. Glenn et al., "The status of music: the multiwavelength sub-millimeter inductance camera," in Millimeter, Submillimeter, and Far-Infrared Detectors and Instrumentation for Astronomy VII, vol. 9153. International Society for Optics and Photonics, 2014, p. 915304.

[5] R. Adam, A. Adane, P. Ade, P. Andre, A. Andrianasolo, H. Aussel, A. Beelen, A. Benoit, A. Bideaud, N. Billot et al., "The nika2 large-field-of-view millimetre continuum camera for the 30-m iram telescope," Astronomy & Astrophysics, vol. 609, p. A115, 2018.

[6] L. Ferrari, O. Yurduseven, N. Llombart, S.J.C. Yates, A.M. Baryshev, J. Bueno, and J. J. A. Baselmans, "Performance verification of a double-slot antenna with an elliptical lens for large format KID arrays," SPIE Astronomical Telescopes + Instrumentation, vol. 9914, pp. 99 142G–99 142G–6, Jul. 2016.

[7] J. Hubmayr, J. Beall, D. Becker, H.-M. Cho, M. Devlin, B. Dober, C. Groppi, G. C. Hilton, K. D. Irwin, D. Li et al., "Photon-noise limited sensitivity in titanium nitride kinetic inductance detectors," Applied Physics Letters, vol. 106, no. 7, p. 073505, 2015.

[8] D. Farrah, K. Ennico Smith, D. Ardila, C. M. Bradford, M. Dipirro, C. Ferkinhoff, J. Glenn, P. Goldsmith, D. Leisawitz, T. Nikola, N. Rangwala, S. A. Rinehart, J. Staguhn, M. Zemcov, J. Zmuidzinas, J. Bartlett, S. Carey, W. J. Fischer, J. Kamenetzky, J. Kartaltepe, M. Lacy, D. C. Lis, E. Lopez-Rodriguez, M. MacGregor, S. H. Moseley, E. J. Murphy, A. Rhodes, M. Richter, D. Rigopoulou, D. Sanders, R. Sankrit, G. Savini, J.-D. Smith, and S. Stierwalt, "Review: Far-Infrared Instrumentation and Technology Development for the Next Decade," ArXiv e-prints, Sep. 2017.

[9] W. Holland, D. Bintley, E. Chapin, A. Chrysostomou, G. Davis, J. Dempsey, W. Duncan, M. Fich, P. Friberg, M. Halpern et al., "Scuba-2: the 10000-pixel bolometer camera on the james clerk maxwell telescope," Monthly Notices of the Royal Astronomical Society, vol. 430, no. 4, pp. 2513–2533, 2013.

[10] J. Fortney, T. Kataria, K. Stevenson, R. Zellem, E. Nielsen, P. Cuartas-Restrepo, E. Gaidos, E. Bergin, M. Meixner, S. Kane et al., "The origins space telescope: Towards an understanding of temperate planetary atmospheres," arXiv preprint arXiv:1803.07730, 2018.

[11] J. Zmuidzinas, "Superconducting Microresonators: Physics and Applications," Annual Review of Condensed Matter Physics, vol. 3, no. 1, pp. 169–214, 2012.

[12] S. Doyle, P. Mauskopf, J. Naylon, A. Porch, and C. Duncombe, "Lumped Element Kinetic Inductance Detectors," Journal of Low Temperature Physics, vol. 151, no. 1, pp. 530–536, Apr. 2008.

[13] M. J. Rosch, F. Mattiocco, T. A. Scherer, M. Siegel, and K.-F. Schuster, "Modeling and Measuring the Optical Coupling of Lumped Element Kinetic Inductance Detectors at 120–180 GHz," IEEE Transactions on Antennas and Propagation, vol. 61, no. 4, pp. 1939–1946, 2013.

[14] M. Roesch, A. Benoit, A. Bideaud, N. Boudou, M. Calvo, A. Cruciani, S. Doyle, H. Leduc, A. Monfardini, L. Swenson et al., "Development of lumped element kinetic inductance detectors for nika," arXiv preprint arXiv:1212.4585, 2012.

[15] M. Roesch, Development of lumped element kinetic inductance detectors for mm-wave astronomy at the IRAM 30 m telescope. KIT Scientific Publishing, 2014, vol. 12.

[16] ANSYS HFSS Software HP: https://www.ansys.com

[17] S. Doyle, "Lumped element kinetic inductance detectors," Ph.D. dissertation, Cardiff University, 2008.

Broadening the IF band of a THz hot electron bolometer mixer by using a magnetic thin film

A. Kawakami[1*], Y. Irimajiri[1], T. Yamashita[2, 3], S. Ochiai[1], Y. Uzawa[4]
[1] National Institute of Information and Communications Technology, Kobe, Hyogo, 651-2492, Japan
[2] Nagoya University, Nagoya, Aichi, 464-8601, Japan
[3] PRESTO, Japan Science and Technology Agency, 4-1-8 Honcho, Kawaguchi, Saitama 332-0012, Japan
[4] National Astronomical Observatory of Japan, Mitaka, Tokyo, 181-8588, Japan
*Contact: kawakami@nict.go.jp

Abstract— To expand the IF band and improve the sensitivity of a hot electron bolometer mixer (HEBM), we examined and proposed a new HEBM structure using a magnetic thin film. We found that it was possible to suppress the superconductivity of the 5-nm thick niobium nitride (NbN) thin film by the addition of a 1.8-nm thick nickel (Ni) thin film. It was also confirmed that the superconductivity disappeared in the Au (70 nm)/Ni (1.8 nm)/NbN (5 nm) tri-layer for forming the electrodes of the HEBM. By using the magnetic thin film for the electrodes, we suggested that the superconductivity of the HEBM strip would be affected and that hot spots near the electrodes would form. This approach is effective for shortening the hot electron drift length and will lead to the expansion of the IF bandwidth. We think that the new structure lowers the required LO power and improves the sensitivity by suppressing the proximity effect under the metal electrode. The HEBMs using a Ni thin film were fabricated and the IF bandwidth was evaluated at 1.9 THz. We confirmed that the IF bandwidth expands, and the evaluated bandwidths were in the range of 5.1–5.7 GHz at 4 K.

I. Introduction

In the terahertz band, there are plenty of emission lines from atmospheric constituents, such as oxygen, atomic oxygen, water vapor, ozone, OH, and CO, which are applicable to observational studies of atmospheric dynamics and chemistry. For such applications, a heterodyne receiver with high frequency resolution is necessary. HEBMs are expected as extremely low noise mixer elements in applications above 1.5 THz. Several reports have already been made on the low noise operation of HEBMs with less than 10 times the quantum noise limit in the terahertz frequency range [1-3]. However, the IF bandwidth of a HEBM is not sufficient compared to a SIS mixer. Consequently, the usable IF band of receivers with HEBM still remain limited to typically 3-5 GHz [4,5]. It may not be sufficient for spectroscopic observations of a variety of atmospheric molecules or efficient wind measurement using several emission lines. Therefore, in recent years, several research efforts have focused on expanding the IF band.

To expand the IF band, efficient cooling of hot electrons is required. In general, there are two cooling mechanisms of HEBM, the heat dissipation process of excited electrons, and they are "lattice cooling" and "diffusion cooling." Lattice cooling releases the excitation energy to the substrate via the lattice, and diffusion cooling diffuses the excited electrons to the metal electrode directly. However, many of the current HEBMs were designed on lattice cooling. Though they are good achievements, we think that there are still problems. One is reduction of the required LO power, and the other is that the operating temperature of low noise performance and wide IF band performance are different.

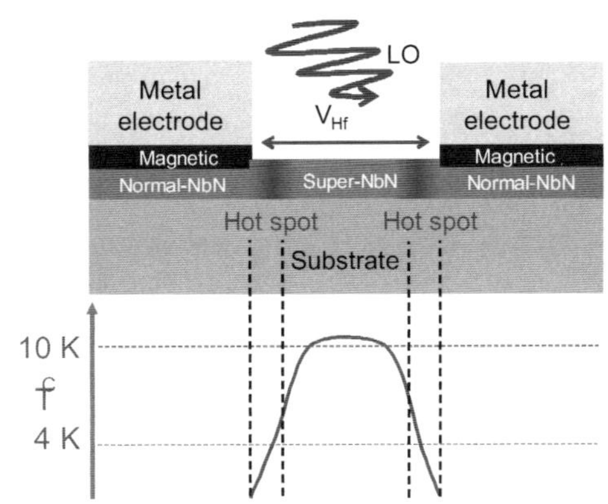

Fig. 1. Schematics of the structures of a new HEBM using a magnetic thin film.

From this background, we have proposed and examined a new HEBM structure using a magnetic thin film. The schematics of the structure of the new HEBM using a magnetic thin film is shown in Figs. 1. The new structure aims to expand the IF band by positively utilizing diffusion cooling. In addition, we are aiming at lowering the required LO power and improving the sensitivity. HEBMs have a structure in which the two metal electrodes are connected by an extremely thin superconducting strip. In general, to ensure good electrical connection, the superconducting strip and both electrodes are usually connected via an overlapped region on the strip. In our proposal, a magnetic thin film is placed between the metal electrode and the superconducting strip in this overlapping region, and the superconductivity of the strip in this region is suppressed by spin electron diffusion from the magnetic thin film. Thus, an HEBM structure with superconductivity only between the two metal electrodes can be realized. Superconductivity of the strip near the electrodes is also

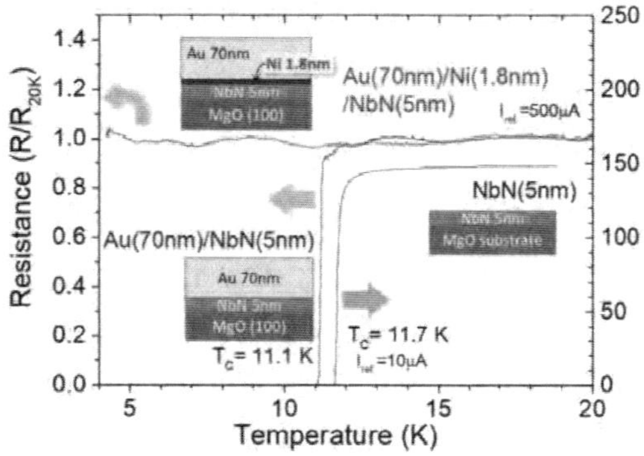

Fig. 2. Suppression of the NbN superconductivity under the HEBM metal electrode by insertion of a Ni thin film.

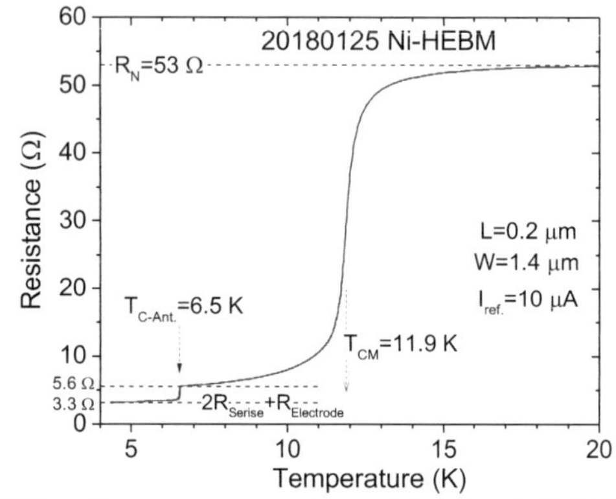

Fig. 3. The resistance–temperature characteristics of a typical Ni-HEBM. The inset shows characteristics near zero voltage.

suppressed by the magnetic thin film. Under LO irradiation, we think that the regions near the electrodes lose the superconductivity first and two hot spots are formed near the electrodes. As a result, the drift length of the hot electrons is shortened and the diffusion efficiency of the hot electrons to the electrodes is improved. Therefore, it is expected that the IF band will be expanded.

We also expect to reduce the required LO power and improve the HEBM sensitivity. In the conventional HEBM structure, the superconductivity under the electrode is remained. During operation of the HEBM, a hot spot is formed at the center between the electrodes with poor cooling efficiency, but it is considered that the superconductivity under both electrodes is maintained by the shielding effect of the metal electrodes against the LO. Therefore, the proximity effect due to the superconducting region under the electrodes is supposed to suppress hot spot formation. Thus, we believe that sensitivity reduction and the required LO power increases, i.e., limiting the superconducting region to only between the electrodes by the addition of a magnetic thin film makes it possible to further minimize the length of the superconducting strip and to realize the widening of the IF band and the reduction of the required LO power. In this paper, Ni thin films were used as the magnetic thin film; therefore, we denote the new HEBM structure as Ni-HEBM.

II. Fabrication of Ni-HEBM

A. Suppression of NbN superconductivity by Ni film

The Ni and Au thin films were deposited by using thermal evaporation methods. The deposition rates of Ni and Au were ~1.8 and ~36 nm/min, respectively, at room temperature. A 5-nm thick NbN thin film was used as the superconducting thin film. The NbN film was deposited by reactive sputtering at room temperature and epitaxially grown by using a magnesium oxide (MgO) single crystal substrate. Details of the NbN film deposition have been published elsewhere [6].

To confirm the superconductivity under the metal electrode of Ni-HEBM, a Au (70 nm)/Ni (1.8 nm)/NbN (5 nm) tri-layer was continuously deposited onto the MgO substrate under high vacuum. Fig. 2 shows the temperature dependence of the resistance of the tri-layer. For comparison, the Au (70 nm)/NbN (5 nm) bilayer, which is used for electrodes in a conventional HEBM, was also fabricated and measured. The Au/Ni/NbN tri-layer film did not show superconductivity until 4.2 K. However, the Au/NbN bilayer film showed a decrease in T_C of ~0.6 K, which is considered to be due to electron diffusion from the metal electrode, but it transited to the superconducting state at 11.1 K. From these results, a Au (100 nm)/Ni (1.8 nm) bilayer was used for the Ni-HEBM metal electrode.

B. Fabrication of Ni-HEBM

A 5-nm thick NbN thin film was used as the superconducting strip. To secure a high T_C, a single crystal MgO (100) substrate was used. NbN thin films were deposited by using DC reactive sputtering, and they showed a T_C of ~12 K. The electrode interval to determine the strip length was set at 0.2 μm and the strip width was set to 0.8–1.4 μm. As a plane antenna, a log spiral antenna was adopted. For both electrodes connected to the superconducting strip, a Au (100 nm)/Ni (1.8 nm) bilayer film was used. The electrode pattern at ~7 μm from the center was drawn by an electron beam lithography system. In this region, the NbN thin film constituting the superconducting strip exists under the Au / Ni bilayer, but the superconductivity was suppressed by the Ni film. On one hand, the antenna pattern of the outer region was formed using a photolithographic process. The antenna was patterned using an Au (150 nm)/Nb (5 nm)/NbN (5 nm) tri-layer by the lift-off method. The tri-layer in this region shows superconductivity. Fig. 3 shows the resistance–temperature characteristics of a typical Ni-HEBM. It was measured by the four-terminal method. The NbN strip length and width are 0.2 and 1.4 μm, respectively. In Fig. 3, the normal resistance (R_N) at 20 K is ~53 Ω and the transition temperature at which the resistance reached half of R_N (T_{CM}) was 11.9 K. The antenna pattern made of the Au/Nb/NbN tri-layer film showed superconductivity at 6.5 K. The residual resistance of 3.3Ω less than 6.5 K is considered as the sum of the NbN strip resistance of the regions near the electrodes which lost superconductivity due to the influence of Ni and the electrode resistances at ~7 μm from the center.

III. Evaluation of the IF bandwidth of Ni-HEBM

The evaluation setup for the IF gain bandwidth is shown in Fig. 4. Here, the IF gain bandwidth of the HEBMs were evaluated at 1.9 THz which is above the NbN gap frequency. As a signal, a stable source, which was generated using a uni-

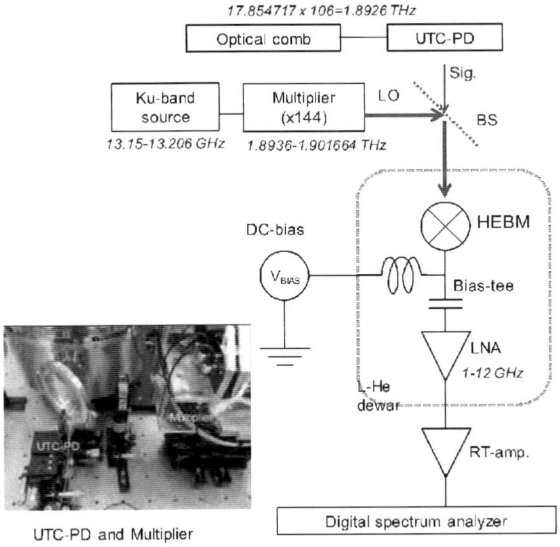

Fig. 4. The IF gain bandwidth evaluation setup of HEBM at 1.9 THz.

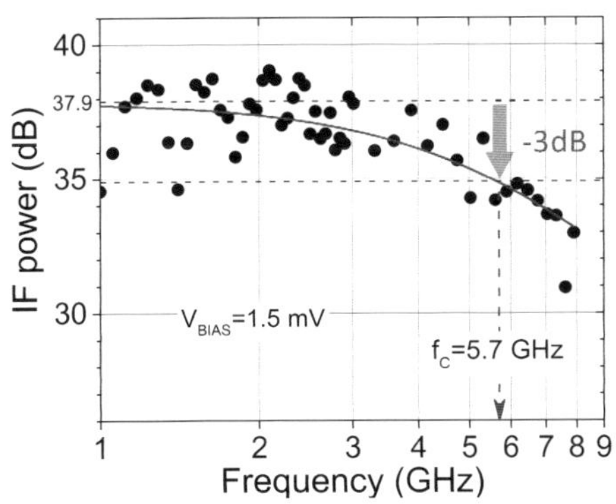

Fig. 5. IF gain bandwidth of the Ni-HEBM. The LO frequency, the measurement temperature, and the bias voltage were 1.9 THz, 4 K, and 1.5 mV, respectively

traveling carrier photodiode (UTC-PD) using the difference frequency component of two optical comb signals, at an appropriate frequency interval of 106 was used. As LO, the output of a multiplier manufactured by VDI was used. A DC bias voltage (V_{BAIS}) was applied through the bias tee to the Ni-HEBM, and the IF output was amplified using a cooled LNA that has a bandwidth of 1–12 GHz. Here, in the IF band measurement, the signal frequency was fixed and the LO frequency was changed. However, the output power changes with respect to the set oscillation frequency. Therefore, by controlling the LO output power with a LO-attenuator so that the bias current flowing in the HEBM becomes constant, the LO power incident on the HEBM was kept constant. The IF output power was measured using a digital spectrum analyzer and the difference between the IF output and the ground level near the IF signal was taken as the IF output value. All of the IF gain bandwidths were evaluated at ~4 K.

For comparison, the IF gain band evaluation of the conventional HEBM was performed. The NbN strip thickness, length, and width were 3.0 nm, 0.4 μm, and 3.2 μm, respectively. Details on the conventional HEBMs have been published elsewhere [7,8]. The IF roll-off frequency was decided by fitting to the equation of the low-pass filter model. As a result, the IF gain bandwidth of the conventional HEBM was ~4.1 GHz when the V_{BIAS} was 1.5 mV. The IF gain bandwidths of HEBM at a V_{BIAS} of 1.0 and 2.0 mV were also measured and they were 3.3 and 4.4 GHz, respectively. We also reported the IF noise bandwidth of the conventional HEBM. The NbN strip thickness, length, and width of the HEBM were 3.0 nm, 0.2 μm, and 2.0 μm, and the IF noise bandwidth was ~3 GHz [8]

Next, the IF gain bandwidths of the Ni-HEBM were evaluated and are shown in Fig. 5. The NbN strip thickness, length, and width of the conventional HEBM were 5.0 nm, 0.2 μm, and 0.8 μm, respectively. The IF bandwidths of the Ni-HEBM were evaluated at a V_{BIAS} of 1.0, 1.5 and 2.0 mV, and they were found to be 5.1, 5.7, and 5.7 GHz, respectively, which were wider than the bandwidth of the conventional HEBM. From these results, it is concluded that the IF bandwidth of the Ni-HEBM was expanded compared to the conventional HEBM.

Conclusions

To increase the IF bandwidth and improve the sensitivity, we proposed a new HEBM structure, Ni-HEBM, which used a magnetic Ni thin film. It was found that it is possible to suppress the superconductivity of the 5-nm thick NbN thin film by using a 1.8-nm thick Ni thin film. It was also confirmed that the superconductivity disappeared in the Au(100 nm)/Ni(1.8 nm)/NbN(5 nm) tri-layer forming the electrodes of the Ni-HEBM. By using these electrodes, hot spots are formed near the electrodes, which are effective for shortening the relaxation time of the hot electrons. The HEBMs using a Ni thin film were fabricated and the IF bandwidth was evaluated at 1.9 THz. We confirmed that the IF bandwidth expands, and the evaluated bandwidths were in the range of 5.1–5.7 GHz at 4 K.

References

[1] J. L. Kloosterman, D. J. Hayton, Y. Ren, T. Y. Kao, J. N. Hovenier, J. R. Gao, T. M. Klapwijk, Q. Hu, C. K. Walker, J. L. Reno, "Hot electron bolometer heterodyne receiver with a 4.7-THz quantum cascade laser as a local oscillator," *Appl., Phys., Lett.*, 102, 011123, 2013.

[2] J. R. Gao, M. Hajenius, Z. Q. Yang, J. J. A. Baselmans, P. Khosropanah, R. Barends, T. M. Klapwijk, "Terahertz Superconducting Hot Electron Bolometer Heterodyne Receivers," *IEEE Trans. on Appl. Supercond.* vol. 17, pp. 252-258, 2007.

[3] F. M. Boussaha, J. H. Kawamura, J. A. Stern, A. Skalare, V. White, "A Low Noise 2.7 THz Waveguide-Based Superconducting Mixer," *IEEE Trans. THz Sci. Technol.*, vol. 2, pp. 284-289, 2012.

[4] K. M. Zhou, W. Miao, S. C. Shi, R. Lefevre, Y. Delorme, "Gain and noise bandwidth of NbN hot-electron bolometric mixers", *URSI AP-RASC.*, pp. 2010-2012, 2016.

[5] D. Meledin, C. E. Tong, R. Blundell, N. Kaurova, K. Smirnov, B. Voronov, G. Goltsman, "Study of the IF bandwidth of NbN HEB mixers based on crystalline quartz substrate with an MgO buffer layer," *IEEE Trans. on Appl. Supercond.* vol. 13 pp. 164-167, 2003.

[6] A. Kawakami, Z. Wang, S. Miki, "Low-loss Epitaxial NbN/MgO/NbN Trilayers for THz Applications," *IEEE Trans. on Appl. Supercond.*, vol. 11, pp. 80 –83, 2001.

[7] A. Kawakami, Y. Irimajiri, S. Tanaka, S. Ochiai, I. Hosako, "Fabrication of 3 THz Quesi-Optical NbN Hot Electron Bolometer Mixer," *TEION KOGAKU*, 49, pp. 433-438, 2014.

[8] Y. Irimajiri, A. Kawakami, I. Morohashi, M. Kumagai, N. Sekine, S. Nagano, S. Ochiai, S. Tanaka, Y. Hanado, Y. Uzawa, I. Hosako, "Development of a Superconducting Low-Noise 3.1-THz Hot Electron Bolometer Receiver," *IEEE Trans. THz Sci. Technol.*, vol. 5, pp. 1154 – 1159, 2015.

SIS Photon Detectors for Terahertz Astronomy

Hajime Ezawa[1*], Hiroshi Matsuo[1], Masahiro Ukibe[2], Go Fujii[2], and Shigetomo Shiki[2]

[1]National Astronomical Observatory of Japan, Tokyo 181-8588, Japan
[2]National Institute of Advanced Industrial Science and Technology, Ibaraki 305-8568, Japan

[*]Contact: h.ezawa@nao.ac.jp

Abstract— Demands of high sensitivity detectors for astronomy and astrophysics are increasing. Observation technologies have advanced extremely in the past decades; large ground base observatory ALMA exhibited a very high resolution and high sensitivity capability at millimeter and submillimeter wavelengths. On the other hand, for space born terahertz missions, photon counting detectors are advantageous to realize high sensitivity and high dynamic range observations. Fast photon counters may significantly improve the capability of future intensity interferometers, aiming for ultra-high resolution imaging.

Considering the observation from space, photon rate from a 1 Jy source, such as stars or active galactic nuclei, is estimated to be in the order of 100 M photons/sec, when observed at a frequency of 1 THz and a bandwidth of 100 GHz with a 10 m telescope. To resolve these photons individually, the detector is to respond as fast as 1 GHz with NEP $< 3\times10^{-17}$ W/ $\sqrt{}$ Hz. We are considering that an SIS junction operated in a photon-assisted tunneling mode may fulfill this requirement, when its leakage current is as low as 1 pA. Our development is undergoing with the CRAVITY facility in AIST. The basic design of our SIS junction consists of thin layers of Nb/Al/AlOx/Al/Nb. As a first step we are developing an SIS junction to demonstrate the fast photon counting capability in lab experiments. For this purpose we utilize a detector with a relatively narrow bandwidth. So far we have successfully fabricated a junction of 3 μm x 3 μm with the leakage current of 1 pA below 0.7 K. The current density was 200 A/cm^2, which leads to the bandwidth of a few GHz. This satisfies our requirement for lab experiments, while developments towards wider bandwidth are necessary to realize future space borne astronomical instruments.

Based on this achievement, we are now designing to integrate this junction into an SIS photon detector. Twin slot antenna optimized for terahertz waves connects to the detector via a coplanar wave guide, where the detector is designed as a parallel connected twin junction (PCTJ). The signal is fed to the readout circuit through a choke filter, similar to conventional SIS mixers. We are considering a GaAs-JFET or a Junction-PHEMT for the first stage FET, followed by a SiGe cryogenic amplifier at 4 K. The development and performance of the SIS junction, as well as the basic design of the detector and its readout circuit will be discussed in the presentation.

Development of a 275-500 GHz waveguide SIS mixer and dual band LO injection system

T. Kojima*[1], M. Kroug[1], A. Gonzalez[1], K. Uemizu[1], K. Kaneko[1], A. Miyachi[1], Y. Kozuki[2], S. Asayama[1]
[1] National Astronomical Observatory of Japan, Mitaka, 181-8588, Japan
[2]Osaka Prefecture University, Sakai, 599-8531, Japan
*Contact: t.kojima@nao.ac.jp

***Abstract*— In this paper, we report on the design and performance of a prototype SIS mixer covering ALMA band-7 (275-373 GHz) and band-8 (385-500 GHz). The waveguide SIS mixer is based on a parallel connected twin-junction circuit with a current density of about 25 kA/cm^2. For the DSB mixer evaluation system, a wideband corrugated horn antenna and a 15-dB RF/LO coupler have been designed and fabricated. LO sources for such a broad frequency coverage are currently not available, therefore, a dual band LO injection system with two LO sources and an LO diplexer have been used. The measured DSB receiver noise temperature ranges from 45 to 90 K and is compliant with existing ALMA band 7 and 8 receiver noise temperature specifications.**

I. Introduction

At NAOJ, we are developing wideband sub-mm SIS mixers based on high-current-density junctions to cover two ALMA bands with the same receiver. In order to benefit from all equipment and know-how acquired during the design and production of ALMA band 8 receivers, we are aiming to cover ALMA band 7 (275-373 GHz) and band 8 (385-500 GHz) simultaneously. The fractional bandwidth of this target band is slightly over 60%. An increase in RF bandwidth of ALMA receivers can lead to the reduction of the total number of receivers to be operated and maintained in the ALMA front end, which would make operations more effective and increase the time devoted to science observations. In addition, the wider spectral range coverage with the same receiver and LO calibrations offers the possibility of new observation modes as described for example in the ALMA band 2+3 science case [1].

The development of a heterodyne receiver with large fractional bandwidths poses technical challenges for most components such as the SIS mixers, local oscillator source and waveguide parts including a corrugated horn and RF/LO coupler. In particular, high critical current density (J_C) junctions yielding low RC products are essential to achieve wide RF band performance [2]-[4]. Moreover, performance of the wideband receiver should not be compromised even if the frequency coverage increases. For example, we should keep the quantum-limited noise performance, and comply with ALMA specifications for both bands 7 and 8 to perform scientific observations in the same level as the current one.

In this paper, we describe the design and performance of the first prototype of a wideband waveguide SIS mixer and evaluation system covering the frequency range from 275 GHz to 500 GHz.

II. Band 7+8 SIS Mixer Design

Our mixer block is similar to the one described in [5]. The mixer block consists of three parts, a main body implementing a waveguide and chip slot, a back piece with the backshort, and an IF module with a 50-Ω microstrip line and a coaxial connector. The SIS mixer chip with dimensions 60 μm×180 μm× 2.2 mm is suspended over the chip slot, and is fixed at both ends with glue as shown in Fig. 1. The mixer chip is designed to cover the very wide RF bandwidth and comprises a waveguide probe in combination with a WR-2.3 (290 μm x 580 μm) waveguide, matching circuits employing high current density junctions, and IF output circuitry incorporating a low pass filter and an IF pad to connect an IF 50 Ω line with bonding wires.

We designed an asymmetric single-side waveguide probe placed across the reduced-height waveguide with dimensions 100 μm x 580 μm. The shape of the waveguide probe was empirically optimized to have nearly constant impedance in the desired RF range. Fig 2 (a) shows the simulation result of the probe impedance. The simulated probe impedance was around 30 Ω and slightly capacitive.

A scanning electron microscope image of the RF matching circuit is shown in Fig.1. The circuit comprises an Nb/SiO_2/Nb microstrip line and parallel-connected Nb/AlN_x/Nb twin junctions and is designed for a current density of around 25 kA/cm^2. The nominal junction diameters are ϕ =0.9 μm. The matching circuit implements a single-section quarter-wave impedance transformer to obtain the wideband performance. Fig. 2 (b) shows the calculated power coupling between feed point of the probe to the SIS junctions. The calculation was performed on the basis of the equivalent circuit shown in the graph in Fig. 2 (b). The design was optimized to maximize the coupling and the bandwidth. The calculated coupling is typically more than 90 % or return loss more than 10 dB. However, the coupling at the lower band edge gets deteriorated. SIS mixers with an updated design applying a multi-section impedance transformer with improved bandwidth coverage will be included in a future publication.

III. Band 7 and 8 Dual Band Evaluation System

Fig. 3 shows our DSB noise temperature measurement system and is similar to of the one described in [5], which focused on the wide IF bandwidth performance of a mixer-

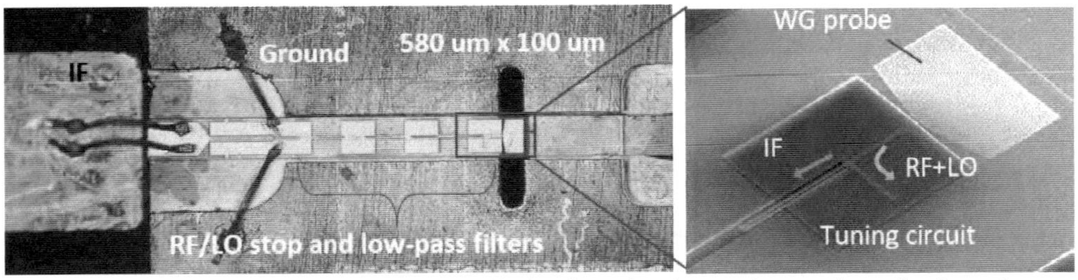

Fig. 1. Optical photograph of the SIS mixer chip mounted on the mixer block and scanning electron microscope image focusing on the waveguide probe and mixer matching circuit.

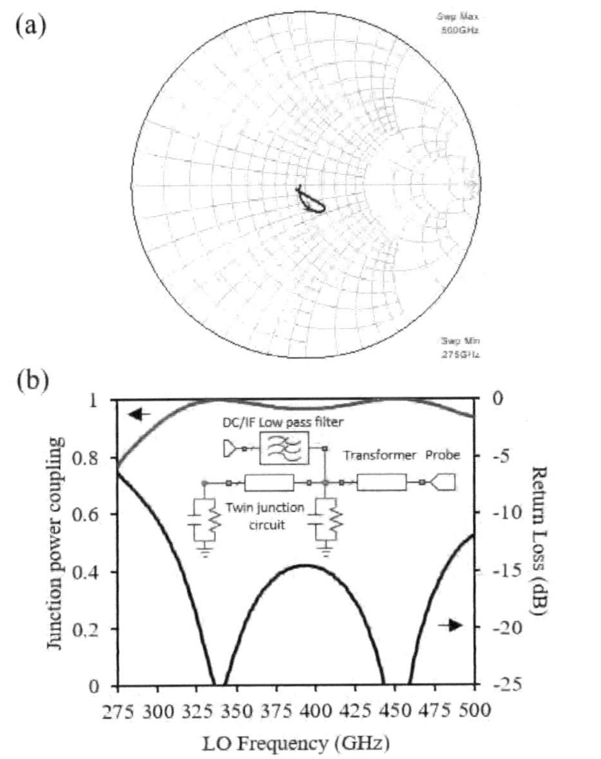

Fig. 2. (a) Simulation result of the designed probe impedance over 275-500 GHz. The Smith chart is normalized to 30 Ω. (b) The calculated junction power coupling and return loss of the designed matching circuit. The calculation is based on the equivalent circuit shown on the graph.

preamplifier module at ALMA band 8 frequencies (385-500 GHz). We modified this system and expanded the RF range to include ALMA band 7 frequencies (275-373 GHz).

On the RF side, the cryostat implements a 12.5-μm-thick polyimide film as a vacuum window and an infrared filter. The reflection loss of the polyimide film has been measured by terahertz time domain spectroscopy and is less than 3 % over the band 7 and 8 frequencies. A corrugated horn antenna [6] attached to a 15-dB RF/LO waveguide coupler, and the SIS mixer block are located on the 10 K stage. The corrugated horn antenna has a profiled shape to achieve the wideband performance in terms of return loss and cross polarization. The measured return loss is higher than 20 dB over the whole band. The RF/LO coupler, fabricated by direct machining, incorporates 3 slots with a 30-μm width between two parallel waveguides of 290 μm x 700 μm. The coupling of the RF/LO coupler ranges from −17 to −14 dB.

The target RF bandwidth is too wide for typical LO sources at these high frequencies, and therefore we applied a dual band LO system incorporating an LO diplexer [7]. The system allows us to evaluate an SIS mixer in one cooling cycle without any change of the configuration in the cryostat. The LO sources use ALMA band-7 and band-8 warm cartridge assemblies, which allow us to electrically control and generate high LO power with low amplitude noise [8], [9]. Cryogenic frequency multipliers x3 and x6 are located on the 10-K stage inside the cryostat and generate the final LO power in the frequency ranges of 283–365 GHz and 393–492 GHz. The frequency diplexer functions as the combiner of the two LO frequency bands. The output LO power from the diplexer is transmitted to the RF/LO coupler through a WR-4.3 oversized waveguide made of CuNi, and then is injected into the SIS mixer. The LO powers to pump SIS junctions are sufficient at the bands 7 and 8 frequencies. The IF chain in this study uses a cryogenic isolator and a cryogenic low-noise amplifier (CLNA) with a typical noise temperature of 2.1 K over the 4-8 GHz range [10]. Output IF power from the cryogenic IF chain is amplified with a room temperature amplifier and measured with a power meter and a spectrum analyzer.

IV. MEASUREMENT RESULT

We have fabricated two wafers with current densities $J_C \sim 26$ kA/cm^2 and $J_C \sim 60$ kA/cm^2, the latter unintentionally resulting in a current density much higher than the design value. The DSB noise temperature measurement was carried out by the standard Y-factor method using liquid-nitrogen cooled and room temperature blackbody loads. Preliminary measurement results for the DSB noise temperatures are 45 K to 90 K over the 283-492 GHz frequency range. Interestingly, both mixers showed similar performance even though their current densities differ by a factor of more than 2, indicating that the circuit design is very tolerant for a wide range of load impedances. The results fully comply with current ALMA band 7 and band 8 receiver noise temperature specifications. Note that the ALMA spec. is divided by 2 to translate from 2SB to DSB specification.

The updated second design of the SIS mixers, its performance and characterization are included in an extended

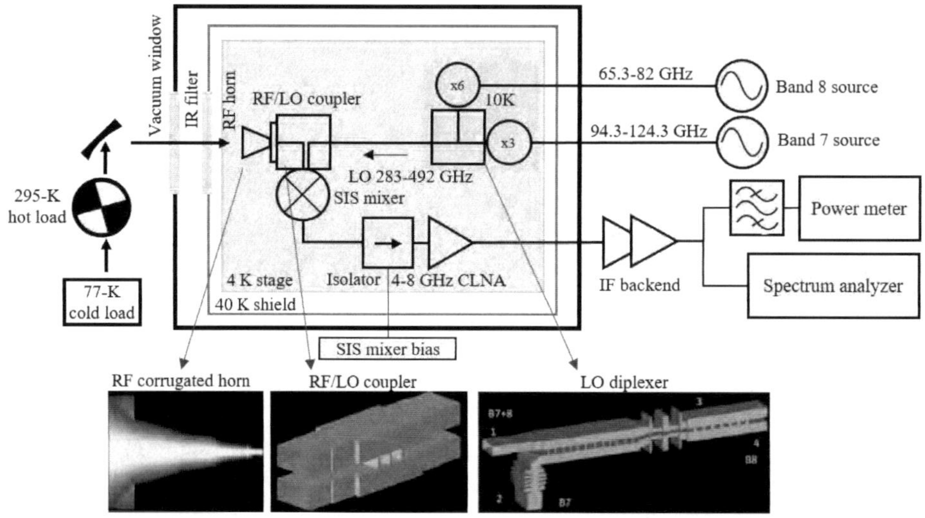

Fig. 3. Schematics of the noise temperature measurement setup with dual band LO injection system. The RF corrugated horn and the LO diplexer have been described in references [6] and [7].

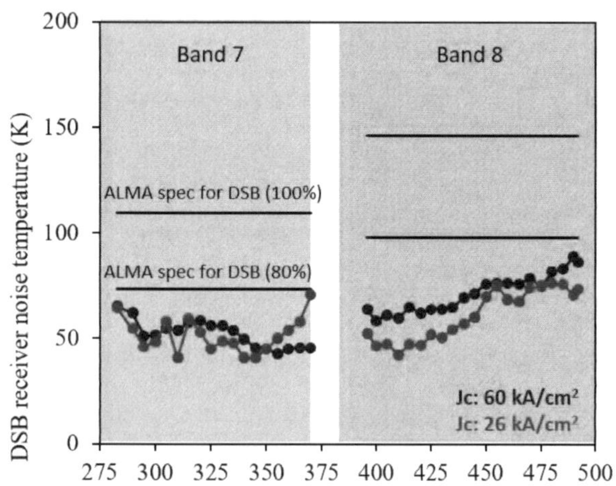

Fig. 4. Measurement results of the DSB receiver noise temperature for two SIS mixers with the critical current density of 26 and 60 kA/cm^2. The ALMA specs. at bands 7 and 8 are divided by 2 to translate from 2SB to DSB specification.

paper which has been submitted to IEEE Transactions on Terahertz Science and Technology.

Acknowledgment

The authors would like to thank Y. Uzawa, W. Shan, Y. Niizeki, S. Ezaki, and Y. Fujii in the NAOJ receiver development team for their support and helpful discussions.

This work was supported in part by JSPS KAKENHI Grant Numbers 15H02074.

References

[1] Fuller, G.A., et al., "The Science Case for ALMA Band 2 and Band 2+3", arXiv:1602.02414.

[2] H. W. Bode, Network Analysis and Feedback Amplifier Design, New York, NY, USA: Van Nostrand, 1945.

[3] R. M. Fano, "Theoretical limitations on the broadband matching of arbitrary impedances," Journal of the Franklin Institute, vol. 249 pp. 57–83, Jan. and pp. 139–154, Feb. 1950

[4] A. R. Kerr, "Some fundamental and practical limits on broadband matching to capacitive devices, and the implications for SIS mixer design," *IEEE Trans. Microw. Theory Techn.*, vol. 43, no. 1, pp. 2-13, Jan 1995.

[5] T. Kojima, M. Kroug, K. Uemizu, Y. Niizeki, H. Takahashi and Y. Uzawa, "Performance and Characterization of a Wide IF SIS-Mixer-Preamplifier Module Employing High-Jc SIS Junctions," *IEEE Trans. THz Sci. Technol.*, vol. 7, no. 6, pp. 694-703, Nov. 2017.

[6] A. Gonzalez, K. Kaneko, and S. Asayama, "Recent work on (Sub-)mm-wave ultra wideband corrugated horns for radio astronomy," 11th European Conference Antennas and Propagation (EuCAP), IEEE, 3327-3331 (2017)

[7] A. Gonzalez, T. Kojima, K. Kaneko, S. Asayama, "275-500 GHz Waveguide Diplexer to Combine Local Oscillators for Different Frequency Bands," *IEEE Trans. THz Sci. Technol.*, vol. 7, no.6, 669–676, Nov. 2017

[8] Y. Sekimoto, Y. Iizuka, N. Satou, T. Ito, K. Kumagai, M. Kamikura, M. Naruse, and W. L. Shan "Development of ALMA Band 8 (385-500 GHz) Cartridge", in *Proc. 19th Int. Symp. on Space THz Technol.*, Groningen, Netherlands, April 2008

[9] E. Bryerton and J. Hesler "Sideband noise screening of multiplier-based sub-millimeter LO chains using a WR-10 Schottky mixer", in *Proc. 19th Int. Symp. on Space THz Technol.*, Groningen, Netherlands, April 2008

[10] Low Noise Factory AB, G¨oteborg, Sweden. [Online]. Available: http://www.lownoisefactory.com/

Contribution of IF Chain Reflections in the Image Rejection Ratio of 2SB SIS receivers

A. Khudchenko[1,*], R. Hesper[1], J. Barkhof[1], and A.M. Baryshev[1]

[1]*NOVA/Kapteyn Astronomical Institute, University of Groningen, Groningen, the Netherlands*
[*]Contact: A.Khudchenko@sron.nl

Abstract—Here we present new results in our study of the Image Rejection Ratio (IRR) of a sideband separating (2SB) receiver based on SIS mixers. In this work we focus specifically on IF part of the receiver. It was found that the reflections from the IF amplifier, or isolator (if present) and from the SIS mixers can give a periodic imbalance in the IF chain, which influences the receiver IRR much more than the imbalance of the IF hybrid. This is clearly seen both in modeling and experiment. The measurements have been done for an ALMA Band 9 2SB receiver (600-720GHz) with different IF configurations: coupling the IF amplifiers to the IF hybrid directly or through an isolator. It was shown that having RF imbalances below a -20dB level we obtain IRR better than 10 dB in the entire receiver band, even when using IF amplifiers with return loss of only -7dB in the worst point (without isolators). Development of a receiver architecture without the isolators is an important step towards implementation of heterodyne arrays.

We explain the mechanism of IF imbalances and show how it influences the final IRR pattern depending on the level of RF imbalance.

Field-Effect Transistor Based Detector for Measuring Power Fluctuations of 4.75-THz Quantum-Cascade Laser-Generated Radiation

J. Zdanevičius[1], D. Čibiraitė[2], K. Ikamas[1,3], M. Bauer[2,4], J. Matukas[1], A. Lisauskas[1*], H. Richter[5], V. Krozer[2,6], H.-W. Hübers[5,7], and H. G. Roskos[2]
[1]Institute of Applied Electrodynamics and Telecommunications, Vilnius University, Vilnius, 10257, Lithuania
[2]Physikalisches Institut, Goethe-University Frankfurt am Main, Frankfurt am Main, 60438, Germany
[3]The General Jonas Žemaitis Military Academy of Lithuania, 10322 Vilnius, Lithuania
[4]Center for Materials Characterization and Testing, Fraunhofer ITWM, Kaiserslautern, 67663, Germany
[5]Institute of Optical Sensor Systems, German Aerospace Center (DLR), 12489 Berlin, Germany
[6]Ferdinand-Braun-Institut, Leibniz Institut für Höchstfrequenztechnik, 12489 Berlin, Germany
[7]Department of Physics, Humboldt-Universität zu Berlin, 12489 Berlin, Germany
*Contact: alvydas.lisauskas@ff.vu.lt

***Abstract*— In order to understand the formation of massive young stars, observation of a major coolant of dense interstellar medium, neutral atomic oxygen which has a fine-structure line at 4.7448 THz, is an important task. It requires high-sensitivity and high-resolution spectroscopy systems in the terahertz (THz) frequency range. One of such systems is in the Stratospheric Observatory For Infrared Astronomy (SOFIA). It essentially requires stable local oscillator (LO) source for heterodyne detection. Therefore, we explore a field-effect transistor based THz detector (TeraFET) with an integrated resonant patch antenna fabricated in commercially available standard 90-nm CMOS technology.**

At 4.75 THz, the detector exhibits an area-normalized minimal noise-equivalent power (NEP) of 404 pW/√Hz and a maximum responsivity of 75 V/W where the effective area is 1750 μm² and directivity reaches 7.4 dBi. These sensitivity and responsivity values contributes to the state of the art for room-temperature electronic detectors operating at around 4.7 THz.

We demonstrate experimentally, that the TeraFET detector can monitor the intensity of THz QCL radiation using only a small fraction of the beam power with S/N ratio of 40 dB and does not require chopping and could be employed in a heterodyne instrument.

I. Introduction

High-resolution and high-sensitivity heterodyne-based spectroscopic observations for astronomy require stable local oscillator (LO) sources operating in terahertz (THz) frequency range. For example, a major coolant of dense interstellar medium, neutral atomic oxygen has a fine-structure line at 4.7448 THz. Monitoring of this line became an important tool in order to understand the formation of massive young stars [1].

One of the monitoring laboratories is the Stratospheric Observatory For Infrared Astronomy (SOFIA). For high-resolution and high-sensitivity measurements at various frequencies, the heterodyne spectrometer GREAT (German REceiver for Astronomy at Terahertz frequencies) employs multiplied microwave sources in all channels except the highest one at 4.7 THz where a solution of this kind is not available. Instead, a quantum cascade laser (QCL) is employed to cover this frequency, which emerged as an alternative THz source solution to the gas lasers for such measurements [2].

It is desirable, that the power of the THz QCL can be continuously monitored. However, commercial devices like pyroelectric detectors [3] require implementation of chopping techniques, that is undesired because of space constrains. Therefore, here started with explorations of practical applicability of compact field-effect transistor based THz detector (TeraFET) [4]. Here we continue with our explorations and show the ability of low-NEP TeraFET detector to monitor the QCL power in continuous wave (CW) mode.

II. TeraFET Design

Reported TeraFETs are fabricated in commercially available standard 90-nm CMOS technology. The dimensions of the patch antenna are 13×13 μm², the antenna is surrounded by a 32×32 μm² metal cup formed using all 10 metal layers. Such a structure is meant to isolate antenna from surrounding wiring and other detectors, although the modeling showed that the directivity of emission (and concomitant effective cross-section of receiving antenna) stays unperturbed.

The FET inside detector has a channel length of 100 nm and a width of 400 nm. The whole chip die is glued and bonded into a ceramic dual-in-line (DIL) package.

III. Experimental Methods

The LO of the GREAT 4.75 THz line is based on a QCL mounted on the cold finger of the two-staged Stirling cryocooler (model Ricor K535) where the second stage is doubled in symmetrical manner to avoid large mechanical vibrations. The default cooling fins of the cryocooler heat sink are changed to the custom ones meeting standards of the SOFIA devices.

The QCL has a temperature sensor and a heater attached to it with a temperature stabilization based on field programmable gate array (FPGA). Such a cooling system maintained the QCL's temperature of 49 K during a whole experiment.

The constant-current source supplies the QCL with a varied current from 450 to 600 mA. The electrical power consumption of the QCL operating in continuous-wave (CW) mode at 600 mA was 4.13 W, which did not overcome the range of the cooling capacity of the Stirling cryocooler. The QCL is placed in vacuum to prevent condensation of water molecules.

Fig. 1. Optical setup. The same as in the SOFIA system. The emitted THz radiation passes through the high-density polyethylene (HDPE) window and is collimated with 25.4 mm focal length off-axis parabolic (OP) mirror. The beam splitter can be changed into various transmission levels. The radiation power is measured by Ge:Ga and TeraFET detectors.

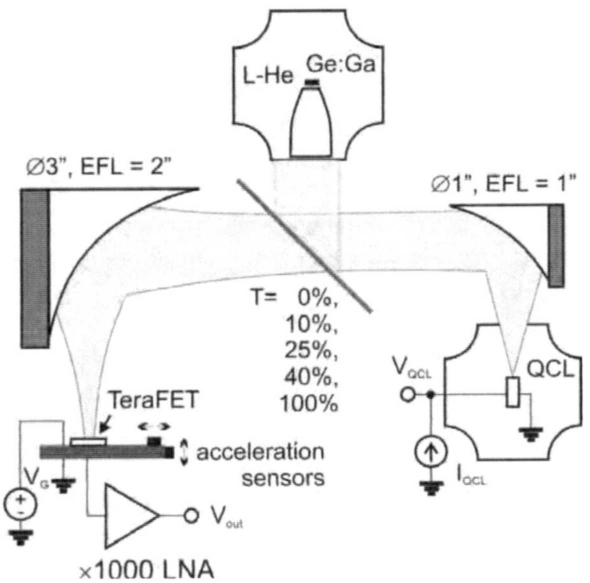

Fig. 1. Optical setup. The same as in the SOFIA system.

IV. TeraFET Application in SOFIA System

During the experiment, physical parameters of LO system, such as temperature, current and voltage of the QCL, power of the heater, and pressure inside the vacuum chamber, were measured and sampled simultaneously along the output voltage of the TeraFET detector which also had acceleration sensors attached.

At 4.75 THz, the detector exhibits an area-normalized minimal noise-equivalent power (NEP) of 404 pW/√Hz and a maximum responsivity of 75 V/W where the effective area is 1750 μm^2 and directivity reaches 7.4 dBi. These values are comparable with state-of-the art Schottky-Barrier Diode (SBD) detector for 4.92 THz based on 130-nm CMOS technology showing the complete device performance of 383 V/W at 300 Hz modulation frequency and shot-noise-limited NEP of 4950 pW/ √Hz [5].

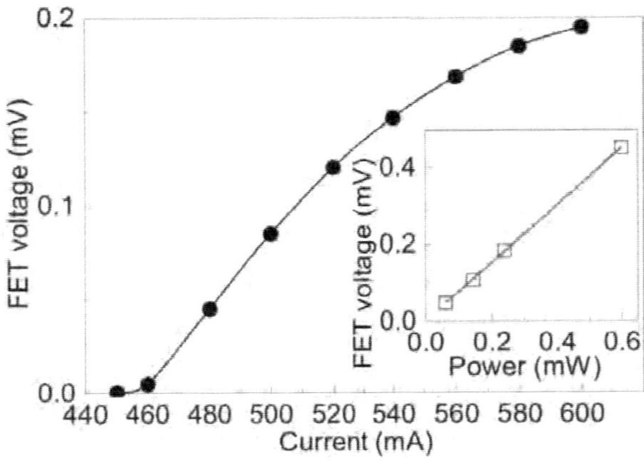

Fig. 2. Measured FET signal response at different QCL currents with 40 % attenuation with the inset of linear response-power dependency for different attenuation levels.

The dependency of the detection response to the incoming power is shown in the inset of Fig. 2. The detector response is linear in a whole power range up to 0.6 mW. The main graph shows the measured TeraFET response to the applied QCL current which controls to the emitted QCL power.

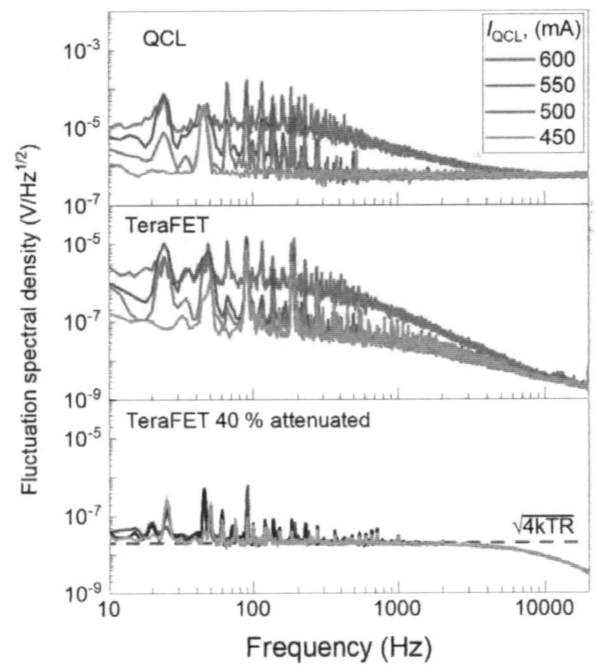

Fig. 3. Spectral voltage noise density of QCL (above), TeraFET (center), and TeraFET with a blocked radiation (bottom). Applied voltage of the QCL for corresponding graphs is shown in the legend and the dashed line shows the thermal noise curve of TeraFET.

Fig. 3. shows evaluated spectral voltage noise density of QCL (above), TeraFET (center), and TeraFET with a 40 % attenuated radiation power (bottom). Spectra exhibit the same frequency modulations caused by the table vibrations transferred through the cables of the setup. Applied current of the QCL for corresponding graphs is shown in the legend and the dashed line shows the thermal noise curve of TeraFET. At higher radiation power, we see a simultaneously increasing noise level in TeraFET and QCL voltage.

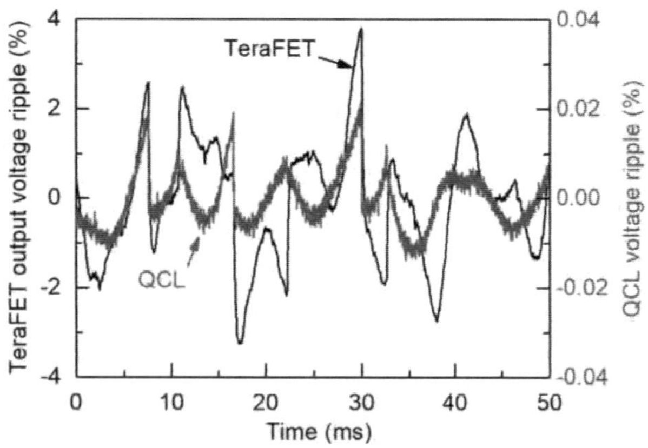

Fig. 4. Detected QCL and TeraFET voltage ripples.

Another important result making the TeraFET a better option for the heterodyne detection than a pyroelectric detector is an extremely low 1/f noise. At the bottom of Fig. 3 we show the TeraFET spectra when THz signal is 40 % attenuated. The dashed line shows the calculated thermal noise of the detector. We can see that the 1/f noise does not influence the spectrum and therefore the detector is very sensitive also for the real-time measurements of QCL power. Moreover, the noise of the detector does not depend on the QCL current and is only limited by the thermal noise to 20 nV/√Hz. And for the effective measurement bandwidth of 1 Hz the S/N ratio is 40 dB what shows the TeraFET ability to monitor the QCL power from a small fraction of the beam power (ca. 250 μW).

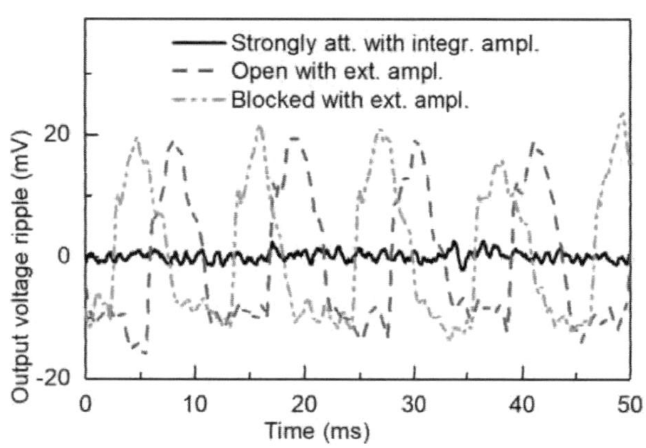

Fig. 5. Detected TeraFET signals with different amplifier configurations.

The measured signals of TeraFET response and QCL voltage are shown in Fig. 4. The QCL voltage ripple of 0.02 % can be monitored as 2 % ripple of the TeraFET signal. This phenomena has been caused by back reflections from the TeraFET. It can be seen that both distortions are equivalent in both noise spectra. We have achieved stable operation by replacing a 2 inch parabolic mirror by a Winston cone (data is not shown here).

The last issue we would like to mention from our experiments is the signal distortion using long coaxial cables and external amplifier. The Fig. 5 shows the time dependent signals measured with an external amplifier by blocked (dash dot line) and present (dash line) THz radiation in comparison to the integrated amplifier by a strongly attenuated THz signal which clearly reduced oscillations originating from cryocooler produced vibrations. Same oscillations are observed in reference Ge:Ga detector.

Conclusions and Discussion

We demonstrate a detector for application at 4.75 THz exhibiting an area-normalized minimal noise-equivalent power (NEP) of 404 pW/√Hz and a maximum responsivity of 75 V/W where the effective area is 1750 μm^2 and directivity reaches 7.4 dBi.

We demonstrate that the TeraFET detector has a linear power dependency in a whole power range of the QCL and can monitor the intensity of THz QCL radiation using a fraction of the beam power (ca. 250 μW) with S/N ratio of 40 dB.

We show a strong cross-correlation between the QCL voltage fluctuations and fluctuations at the detector output induced by a strong back-reflection. This fact directly proves that TeraFETs allow tracking of the intensity fluctuations on a millisecond time scale.

The reflection-induced intensity fluctuations in emitted QCL radiation can be improved by the use of attenuator, which reduce back-reflection or a Winston cone.

Since the TeraFET does not require chopping, it has an advantage over alternative room-temperature detectors (such like a pyroelectric sensor) and can be employed in a heterodyne instrument as a reference detector for real-time QCL power monitoring.

Acknowledgment

The Vilnius team is grateful the Research Council of Lithuania for the financial support under Contract No. S-LAT-17-3.

The Frankfurt team acknowledges for partial financial support in the frame of the Marie Curie European Training Networks ITN CELTA, Contract No. 675683.

References

[1] R. Boreiko and A. L. Betz, "Heterodyne spectroscopy of the 63 um OI line in M42," Astrophys. J., vol. 464, no. L83L86, 1996.
[2] H. Richter et al., "4.7-THz Local Oscillator for the GREAT Heterodyne Spectrometer on SOFIA," IEEE Transactions on Terahertz Science and, vol. 5, no. 4, pp. 539-545, 2015.
[3] Z. Liang et al., "High performance THz detector based on ultra-thin LiTaO3 crystal," in Infrared, Millimeter, and Terahertz waves (IRMMW-THz), 2015 40th International Conference on, Hong Kong, 2015.
[4] J. Zdanevicius et al., "TeraFET detector for measuring power fluctuations of 4.75-THz QCL-generated radiation," In Proceedings of ICNF 2017, Vilnius, pp. 1–4, 2017.
[5] Z. Ahmad et al., "9.74-THz electronic Far-Infrared detection using Schottky barrier diodes in CMOS," in International Electron Devices Meeting. San Francisco, CA: IEEE, Dec. 2014, pp. 4.4.1–4.4.4.

THz performance of MgB_2 HEB mixer with non-uniform thickness profile

Daniel Cunnane[1], Narendra Acharya[2], Wenura K. Withanage[2], Xiaoxing Xi[2], and Boris S. Karasik[*1]

[1]Jet Propulsion Laboratory, California Institute of Technology, Pasadena, CA 91109, USA
[2]Temple University, Philadelphia, PA 19122, USA
[*]Contact: boris.s.karasik@jpl.nasa.gov

Abstract—MgB_2 based HEB mixers have been under development for the past several years with the goal to achieve an adequate broadband heterodyne detector for high-resolution spectroscopy above 3 THz. Here, certain molecular lines in our galaxy may have radial velocity spread reaching several hundred km/s. This will require an intermediate frequency (IF) bandwidth up to 8 GHz at around 5 THz. Such a large bandwidth has been currently demonstrated, however other characteristics of the MgB_2 HEB mixer need further investigation and improvement.

The principal technique for make thin MgB2 films as thin as 5 nm with high critical temperature > 30 K is Hybrid Chemical-Physical Vapor Deposition. As a result, MgB_2 HEB devices of the small size similar to that of NbN HEB mixers have become feasible. However, the local oscillator (LO) power required for pumping such devices is substantially larger (~ 10 μW) than that needed for pumping the NbN HEB (~ 100 nW). This is quite expected given the large IF bandwidth and high electron temperature (electron heat capacity) in MgB_2 devices. Reduction of the required LO power is very desirable for enabling large heterodyne arrays. This can, in principle, be achieved by decreasing the device area to ≈ 100 nm × 100 nm since the film normal sheet resistance is several tens of Ohm (compared to ~ 1000 Ohm in NbN film).

This presentation will describe the operation of small (submicron size) MgB_2 HEB where the device area reduction was achieved in an unconventional way, using a postfabrication milling of the device with Ar-ion beam. The original devices made from a 40-nm thick film and integrated with planar log-spiral antennas are milled down to achieve smaller thickness and higher sheet resistance. However, the milling process of the MgB_2 film patch confined between tall gold walls (contacts to integrated antenna) is non-uniform and leads to the arched thickness profile. In the experiment, the thinnest central part of the device behaves as a very small HEB which is possible to pump with an LO power of 70-100 nW. These non-uniform devices still demonstrate very robust noise performance with a double-side band (DSB) noise temperature ~ 2,000 K in the 0.6-4.3 THz range using both molecular gas laser and quantum cascade laser LOs. Longer ion mill time results in creation of a weak-link Josephson junction with good sensitivity up to 2 THz.

Despite the difficulties in reproduction of such devices, this method deserves attention given the important benefits associated with small LO power requirement. The latter is critical for achieving large scale (~ 100 pixels) heterodyne cameras. An alternative approach may include the Focused Ion Beam technique in order to define better the geometry of the small thickness sub-HEB.

A Tunable Antenna-coupled Intersubband Terahertz Detector

C. Yoo[1], M. Huang[1], J. Kawamura[2],B. S. Karasik[2], L. Pfeiffer [3]and M. S. Sherwin[1*]

[1]*Physics Department and Institute for Terahertz Science and Technology, University of California at Santa Barbara, 93106, Santa Barbara, CA*
[2]*Jet Propulsion Laboratory, California Institute of Technology, 91109, Pasadena, CA*
[3]*Department of Electrical Engineering, Princeton University, 08544, Princeton, NJ*
[*]Contact: sherwin@ucsb.edu

Abstract— We are developing a 2.5 THz electrically tunable heterodyne detector for molecular line spectroscopy, operating at 50 K to 70 K. Named as Tunable Antenna-Coupled Intersubband Terahertz (TACIT) detectors, the detectors promise noise performance <10 times the quantum limit at frequencies above 1 THz while operating at temperatures accessible with passive cooling to 50-60 K in space. The detection frequency of a TACIT device can be electrically tuned in the THz frequency during a mission, adding the flexibility to search for a range of species with spectral signatures in different frequency bands. Compared to Schottky-diode mixers, TACIT mixers would require much smaller local oscillator (LO) power (~ μW) which enable operation in the 1-5 THz range where powerful LO sources are not readily available.

The TACIT detector can be approximated as a hot-electron bolometer based on the intersubband transition of two-dimensional electron gas (2DEG) in a 40 nm GaAs/AlGaAs quantum well (QW) heterostructure. The detector is fabricated from a high mobility (>10^6 cm^2/V-s at 4 K) 2DEG where the device resistance is highly sensitive to temperature in the temperature range between 50K and 70K. The designed TACIT detector is a 4-terminal device with source, drain, top gate and back gate where the source and the drain are Ohmic contacts to the 2DEG and the top and the bottom gates are used to tune the resonant frequency of the intersubband transition of the 2DEG. Indium Bonding And Stop Etch (IBASE) method is designed to process the backside of the device where a planar antenna structure will be deposited to efficiently couple the THz radiation to the active region of the device. When THz radiation near the resonant frequency of the intersubband transition impinges on the device, electrons in the 2DEG are excited to the next subband, leading to an increase in the electron temperature. In the presence of a local oscillator signal, the intermediate frequency (IF) signal can be detected between the source and the drain.

Currently, we have optimized the quantum well structure to have electron density of ~10^{11} cm^{-2} and high electron mobility of 10^6 cm^2/V-s at 4 K. Our preliminary devices retain the required electron density and high mobility throughout all the microfabrication steps up to the IBASE step. Our initial trial with the IBASE step show that a smaller sample size (~ 3 mm by 3 mm) and fewer number of bonding pads (~3) would be beneficial for the process. Optical measurements of the test structure using a 2.5 THz source show the tunability of the response by varying the gate voltage.

We will report the results of characterization of complete TACIT devices during the meeting.

An array scalable far-IR detector with NEP < 10^{-20} W/Hz$^{1/2}$

Boris S. Karasik

Jet Propulsion Laboratory, California Institute of Technology, Pasadena, CA 91109, USA
Contact: boris.s.karasik@jpl.nasa.gov

Abstract**—** As more powerful instruments are being planned for the next generation of submillimeter telescopes (e.g., Origins Space Telescope), the need for better detectors is becoming more urgent. Several advanced concepts have been pursued in the recent years with the goal to achieve a detector Noise Equivalent Power (NEP) of the order of 10^{-20} W/Hz$^{1/2}$ that corresponds to the photon noise limited operation of the future space borne far-IR spectrometers under an optical load ~ 10^{-19} W. Our recent work was focusing on the hot-electron nanobolometer (nano-HEB), a Transition-Edge Sensor (TES) where a very low thermal conductance was achieved due to the weak electron-phonon coupling in a micron- or submicron-size device. Using this approach, the targeted low NEP values have been confirmed via direct optical measurements. The kinetic inductance detector and the quantum capacitance detector also demonstrated recently high sensitivity. The next challenge is to arrive to a sizable array (1000s pixels) of such detectors. Many hurdles associated with electrical and thermal crosstalks, wire management, and readout noise should be addressed.

A novel hot-electron bolometric direct detector presented here uses a submicron size normal metal (e.g., non-superconducting Ti) patch coupled to a planar microantenna. The detector does not require any bias (dc or rf). The Johnson Noise Thermometry using a quantum noise limited microwave amplifier (LNA) allows for the direct read of an increase of electron temperate caused by the absorbed far-IR radiation. At 50 mK, the Noise Equivalent Power (NEP) is less than 10^{-20} W/Hz$^{1/2}$ whereas the dynamic range is 60-100 dB. Multiplexing of a 1000-pixel array is feasible using a single LNA with a bank of narrowband bandpass filters.

The NEP depends on the noise temperature of the readout amplifier and is not very sensitive to the electro-phonon coupling strength. However, materials with weak electron-phonon coupling (Bi, graphene) may provide a larger usable microwave bandwidth and thus will allow for more pixels per a single amplifier.

Although, the primary application is envisioned on a telescope with cryogenically cooled mirror (Origins Space Telescope), this detector can be employed on high-background suborbital and ground based platforms since the detector does not exhibit hard saturation with an increase of the radiation power. This can be important for gradual advancing the technology towards space applications. For such less sensitivity demanding applications, various simplifications of the detector array architecture are possible. For example, an NEP < 10^{-19} W/Hz$^{1/2}$ can be achieved with commercially available LNA (T_A ~ 2 K). Also, operating temperature can be increased for the use of the detector in photometers and polarimeters where the background is higher (NEP = 10^{-18}- 10^{-16} W/Hz$^{1/2}$).

In this paper, we will present an initial experimental study of a normal metal HEB made from a 1-square-micron normal metal Ti patch coupled to a planar twin-slot microantenna. A SQUID-rf LNA with the noise temperature $T_A < 1$ K followed by a HEMT LNA with $T_A \approx 5$ K and a large gain were used for readout. Electrical NEP was measured by sending a dc current through the device and measuring a change of output noise power caused by the heating. At 50 mK, the NEP was ~ 10^{-19} W/Hz$^{1/2}$. The data were obtained as function of bath temperature and the filter passband and compared with the model. An on-going effort to design a 1000-element bank of narrowband (few MHz) filters needed to array multiplexing will be presented too. We will also discuss various options of LNA (HEMT, parametric superconducting amplifiers, etc.) and associated sensitivity and dynamic range trade-offs.

Noise Performance of ALMA Band10 Receivers Employing High-j_c SIS Mixers

M. Kroug[*1], T. Kojima[1], Y. Fujii[1], K. Ohtawara[1], A. Miyachi[1], and Y. Uzawa[2]

[1] *National Astronomical Observatory of Japan, Mitaka, 181-8588, Japan*
[2]*National Institute of Information and Communications Technology, Koganei, 184-0015, Japan*
[*]Contact: matthias.kroug@nao.ac.jp

Abstract—Current ALMA Band10 cartridges covering the frequency range 790 – 950 GHz are equipped with SIS mixers based on standard Nb/AlO_x/Nb junctions integrated with NbTiN/SiO_2/Al strip lines for the embedding circuit. The current density j_c of the AlO_x-barrier type junctions is 10 – 14 kA/cm^2. While the sensitivity performance complies with ALMA receiver noise specifications, tolerance margins are small because of the rather narrow RF bandwidth of the devices [1]. We have fabricated and tested a new set of mixer chips using Nb/AlN_x/Nb SIS tri-layers with current density j_c ~30 kA/cm^2 and low leakage. Circuit design is unchanged except for small modifications in the geometry of the matching circuit to accommodate higher-j_c junctions. Compared with junctions in all-Nb SIS circuits, these junctions usually display a lower gap voltage and higher sub-gap leakage, an observation already made with AlO_x type junctions. This is likely because of degraded quality of the tri-layer when grown onto the NbTiN film instead of a blank quartz substrate. The high-j_c mixers under test have a gap voltage of V_{gap}=2.65 mV and sub-gap to normal state resistance ratios $q := R_{2mV}/R_n$ = 16 – 17 (all-Nb devices: V_{gap} ~2.8 mV, q = 20 or above). For noise performance characterization mixers are mounted into a Band10 cartridge type test set-up. Measured DSB noise temperatures are below 150 K up to 900 GHz and then increase moderately to T_{rx} = 175 K at 940 GHz. This fairly flat frequency dependence is a significant improvement over the low-j_c mixers.
However, we encountered several difficulties in finding optimum bias conditions for low noise performance *and* linearity. Depending on the position on the mixer`s power voltage curve, we observe points with gain expansion or gain compression, a phenomena described in [2]. This is directly related to the voltage swing in the IF output power around $V_{gap}/2$ and seems to be more pronounced for mixers based on high-j_c junctions. Details of the measurement method and analysis of the results will be presented at the conference.

[1] Y. Uzawa, "Development and Testing of Band10 Receivers for the ALMA Project," *Physica C,* vol. 494, 2013
[2] C.E. Tong, "Gain Expansion and Compression of SIS Mixers," *IEEE Transactions on Applied Superconductivity,* vol. 19 (3), 2009

Investigation of the Performance of an SIS Mixer with Nb-AlN-NbN Tunnel Junctions in the 780–950 GHz Frequency Band

Boon-Kok Tan*¶, Sumedh Mahashabde*, Andre Hector*, Ghassan Yassin*, Andrey Khudchenko‡, Ronald Hesper‡, Andrey M. Baryshev‡, Pavel Dmitriev¶, Kirill Rudakov‡§†† and Valery P. Koshelets¶
*Department of Physics (Astrophysics), University of Oxford, Keble Road, Oxford OX1 3RH, UK
‡University of Groningen, Kapteyn Astronomical Institute, 9747 AD, Groningen, The Netherlands
§Kotel'nikov Institute of Radio Engineering and Electronics RAS, 125009 Moscow, Russia
††Moscow Institute of Physics and Technology, 141700 Dolgoprudny, Russia
¶Contact: boonkok.tan@physics.ox.ac.uk, phone +44 1865 273 303

***Abstract*—In this paper, we present preliminary measured performance of an SIS mixer employing a Nb/AlN/NbN tunnel junction in the frequency range of 780–950 GHz range. The mixer design is an upgrade of the Carbon Heterodyne Array of the Max-Planck-Institute Plus (CHAMP+) mixer, coupled with an easy to fabricate smooth-walled horn. The noise temperature of the mixer is measured using the standard Y-factor method, but all the RF optics is enclosed in the cryostat. We use a rotating mirror in the cryostat to switch between a room temperature load and a 4 K blackbody load. With this method, we have measured a noise temperature of 330 K around 850 GHz, corrected for a mismatch between a reduced height rectangular waveguide at the input of the mixer block and a full height waveguide at the output of the horn. To remove this mismatch we now plan to redesign a new mixer chip with a full-height waveguide back-piece. The expected performance of the new mixer chip is also reported.**

I. Introduction

The Carbon Heterodyne Array of the Max-Planck Institute Plus (CHAMP+) installed in the Atacama Pathfinder EXperiment (APEX) telescope consists of two 7-pixel arrays: the low frequency band array operating from 620–720 GHz and the high frequency band array from 790–950 GHz [1]. The low frequency band array have shown state-of-the-art sensitivity, but the high frequency band array has struggled to achieve the similar quantum-limited performance [2], [3]. We have therefore designed and fabricated a new set of high-gap superconductor-insulator-superconductor (SIS) mixers to investigate the performance of the CHAMP+ high band array devices [4], [5]. In order to allow for the potential fabrication of a larger heterodyne array on APEX at such high frequencies, we have replaced the original corrugated horn with the Oxford-pioneered easy-to-fabricate smooth-walled horn designed to operate in the same frequency region. The integration of the smooth-walled horn with the high-gap SIS mixer also allows us to investigate the feasibility of constructing a large focal plane SIS mixer array near terahertz (THz) frequencies for future astronomical projects such as the Origin Space Telescope (OST) [6] and the Millimetron [7] project. Another key component of our project is the inclusion of the optics and the RF loads inside the cryostat. This will hopefully reduce the optical losses, allows more accurate measurement of the Y-factor by enclosing the RF loads in the vacuum environment, and utilise the full power of the local oscillator (LO) by reducing the optical path of the pumping signal.

Sumedh Mahashabde is now with the Department of Microtechnology & Nanoscience, MC2, Chalmers University of Technology, Kemivägen 9, 412 58 Göteborg, Sweden.

II. SIS Mixer & Experimental Setup

The design of the high-gap SIS mixer used here has been reported before in [4], hence will only be briefly summarised here. We used two high current density (30 kA/cm^2) 0.5 μm^2 SIS junctions separated by a quarter-wavelength microstrip [8], [9] to tune out the parasitic junction capacitances for broadband performance. The tunnel junctions comprise of Niobium (Nb) ground and Niobium Nitride (NbN) top electrode, both 100 nm thick, sandwiching a thin Aluminium Nitride (AlN) insulation layer. The junctions are fabricated on top of a 300 nm thick Niobium Titanium Nitride (NbTiN) ground plane, with a 250 nm Silicon Dioxide (SiO_2) dielectric layer and a 500 nm thick Aluminium (Al) overlaid on top to form the microstrip transmission line, connecting the junctions to the rest of the superconducting circuits. The whole structure is

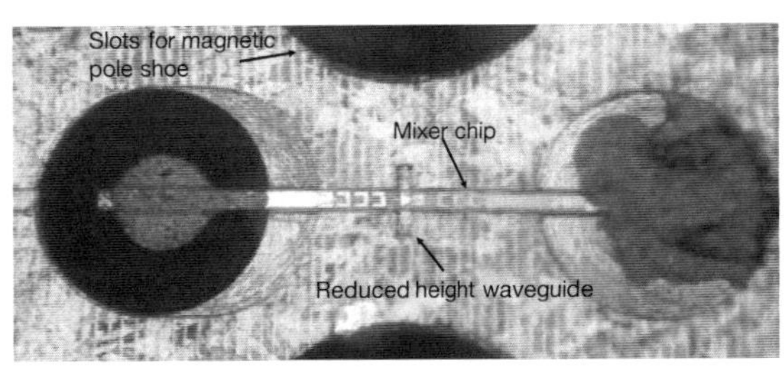

Fig. 1. An image of the mixer block's back-piece with the SIS mixer chip mounted on a 60 μm deep groove milled into the surface of the back-piece, with the triangular probe antenna positioned along the E-plane of a reduced height rectangular waveguide to couple the RF signal to the mixer.

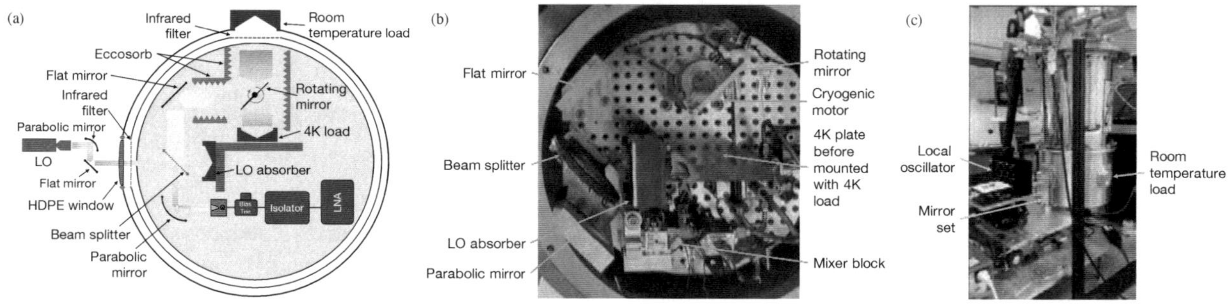

Fig. 2. (a) Schematic layout of the internal optics of the cryostat. (b) Image showing the actual optical setup. (c) A photo showing the external LO setup along with the PTC cryostat.

supported by a 40 μm thick Quartz substrate, positioned along the E-plane axis of a reduced height rectangular waveguide fabricated on the back-piece of the mixer block, with the normal vector to the chip aligned to the direction of the propagation of the incoming RF wave along the waveguide, as shown in Fig. 1.

The back-piece carrying the mixer chip was aligned to the rectangular waveguide output of the feed horn and held in place using a threaded centring ring. The whole assembly (horn + mixer) is then slotted into an Atacama Large Millimetre/Sub-Millimetre Array (ALMA)-style mixer block, which contains a temperature sensor, magnetic coils and the DC/IF SMA connector. This assembly was designed such that a separate back-piece could be bolted to the same mixer block for quick turn-around testing. The SIS device-loaded back-piece and part of the mixer block were supplied by the Kapteyn Astronomical Institute (RUG) group to Oxford, where the rest of the mixer block including the feed-horn was fabricated.

In the original CHAMP+ mixer assembly, the feed horn block was electro-formed with the corrugated horn and the circular-to-rectangular waveguide fabricated in a single piece. However, to avoid the need for the complex electro-forming, we fabricated a smooth-walled horn by direct drilling with a circular to full-height rectangular waveguide transition. The performance of the smooth-walled horn has been reported previously [10], and has shown to have comparable high quality as the corrugated horn in the frequency range.

Fig 2 shows the optical arrangement of the experimental setup. The mixer test was performed in a Janis pulse-tube cooler (PTC) system at Oxford, with the optics that bring the RF signal to the mixer mounted inside the cryostat, including the hot and cold loads for Y-factor measurements. This should reduce the optical path of the LO signal in air and eliminate losses from the cryostat window. The room temperature load was mounted on the outer wall of the cryostat (via a window socket), while the 4 K load was attached to a 4 K plate anchor mounted on the cold plate. We used a rotating mirror mounted on top of a cryogenic motor to sweep the mirror between the hot and cold load for Y-factor measurements. LO power was coupled from outside the cryostat to the mixer using a Gaussian telescope arrangement via another optical window. This arrangement ensures that the effect of the water absorption at high frequencies is minimised, thus allowing us to access the actual performance of the mixer near THz frequencies.

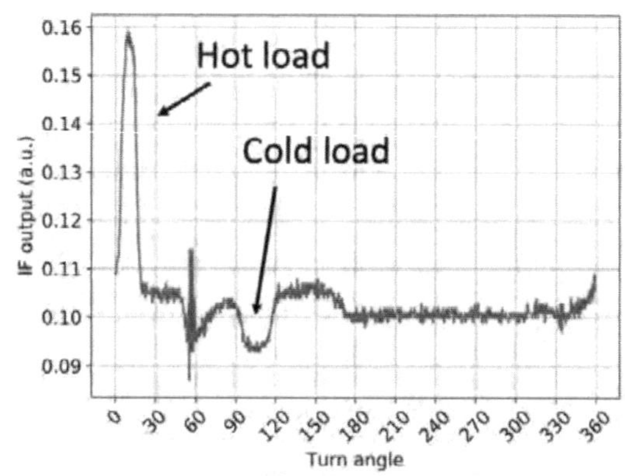

(a)

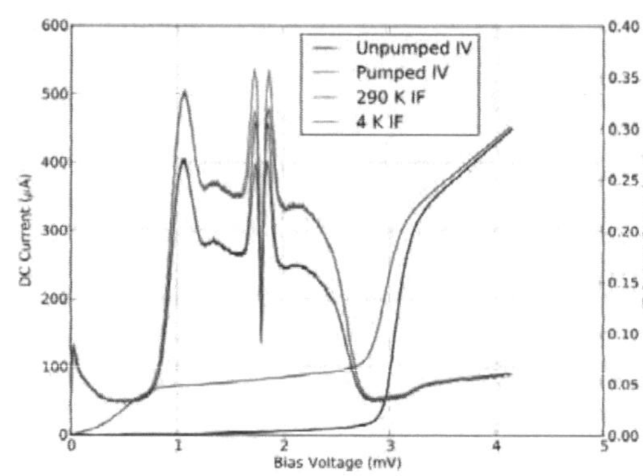

(b)

Fig. 3. (a) The measured IF response of the mixer when the mirror is swept along its rotational axis, pointing the mixer beam towards different surfaces within the cryostat, including the room temperature hot and 4 K cold load. The mixer was optimally biased at 2.4 mV and the LO set at 850 GHz. (b) The measured pumped and unpumped IV curves, along with the hot and cold IF responses, at 850 GHz. All plots are corrected for a 3.1Ω series resistance in the device.

III. Preliminary Measured Results

Fig. 3 (a) shows the intermediate frequencies (IF) response of the SIS mixer when the rotating mirror was swept around in a full rotation cycle. We can clearly see the difference in the IF responses when the mixer is illuminated by a hot (room temperature) or a cold load (4 K) loads. Fig 3 (b) shows the pumped and unpumped current-voltage (IV) curves, along with the IF responses when the mirror was fixed to face either the hot or the cold blackbody loads. When the LO was set at 850 GHz, we measured an uncorrected noise temperature of about 600 K (double side-bands, DSB), and the noise temperature remains at a similar level from about 820 to 880 GHz, as shown in Fig 4. The sensitivity has then started to deteriorate and the cause of this increase in noise temperature at the high frequency end is still under investigation.

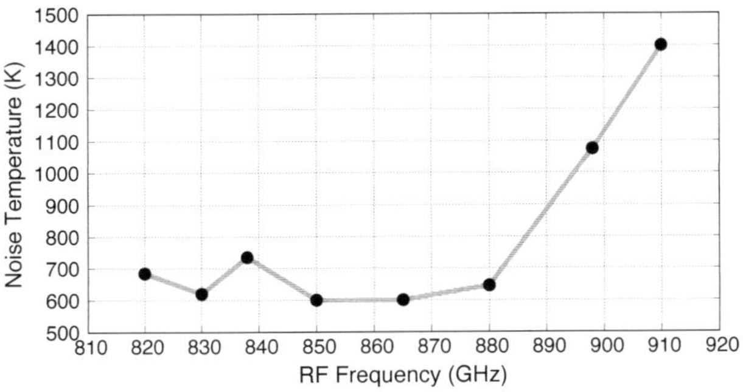

Fig. 4. The measured uncorrected noise temperature from 820–910 GHz.

We would like to emphasise that the relatively high value of 600 K noise temperature was caused by the rectangular waveguide mismatch between the output of the feed horn and the back-piece where the mixer is mounted. As shown in detail in Fig. 5, the rectangular waveguide port of the feed horn is a full-height waveguide with the narrow wall of 150 μm, in comparison to the reduced height back-piece waveguide that is half the size. We estimated that this costs about 16–20% losses of RF power coupling to the SIS mixer, and the situation is probably more severe when they are compounded by the conversion loss of the mixer itself. Furthermore, the dielectric beam splitter we used in our setup has a reflection coefficient of about 30% across the band, which was necessary due to the low output power of the LO source at these frequencies. By taking into account these two effects and other optical losses such as the infrared filter and the dielectric window, we estimated that the noise temperature measured is in the range of 250–400 K, which is similar to the values reported by [4].

IV. Improved Mixer Chip Design

In order to improve the performance of the SIS receiver, we have since started to modify the SIS mixer design such that it can be used along with a full-height waveguide back-piece, eliminating the waveguide mismatch effect. Furthermore, as shown in Fig. 3, the current SIS mixer suffers from the junction heating effect. This is probably caused by the resistive properties of the mixer, and partly due to the fact that the bottom electrode of the tunnel junction, which was made out of Nb is being sandwiched between the high gap NbTiN ground plane and the NbN top electrode. Although the junction heating may not be critical for the mixer operation, we aim to

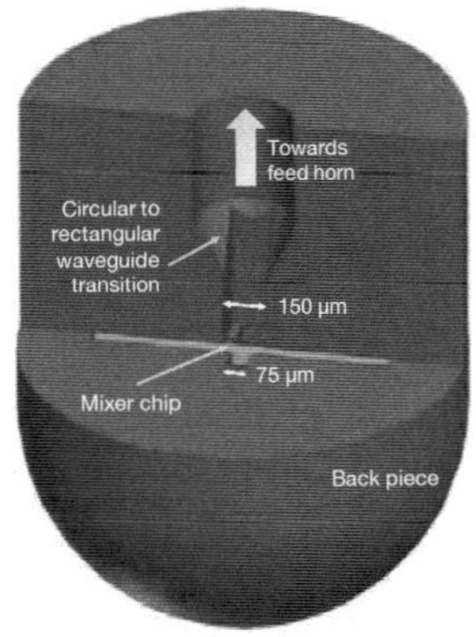

Fig. 5. Illustration showing the rectangular waveguide mismatch between the reduced-height waveguide back-piece with the full-height circular-to-rectangular waveguide transition.

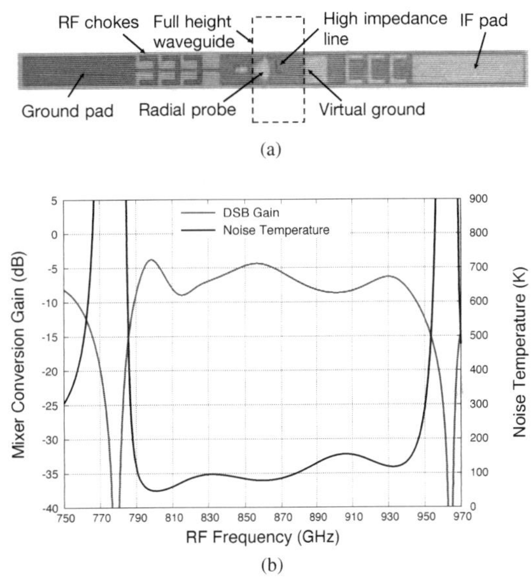

Fig. 6. (a) Preliminary design of an SIS mixer chip with radial probe coupled to a full-height rectangular waveguide. (b) Simulated noise temperature performance of the SIS mixer chip using High Frequency Structure Simulator (HFSS) and SuperMix [11] software package.

resolve this problem in the new mixer design, by replacing the NbTiN ground plane with an aluminium (Al) layer, so that the hot electrons can escape via the normal metal. We will also replace the top wiring layer with NbN, so that it has the same superconducting material as the top electrode of the tunnel junction. The tunnel junctions are expected to be 1 μm^2 in area, with an AlO_x insulation layer.

Fig. 6 (a) shows the preliminary design of a new mixer chip. The design is similar to the existing mixer chip, except that we replaced the triangular probe with a radial probe to cover broader bandwidth. The probe was also extended to about 35% of the waveguide height to achieve maximum coupling. A high impedance line is used to connect the radial probe to the virtual ground pad, along with the RF chokes and the IF bonding pads, forming an integrated bias-tee arrangement [12]. We expect a good noise temperature performance from 790–940 GHz, as shown in Fig. 6 (b). However, the simulated double sideband (DSB) gain is not ideal yet, at the level of about –5 to –10 dB level across the band. Hence, we are currently in the process of optimising the mixer chip design, to cover the full 780–950 GHz, and to improve the conversion gain performance as well.

V. Conclusion

We have presented preliminary measured performance of an SIS mixer operating in the ALMA Band 10 frequency range. The mixer was coupled with an easy to fabricate smooth-walled horn operating in the same frequency region. The hot and cold load for performing the noise temperature measurement, along with other optical components, are enclosed within the vacuum environment of the test cryostat to reduce the water vapour contaminations. Using this arrangement, we measured a preliminary noise temperatures of about 330 K, corrected for the waveguide mismatch and the loss from a thick beam splitter. To overcome this mismatch and to include other design improvements, we are currently in the process of designing a new mixer chip to further improve the performance of our THz SIS receiver.

References

[1] C. Kasemann, R. Güsten, S. Heyminck, B. Klein, T. Klein, S. Philipp, A. Korn, G. Schneider, A. Henseler, A. Baryshev, *et al.*, "CHAMP+:

[11] J. Ward, F. Rice, G. Chattopadhyay, and J. Zmuidzinas, "SuperMix: A flexible software library for high-frequency circuit simulation, including SIS mixers and superconducting elements," in *Proceedings, Tenth International Symposium on Space Terahertz Technology*, pp. 269–281, 1999.

a powerful array receiver for APEX," in *Millimeter and Submillimeter Detectors and Instrumentation for Astronomy III*, vol. 6275, p. 62750N, International Society for Optics and Photonics, 2006.

[2] Y. Uzawa, Y. Fujii, A. Gonzalez, K. Kaneko, M. Kroug, T. Kojima, A. Miyachi, K. Makise, S. Saito, H. Terai, *et al.*, "Tuning circuit material for mass-produced terahertz SIS receivers," *IEEE Transactions on Applied Superconductivity*, vol. 25, no. 3, pp. 1–5, 2015.

[3] G. De Lange, B. Jackson, M. Jochemsen, W. Laauwen, L. De Jong, M. Kroug, T. Zijlstra, and T. Klapwijk, "Performance of the flight model HIFI band 3 and 4 mixer units," in *Millimeter and Submillimeter Detectors and Instrumentation for Astronomy III*, vol. 6275, p. 627517, International Society for Optics and Photonics, 2006.

[4] A. Khudchenko, A. M. Baryshev, K. I. Rudakov, P. M. Dmitriev, R. Hesper, L. de Jong, and V. P. Koshelets, "High-gap Nb-AlN-NbN SIS junctions for frequency band 790–950 GHz," *IEEE Transactions on Terahertz Science and Technology*, vol. 6, no. 1, pp. 127–132, 2016.

[5] A. Khudchenko, A. M. Baryshev, R. Hesper, M. E. Bekema, R. de Haan Stijkel, J. Barkhof, D. van Nguyen, V. P. Koshelets, P. Dmitriev, M. Fominsky, *et al.*, "Performance of sis mixers for upgrade of champ+ 7-pixel arrays," in *Proceedings, Twenty Eighth International Symposium on Space Terahertz Technology*, 2017.

[6] "Origin Space Telescope (OST)." https://asd.gsfc.nasa.gov/firs/.

[7] A. Smirnov, A. Baryshev, P. de Bernardis, V. Vdovin, G. Gol'tsman, N. Kardashev, L. Kuz'min, V. Koshelets, A. Vystavkin, Y. V. Lobanov, *et al.*, "The current stage of development of the receiving complex of the Millimetron space observatory," *Radiophysics and quantum electronics*, vol. 54, no. 8-9, pp. 557–568, 2012.

[8] V. Y. Belitsky, S. Jacobsson, L. Filippenko, and E. Kollberg, "Broadband twin-junction tuning circuit for submillimeter SIS mixers," *Microwave and Optical Technology Letters*, vol. 10, no. 2, pp. 74–78, 1995.

[9] J. Zmuidzinas, H. G. LeDuc, J. A. Stern, and S. R. Cypher, "Two-junction tuning circuits for submillimeter SIS mixers," *IEEE transactions on microwave theory and techniques*, vol. 42, no. 4, pp. 698–706, 1994.

[10] A. Hector, G. Yassin, B. Ellison, M. Tacon, and A. Stokes, "A Smooth-Walled Feed Horn Operating in the Terahertz Regime," in preparation.

[12] C. Risacher, V. Vassilev, A. Pavolotsky, and V. Belitsky, "Waveguide-to-microstrip transition with integrated bias-T," *IEEE Microwave and Wireless Components Letters*, vol. 13, no. 7, pp. 262–264, 2003.

Superconducting Parametric Amplifiers: the Next Big Thing in (Sub)Millimeter-wave Receivers

O. Noroozian*[1, 2, 3], P. K. Day[4], H.G. Leduc[4], D. Woody[5], J. Zmuidzinas[5], A. R. Kerr[1], J. G. Mangum[1], M. Cyberey[3], A. W. Lichtenberger[3]

[1]*National Radio Astronomy Observatory, 22903, Charlottesville, VA,USA*
[2]*NASA Goddard Space Flight Center, 20771, Greenbelt, MD, USA*
[3]*University of Virginia, 22903, Charlottesville, VA, USA*
[4]*Jet Propulsion Laboratory, NASA, Pasadena, CA, USA*
[5]*California Institute of Technology, Pasadena, CA, USA*

*Contact: onorooz@nrao.edu

Abstract—Superconducting paramps developed for quantum computing have narrow bandwidths, low dynamic range, and operate at sub-Kelvin temperatures. Our collaborators at JPL/Caltech recently demonstrated a microwave Travelling-Wave Kinetic Inductance Parametric (TKIP) amplifier with several GHz of bandwidth, and orders of magnitude larger dynamic range with near quantum-limited noise performance up to an operating temperature of 3K. Similar amplifiers but at higher frequencies up to 1 THz could be designed since the operation principle and physics remain largely unchanged. These qualities make TKIPs suitable candidates for ground-based astronomy with instruments such as ALMA.

In a 2-year study at NRAO we are looking into the feasibility of a high-frequency TKIP demonstration prototype as a front-end replacement for the ALMA band-3 SIS mixer receivers. Our collaborators at Caltech have designed and fabricated a TKIP chip that covers a very wide bandwidth of 55-175 GHz, and we are preparing to test this device at NRAO-CDL. We believe that ultimately the main challenge will be in further optimizing the superconducting thin-film materials to retain their desired properties such as ultra-low loss at temperatures close to 4K.

The enhanced observational capabilities that would be enabled by a RF TKIP amplifier front-end would benefit ALMA science across all bands. For example the Band-3 improved signal-to-noise would be a factor of ~ 5 measured at the receiver input. Including the loss of atmosphere this translates to a doubling of system sensitivity and a factor of ~ 4 increase in array efficiency (speed) enabling the detection of weaker spectral lines and continuum sources and mapping larger fields. The increased sensitivity from the RF front-end relaxes the requirements on IF amplifiers and allows for tradeoff with bandwidth to increase the instantaneous IF bandwidth from the current 4 GHz per sideband per polarization to ~ 10 GHz. For continuum observations, this provides a greater than factor of two increase in efficiency (speed), which combined with the increased RF efficiency would result in a factor of ~ 8 increase in observation efficiency (speed). For spectral observations such a wide bandwidth also enables the detection of various spectral lines simultaneously, removing the need of multiple observations at different LO frequencies to cover the whole band.

These amplifiers are not only interesting for ALMA, but are also very useful for direct detection astronomy (e.g. MKID detectors, TES detectors) for amplifying or multiplexing signals from large focal-plane arrays of photon detectors for space telescopes such as NASA's Origins Space Telescope and X-ray telescopes. Therefore, this development activity is highly synergistic with development of future generations of direct detectors on space and ground platforms.

Graphene Field Effect Transistors for Microwave and mm-Wave Applications

Jake A. Connors*, Paul K. Grimes[2], Raymond Blundell[2], Young Jae Shin[1], and Philip Kim[1]

[1]*Department of Physics, Harvard University, Cambridge, MA, 02138*
[2]*Smithsonian Astrophysical Observatory, Cambridge, MA, 02138*
*Contact: jconnors@cfa.harvard,edu

Abstract—In order to operate at microwave frequencies and higher, the channel length of a field effect transistor must be made very short (~25nm at 1THz) to minimize input capacitance and the drift time of carriers through the channel. This scaling cannot continue indefinitely however, as short channel effects limit FET transconductance at very short gate lengths. These effects are largely due to a degradation in the ability of the gate to electrostatically control channel doping as the gate-to-channel distance and the finite thickness of the 2DEG channel become comparable to the channel length. This degradation can be avoided by utilizing graphene, a zero-bandgap two-dimensional material consisting of carbon atoms on a hexagonal lattice, as a channel material due to its high carrier mobility, truly atomic thickness and ability to be integrated with other Van der Waals materials for ultra-thin gate dielectrics. Presently however, the difficulty of producing low-resistance electrical contacts to graphene and the absence of saturating behavior in graphene transistors lead to poor RF performance especially for short channel devices. By utilizing current annealing to reduce contact resistance and biasing in a high-field regime with saturated carrier velocities, graphene field effect transistors (GFETs) with improved RF performance can be realized. Here, we present fabrication techniques to create edge-contacted, exfoliated, hexagonal Boron-Nitride encapsulated graphene field effect transistors (GFET), including self-aligned processes for producing ultra-short gate lengths scalable to dimensions enabling THz and higher transistors. We discuss measurements of the DC and AC linear response of several GFETs of gate lengths between 1um and 100nm, as well as potential for future improvements. Notably an f_{max} of 59GHz is achieved in a 1µm gate length transistor with softly saturating DC characteristics. In contrast to the high-field operation of GFET amplifiers, when biased near the zero density of states Dirac point, the low field conductivity of graphene exhibits a distinctly non-linear behavior with respect to gate voltage. Utilizing this non-linearity to rectify incident voltage waves, a GFET can be made to be a very broadband power detector. In a 1µm gate length GFET at 30K, a responsivity from 1-20GHz of 55V/W is achieved with a noise equivalent power limited by the room temperature readout to 1nW/√Hz.

MKID large format array testbed

L. Ferrari[1*], S. J. C. Yates[1], M. Eggens[1], A. M. Baryshev[2], and J. J. A. Baselmans[3,4]

[1]SRON Netherlands Institute for Space Research, Groningen, 9747AD, The Netherlands
[2]Kapteyn Institute, University of Groningen, Groningen, 9747 AD The Netherlands
[3]SRON Netherlands Institute for Space Research, Utrecht, 3584CA, The Netherlands
[4] Terahertz Sensing Group, Delft University of Technology, Delft 2628CD, The Netherlands
*Contact: l.ferrari@sron.nl

***Abstract*— The development of an astrophysics kilo-pixel imaging instrument requires a dedicated cryogenics and optics setup to measure the expected performance of the detector arrays in terms of sensitivity and optical response.**
We have developed a testbed for the characterization of MKID large format arrays (6 cm x 6 cm) in the 350 GHz and 850 GHz band. The testbed is a wide field camera, allowing full optical tests in a realistic environment. The cryostat is based upon a commercial pulse-tube cooled 3K system with a He10 sorption cooler to reach base temperature below 250 mK. We will describe the thermo-mechanical solutions implemented in our system to minimize the thermal loading on the cold stage allowing a base temperature of 250 mK in combination with a large focal plane. The optics design and the stray light philosophy will also be presented.

I. Introduction

Microwave Kinetic Inductance Detectors (MKIDs) have matured in the last decade and have become an excellent detector choice for future submm and mm astronomical instruments. Their intrinsic multiplexing capability makes the readout of kilopixels arrays relatively cheap and easy compared to other technologies [2] [3]. At SRON we are developing antenna coupled KIDs for a KID camera called AMKID to be deployed at the APEX telescope in Chile [4]. Our design is targeting two of the atmospheric windows available at the telescope centered at 350 and at 85O GHz [5][6] . The detector arrays before deployment to the telescope need a full optical and electrical characterization. The array performance verification can only be realized using a wide field camera which simulates the thermal and optical properties of the telescope instrument. For this goal we have designed and built such testbed. The cryostat requirement specification derived from the telescope instrument and the detector design are:

1) 2 bands of operation: 350 GHz and 850 GHz;
2) Use of reimaging optics with F#=2 at the detector, Strehl ratio better than 95%, near telecentric behavior and a band transmission better than 30%;
3) Stray light and spillover rejection better than 60 dB;
4) Base temperature of 250 mK or lower;
5) A turn-around of roughly a week.

In this paper we will describe the thermal, optical and mechanical solutions implemented to meet the specified requirement.

II. Camera design

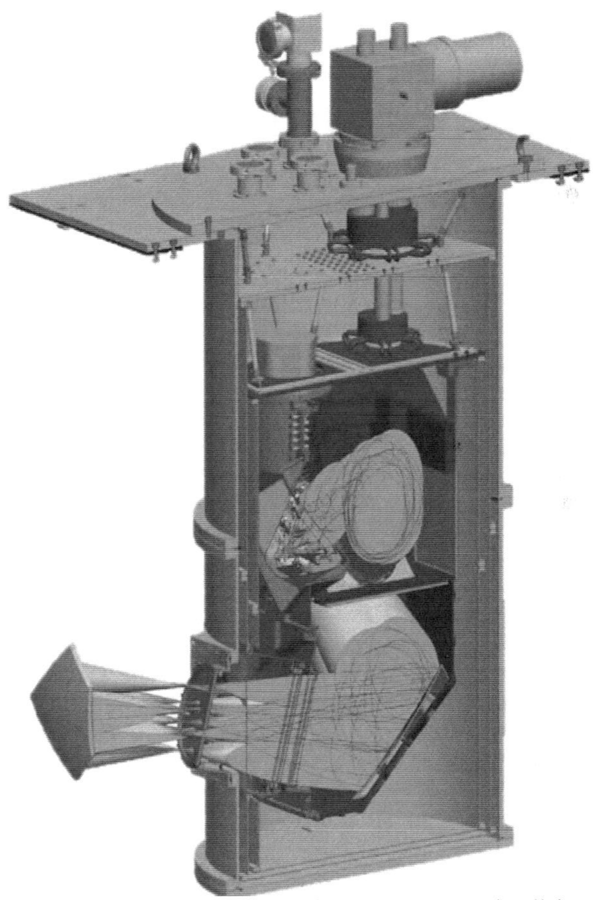

Fig. 1. Schematic of our wide field camera. For simplicity the warm optics is omitted. The light enters via the cryostat window and is reimaged via the optics onto the array, mounted on a thermal suspension unit. The cold optics is mounted inside an enclosed box connected to the 4 K plate via spring leafs. The detector array is cooled down to base temperature via a He10 cooling unit.

The camera design is shown in Fig. 2. The cryostat cooling unit is based on a commercial Pulse tube cooler (PTC) and an extra He3-He3-He4 sorption cooler [7] used to reach the base temperature at the detector array. A liquid nitrogen LN2 precooling unit allows fast turnaround of the system, speeding up the first part of the cool-down to roughly 80 K. The radiation enters the cryostat via an HDPE window and travels via the optics to the detector array mounted inside a thermal suspension unit. The cold optics are enclosed in a box mounted on the 4 K cryostat plate via spring-leaves.

In the following subsections each component is described in more detail.

A. *Re-imaging optics and optics box*

The camera optics create an image of the detector array at a warm focal plane outside the cryostat using a seven mirror system with a total magnification of 3. The optical design is based on two parabolic relays, one in the cold at 4 K and one outside the cryostat. The optical design is based on aberration compensation [8] that cancels out the aberrations and cross-polarization of the optics near the optical axis. All mirrors were modelled as biconics, allowing different curvature and conic constants in the x and y direction. An angular limiting aperturein the pupil limits the beam to F#=2 and is placed between the 4 K active mirrors where all the different pixel beams overlap. The design was simulated using the Zemax software [9] (see Fig. 2 for the final design). The cold optics are mounted inside an enclosed 3K box (800 X 425 X 400 mm^3) using three mounting pads for each mirror that allow fine positioning. The alignment of all mirrors was verified by measuring their position with a 3D measuring machine and found to be within tolerances. The optics box provides also the mounting interface for the He10 and the detector thermal suspension unit and baffles. The inside of the box is coated with radiation absorbing coating made of a mixture of 1 mm SiC grains and Stycast 2850 [10]. The box is equipped with flexible thermal straps to ensure a good thermal contact and evenly cool down the big structure.

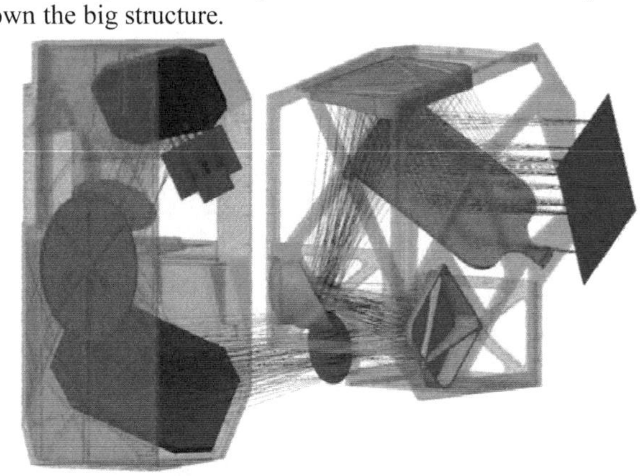

Fig. 2. Scheme of the cold and warm optics. The cold optics is mounted inside an enclosed 4 K box. The warm optics is mounted outside the cryostat.

B. *Radiation filtering*

The optical measurement band is defined by a set of quasi optical filters [11]. More than 60 dB out of band rejection is needed to suppress the out of band 300 K blackbody spectrum and self-emission of the windows and filters. A known problem is the filter heating due to the fact that the filters are large and made of plastic (mostly mylar) so poorly thermalized [12]. This issues requires additional filters to block the reradiated heat. In Fig. 3 the spectral transmission of the filters is shown and in Table I the filter stack overview for the 350 GHz band is summarized including the main task of each filter.

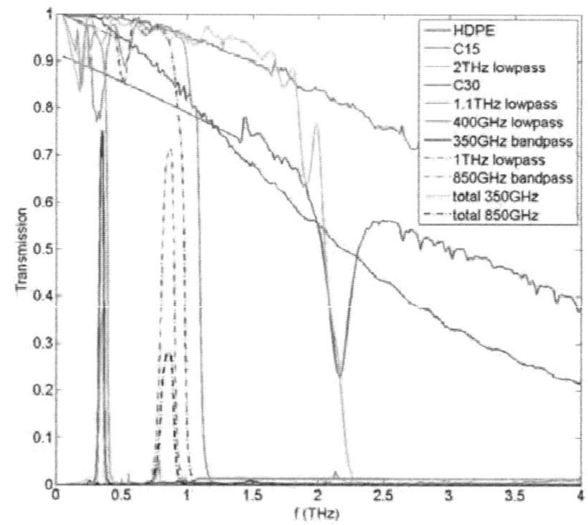

Fig. 3 Spectral transmission of the filters used in our system.

TABLE I
FILTER STACK OVERVIEW FOR THE 350 GHZ BAND

Name	Position and minimal T	Task
HDPE window	Window 300 K	Absorbs IR
Scatterer	300 K	Scatters near IR radiation, reduce condensation on the window
Shader 15 μm	50 K	Reflects near to mid -IR power back
Scatterer	50 K	Scatters near IR radiation, reduce condensation on the window
Shader 15 μm	50 K	Reflects power back
LP 3 THz	50 K	
Shader (2x) 30 μm	Window 4 K	Reflects near to mid-IR power back
LP 1.1 THz	Window 4 K	Reflects far IR and out of band mm wave but absorbs mid-IR
Goretex sheet	Pupil 4 K	Absorbs IR
LP 400 GHz	850 mK	Reflects far IR and out of band mm wave but absorbs mid-IR
BP 350 GHz	250 mK	Defines measurement band

C. MKID Readout

The MKID readout is fed into the cryostat by standard coaxial cables with SMA connectors with double DC blocks at the input to prevent ground loops. The signals runs from 300 K to the detector array and back. From 300 K to 4 K we use CuNi coax cables with Ag cladding on the central conductor [13]. From 4 K to the optics box we use copper coax and from the inner side of the box and in the suspension unit we use 0.86 mm CuNi coax cables with Ag cladding. The signal is attenuated on its way to the detector array and on the way back is amplified by 4-8 GHz Low Noise Amplifier. The loading of the cables has been designed to be well below the cooling power of the He10 unit.

D. Thermo-mechanical suspension unit

The detector array is mounted inside a suspension unit (see Fig. 4) which provides the thermal interface to the optics box and the intermediate thermal steps going from 3K to the 250 mK base temperature. The temperature levels are provided by the He10 cooler via flexible thermal straps. All temperature stages consist of a square annular deck, that are separated using three Vespel SP1 supports. The inner side has baffles coated with absorbing coating. The coaxial harness is thermally coupled to each deck to reduce the thermal load at the detector unit. The attenuators and Dc blocks and SMA connectors are used to rigidly clamp the coaxial harness. To attenuate magnetic fields the whole thermal suspension unit is surrounded by a niobium superconducting shield.

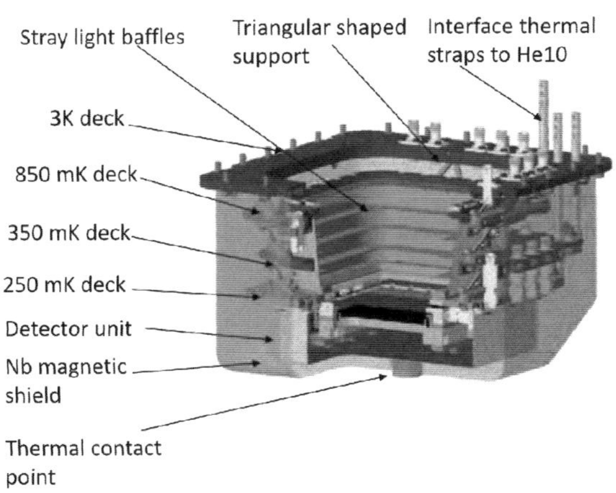

Fig. 4 Schematic of the thermo-mechanical suspension unit.

III. VERIFICATION

The thermal operation of the cryostat shows excellent performance: the base temperature of 250 mK is reached in roughly two days and an the hold time is about 30 hours with an addition recharge of the He4 buffer head of the He10 cooler.

The detectors show photon noise background operation and the optics quality is verified by the comparison of the simulated beam pattern using the Zemax physics optics tool and the measured beam patterns [14]. The agreement is excellent.

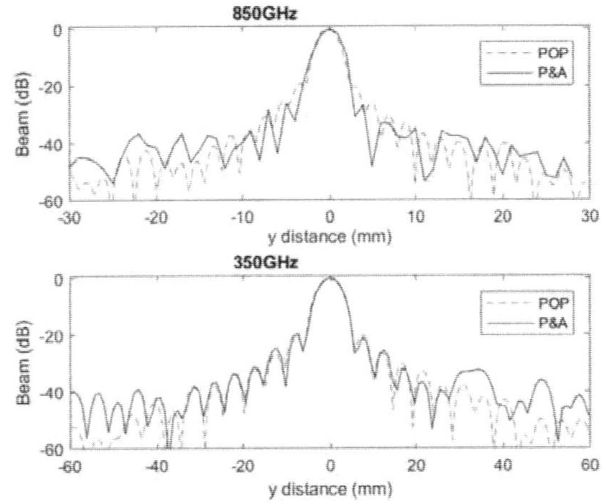

Fig. 5 Comparison between simulated and measured beam patterns.

CONCLUSIONS

We have successfully designed, build and verified a wide field camera testbed for the development of large format KID arrays. It mimics the conditions found in typical submm telescope instruments (APEX).

REFERENCES

[1] P. K. Day, H. G. Leduc, B. A. Mazin, A. Vayonokis and J. Zmuidzinas, "A broadband superconducting detector suitable for large arrays", *Nature*, vol. 425, 817, 2003.

[2] W. S. Holland et al, "SCUBA-2: the 10 000 pixel bolometer camera on the James Clerk Maxwell Telescope", *MNRAS*, vol. 430, Issue 4, pp..2513-2533, 2013.

[3] V. Reveret et al, "The ArTéMiS wide-field submillimeter camera: preliminary on-sky performances at 350 microns", Proc. SPIE 9153, doi:10.1117/12.2055985, June 2014.

[4] APEX telescope [Online]. Available:http://www.apex-telescope.org/.

[5] R. M. J. Janssen, J. J. A. Baselmans, A. Endo, L. Ferrari, S. J. C. Yates, A. M. Baryshev and T. M. Klapwijk,, "High optical efficiency and photon noise limited sensitivity of microwave kinetic inductance detectors using phase readout", *Appl. Phys. Letter*, 103, 203503, 2013.

[6] L. Ferrari, O. Yurduseven, N. Llombart, S. J. C. Yates, J. Bueno, V. Murugesan, D. J. Thoen, A. Endo, A. M. Baryshev, J. J. A. Baselmans, Äntenna coupled MKID Performance verification at 850 GHz for large Format Astrophysics Array", *IEEE Transactions on Terahertz Science and Technology*, Volume 8, Issue 1, pp. 127-139, Jan. 2018.

[7] Chase Research Cryogenics [Online]. Available: http://www.chasecryogenics.com/

[8] J. A. Murphy, "Distortion of a simple Gaussian beam on reflection from off-axis ellipsoidal mirrors", *Int. J. Infrared and Milli. Waves*, Vol. 8, no. 9, pp. 1165-1187, Sep. 1987.

[9] Zemax LLC, "Zemax optical studio", 2013 . [Online] Available: www.zemax.com.

[10] J. J. A. Baselmans, S. J. C. Yates, P. Diener, P. J. de Visser, *J. Low Temp. Phys.* , vol. 167, 360, 2012.

[11] QMC Instruments Ltd, 2008. [Online] Available: www.terahertz.co.uk/qmc-instruments-ltd

[12] C. E. Tucker, P. A. R. Ade, "thermal filtering for large aperture cryogenic detector arrays", *Proc. SPIE.* , vol. 6275, 2006.

[13] COAX Co. Ltd., [Online] Available: www.coax-co.jp/en

[14] K. K. Davis, W. Jellema, S. J. C. Yates, L. Ferrari, J. J. A. Baselmans, K. Kohno, D. Thoen, V. Murugesan and A. M. Baryshev, "Proof of concept demonstration of vector beam pattern measurements of kinetic inductance detectors", *IEEE Trans. THz Sci. Technology*, vol. 7, no. 1, pp. 98-106, Jan. 2017.

Development of Parallel Capacitor Based Kinetic Inductance Detectors (KIDs)

Faouzi Boussaha*, Samir Beldi, Christine Chaumont, Piercarlo Bonifacio

Galaxies, Etoiles, Physique et Instrumentation (GEPI),
Observatoire de Paris, 77 avenue Denfert-Rochereau, 75014 Paris – France
*Contact: faouzi.boussaha@obspm.fr

Abstract—We are developing low TLS (Two-Level System) noise Lumped Element KIDs. It is well established that KIDs feature an excess noise due to TLSs generated in dielectrics, particularly the amorphous ones. TLSs are generated in the whole LC circuit but it would be predominant in the capacitive portion [1]. In this study, we propose to replace the standard interdigitated capacitance by a parallel plate capacitor by means of which it is possible to reduce the TLS in two ways. One way consists in the use of very thin films featuring a high dielectric constant (ε_r=9-11). In this case, according to the theoretical model [2], TLSs can be lowered thanks to the high electric field that can be set inside the capacitor through the power driving the resonators. The second approach consists in freeing the capacitor from dielectric by implementing vacuum gap capacitors. These concepts can be applied for a wide range of wavelengths from Submillimeter/THz through to near IR. We will present designs, simulations and the first detector realizations.

[1] P. Noroozian et al., "Two-level system noise reduction for Microwave Kinetic Inductance Detectors", AIP Conference Proceedings 1185, 148-151 (2009).
[2] J. Gao et al., "Experimental evidence for a surface distribution of two-level systems in superconducting lithographed microwave resonators", Appl. Phys. Lett. 92, 152505, 2008.

The SAFARI grating spectrometer for SPICA

Gert de Lange[1,2], Peter Roelfsema[1,2], Willem Jellema[1,2], Martin Giard[3], Francisco Najarro[4], KeesWafelbakker[1], Brian Jackson[1], Lee Armus[5], Marc Audard[6], Matt Bradford[7], YasuoDoi[8], Matt Griffin[9], Frank Helmich[1,2], Inga Kamp[2], Franz Kerschbaum[10], Oliver Krause[11], Bengt Larsson[12], Suzanne Madden[13], David Naylor[14], Takahashi Onaka[8], Luigi Spinoglio[15], Floris van der Tak [1,2]

[1]SRON Netherlands Institute for Space Research, Groningen, The Netherlands; [2]Kapteyn Astronomical Institute, Groningen, The Netherlands; [3]IRAP, Toulouse, France; [4]CAB-INTA, Madrid, Spain; [5]IPAC/Caltech, USA; [6]ISDC, Geneva, Switzerland; [7]JPL/Caltech, USA; [8]University of Tokyo, Tokyo, Japan; [9]University of Cardiff, Cardiff, United Kingdom; [10]University of Vienna, Vienna, Austria; [11]MPIA, Heidelberg, Germany; ***12**Stockholm university, Stockholm, Sweden; 13CEA, Saclay, France; 14University of Lethbridge, Lethbridge, Canada; 15IAPS-INAF, Rome, Italy;*

*Contact: g.de.lange@sron.nl

Abstract— The far-infrared spectrograph SAFARI, on the joint European-Japanese space telescope SPICA (proposed in the ESA M5 Call), will provide the most sensitive view ever of the cool, obscured universe. By cooling the SPICA telescope to below 8 K its thermal emission is decreased to such low levels that the latest generation of ultra-sensitive Transition Edge Sensor (TES) detectors can be utilized to their full potential. With this combination of low background and extreme detector sensitivity SPICA will be able to look far deeper into space than was possible with any of its predecessors. SPICA/SAFARI is the only facility that will fill the gap in the wavelength domain between the other great observatories, and is as sensitive as both the James Webb Space telescope and the ALMA radio observatory – only with SPICA/SAFARI we will complete the view on the star-formation history of our universe.

The current baseline SAFARI design uses a beam steering mirror (BSM) that forwards the incoming signal to the dispersing and detection optics. The BSM is used to select sky or calibration signals and forward that to a nominal R~300 (low) resolution optics chain or to a R~11000 (high) resolution optics chain. The low resolution is obtained by dispersion through a diffraction grating illuminating a line of TES detectors. For the high-resolution mode, the signal is first pre-dispersed using a Martin-Puplett interferometer before entering the grating. The full 35-230 μm wavelength range is split in to several different bands, each with its own grating and TES detectors. The baseline design has for each of the bands three separate spatial pixels, to provide background reference measurements, but also to provide some imaging capability.

With SPICA's cold, 2.5-meter telescope and the baseline TES NEP of $2x10^{-19}$ W/ $\sqrt{Hz}$, for the new grating based SAFARI the sensitivity of the R~300 mode will be about 5 x 10^{-20} W/m^2 (5 σ , 1hr). With this high sensitivity astronomers will e.g. be able to detect the [OIV] line in relatively average galaxies out to a redshift z~3. Thus, the evolution of galaxies can be followed through their most active periods in cosmic time from about 10 billion years ago to what they look like today. Additionally, we will be able to observe dust features from even earlier epochs, out to redshifts of z~7-8, thus providing insight into dust formation in the very early phases of the universe.

Wide bandwidth measurements of microwave and millimeter wave impedance in MgB_2 HEB mixers

Narendra Acharya, Usman Ul-Haq, Sergey Cherednichenko*

Terahertz and Millimetre Wave Laboratory, Department of Microtechnology and Nanoscience, Chalmers University of Technology, SE-412 96 Gothenburg, Sweden
*Contact: serguei@chalmers.se

Abstract—Superconducting Hot electron Bolometer (HEB) mixers are used in many terahertz astronomical receivers. At frequencies above 1THz, HEB mixers have noise temperature lower than any other mixers working at that range. There have been some studies conducted to investigate THz performance for HEB mixers made of both NbN and MgB_2 thin films. However, experimental data for HEB mixer characteristics at microwave (intermediate frequency) frequencies are very scarce and noisy. There are no accurate and systematic IF impedance data. The problem is explained by difficulty of system calibration at cryogenic environment, and device fixture de-embedding. From theoretical models and some experiments, it is expected that IF impedance varies strongly with IF, with a significant imaginary part. With the lack of such data, accurate HEB-to-LNA impedance matching is rarely conducted. As a consequence, electrical standing waves are observed, leading to ripples in the receiver gain and the system noise temperature.

In this work we studied IF impedance of MgB_2 HEB mixers from 20MHz to 67GHz in a continuous sweep, using an advanced cryogenic microwave/millimeter wave probing station and Vector Network Analyzer. HEB devices were fabricated from high quality 5-10nm MgB_2 thin films with many bridge dimensions, integrated with one-port CPW contacts. Device temperature was adjusted from 5K to 40K (normal state). On-wafer calibration kit was mounted next to HEB wafers, and allowed for accurate system calibration up to 67GHz using a 100μm Ground-Signal-Ground (GSG) probe. This way, impedance was measured directly, without any need for de-embedding, in a frequency range far exceeding any other previous studies.

HEB devices were set on different temperatures, and different bias voltages were applied. We clearly observe correlation of the measured impedance with dc characteristics of HEB devices. Using measured complex-S11 data, we analyze scenarios for HEB-LNA matching.

Terahertz Transition-Edge Sensor with a Kinetic-Inductance Amplifier at 4.2 K.

Artem Kuzmin[1,*], Steffen Doerner[1], Stefan Singer[1], Ilya Charaev[1], Konstantin Ilin[1], Stefan Wuensch[1] and Michael Siegel[1]

[1]Institut für Mikro- und Nanoelektronische Systeme, Karlsruhe Institute of Technology, Karlsruhe, 76187, Germany

*Contact: artem.kuzmin@kit.edu

***Abstract*— Different terrestrial terahertz applications would benefit from large-format arrays, operating in compact and inexpensive cryocoolers at 4.2 K with sensitivity, limited by the 300-K background radiation only. A voltage-biased Transition-Edge Sensor (TES) as a THz detector can have sufficient sensitivity and has a number of advantages important for real applications, however it requires a low-noise current readout. Usually, a current amplifier based on Superconducting Quantum-Interference Device (SQUID) is used for readout, but the scalability of this approach is limited due to complexity of the operation and fabrication. Recently, it has been shown that instead of SQUID it is possible to use a current sensor, which is based on the nonlinearity of the kinetic inductance of a current-carrying superconducting stripe. Here, we demonstrate the operation of a voltage-biased TES with a microwave kinetic-inductance current amplifier at temperature 4.2 K. We measured the expected intrinsic Noise-Equivalent Power NEP ~ 5×10^{-14} W/Hz$^{1/2}$ and confirmed that a sufficient sensitivity of the readout can be reached in conjunction with a real TES operation. The construction of an array with the improved sensitivity ~ 10^{-15} W/Hz$^{1/2}$ at 4.2 K could be realized using a combination of the new current amplifier and already existing TES detectors with improved thermal isolation.**

I. Introduction

Large-format arrays of sensitive THz detectors are required today in different terrestrial applications. Among them is THz passive security scanning for concealed hazardous objects, THz imaging for non-destructive testing (NDT) in production lines and imaging far-infrared Fourier-Transform Spectroscopy for material research and atmospheric studies [1-5]. Estimations show that the sensitivity of a diffraction-limited single-pixel THz detector with Noise Equivalent Power NEP ~ 10^{-15} W/Hz$^{1/2}$ would be sufficient for applications with a 300-K background [6].

Cryogenic detectors can achieve the required sensitivity, but construction of large arrays is challenging. A practicable system should operate at temperatures not much lower than 4.2 K. A superconducting Transition-Edge Sensor (TES), together with a low-noise current amplifier can have sufficient sensitivity at moderate temperatures [7]. Usually, a Superconducting Quantum Interference Device (SQUID-amplifier) is used for measurements of the current response of a TES. However, readout with many SQUID amplifiers becomes expensive and difficult in fabrication and operation.

A microwave kinetic-inductance detector (MKID) is an example of a scalable detector, which can be operated in the THz spectral range. It is based on high-Q superconducting resonators and a Frequency-Division Multiplexing (FDM) in GHz-range [8]. However, for a pair-breaking process to be effective in the lower THz range, the superconducting energy gap should be small. At 4.2 K this will lead to a low Q factor of the resonator and high generation-recombination noise and thus high NEP.

Luomahaara *et al.* demonstrated a kinetic-inductance magnetometer, which is based on the nonlinearity of kinetic inductance of a small NbN stripe, imbedded in a high-Q resonator [9]. Kher *et al.* demonstrated a current amplifier based on the same principle with a sensitivity ~5 pA/Hz$^{1/2}$ [10].

Here, we demonstrate an operation of a voltage-biased THz TES detector together with an array-scalable current amplifier based on nonlinear-kinetic inductance at 4.2 K. The TES is an antenna-coupled superconducting nano-bolometer, and the current-sensitive inductor is a superconducting nanowire embedded in a high-Q microwave resonant circuit. The resonator is inductively coupled to a coplanar-waveguide transmission line (CPW). It allows for simultaneous injection of the TES current and measurement of the CPW transmission. We called this device a Microwave Kinetic-Inductance Nanowire Galvanometer (MKING). Previously, the MKING achieved a current sensitivity of about 10 pA/Hz$^{1/2}$ [11], which should be sufficient to reach a NEP ~ 10^{-15} W/Hz$^{1/2}$ with a well thermally-isolated THz TES at 4.2 K. This opens a way for building large and sensitive imaging arrays in a compact and inexpensive cryogenic system.

II. Detector system at 4.2 K

A detector system for demonstration of functionality consists of a single TES and a kinetic-inductance current amplifier (MKING) in separate housings. We performed measurements in a liquid-helium-bath cryostat with a THz-transparent window (Fig. 1).

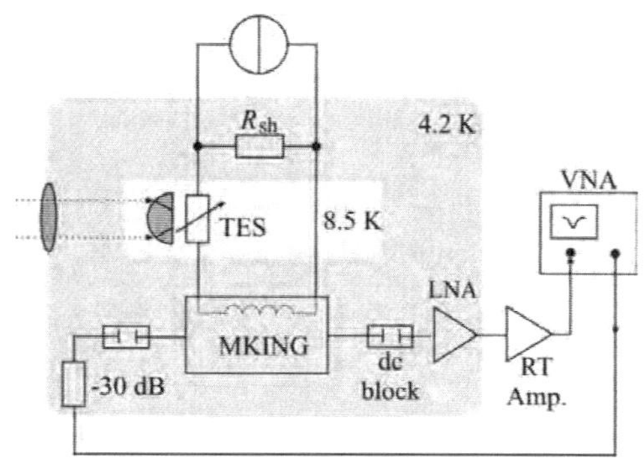

Fig. 1. The schematics of the experimental setup. The gray area is the cold stage of the 4.2 K-cryostat with THz-transparent window; the yellow area is the detector block with antenna-coupled TES on a lens at 8.5 K.

The voltage bias of the TES is realized using a 2-Ω Manganin shunt resistor R_{sh}, which is biased with a current from a room-temperature source. The current inputs of MKING are connected in series with TES. Changes of the microwave transmission of CPW in the MKING are measured using a Vector Network Analyzer (VNA) and a cryogenic microwave low-noise high-electron-mobility transistor (HEMT) amplifier. The equivalent noise temperature Tn of the complete setup is about 25 K and is dominated by the VNA.

A. Transition-Edge Sensor

We have fabricated the THz TES which is an antenna-coupled superconducting nano-bolometer from 5-nm-thin NbN film on a substrate from highly-resistive silicon. The antenna is a bi-layer structure of in-situ magnetron sputtered 20-nm NbN buffer and 200-nm gold layers. The size of the nano-bolometer: length × width = 200 nm × 700 nm. Fig. 2 shows the dependence of the resistance of a nano-bolometer on temperature (R-T curve). The superconducting transitions is at $T_c \approx 9.4$ K. The steepness parameter of R-T curve $\alpha = d(\log R)/d(\log T) \approx 40$. We also measured a non-hysteretic current-voltage characteristic (I-V curve) at 8.5 K, where electrical bias can provide a stable and uniform Joule heating of the nano-bolometer close to T_{c2} (inset in Fig. 2). The estimated thermal conductance of the nano-bolometer $G_{th} \approx 25$ nW/K and a loop gain for the negative electro-thermal feedback $L = \alpha I V_b / G_{th} T \approx 4$ for a working point $V_b = 1$ mV and $T = 9.4$ K. Using these values, we calculated the current responsivity $\Re_I = (L/L+1)/V_b \approx 800$ A/W, a minimum phonon-limited $\mathrm{NEP_{ph}} = \sqrt{4kT^2G_{th}} \approx 10^{-14}$ W/Hz$^{1/2}$, and a minimum noise current $\delta I \approx 8$ pA/Hz$^{1/2}$ of the nano-bolometer. The TES chip with the size of 3 mm × 3 mm was glued in the focus of the hyper-hemispherical silicon lens with 12-mm diameter and without anti-reflection coating. The lens with the chip were mounted in a copper block. The temperature sensor and a heater were integrated into a detector block for the control of the TES operation temperature.

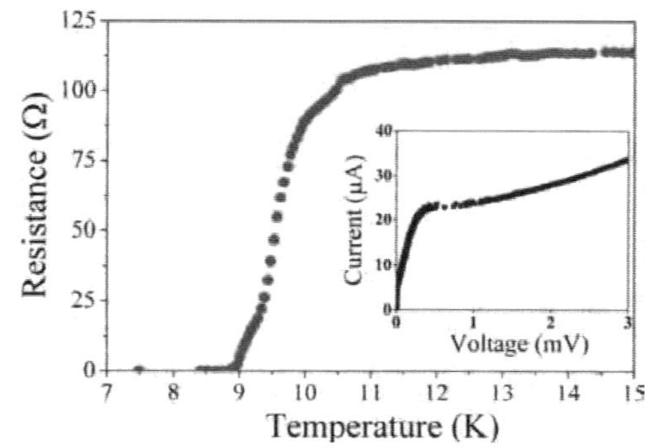

Fig. 2. The R-T curve of the nano-bolometer. Inset: the I-V curve of the nano-bolometer measured at $T = 8.5$ K in a current-bias mode.

B. Kinetic-Inductance current amplifier

To avoid in the future a limited scalability and complexity of readout based on SQUID-amplifier, we replaced it with a superconducting current amplifier, MKING [11]. The resonance frequency of the device is about. The resonance frequency of the device is about 4.7 GHz. In our experiment we used phase response of the device with a fixed frequency of the microwave probe. The single chip of MKING device with a size of 3 mm × 3 mm was mounted into a separeate block with an adapter plate, two SMA connectors for CPW feed line and a DC input port. The particular MKING device had a noise current of about 40 pA/Hz$^{1/2}$ (Fig. 3), which is higher than the value that was demonstrated with the previous 3.8-GHz device.

III. Measurements of Responsivity and NEP

To measure the optical response, we applied a THz signal from a calibrated 0.65-THz quasi-optical source through the cryostat window. The measured optical responsivity of 3.3 A/W is much lower than the electrical one due to the low coupling efficiency $\eta = 0.5\%$, since we have a not optimized matching of Gaussian beams from the THz source and TES antenna. Moreover, antenna was not designed for frequencies below 1 THz and its performance at 0.65 THz might be low. For measurement of a small-signal response, we applied a chopped THz signal, which corresponds to an absorbed power of $P_{abs} = 160$ pW (inset in Fig. 4).

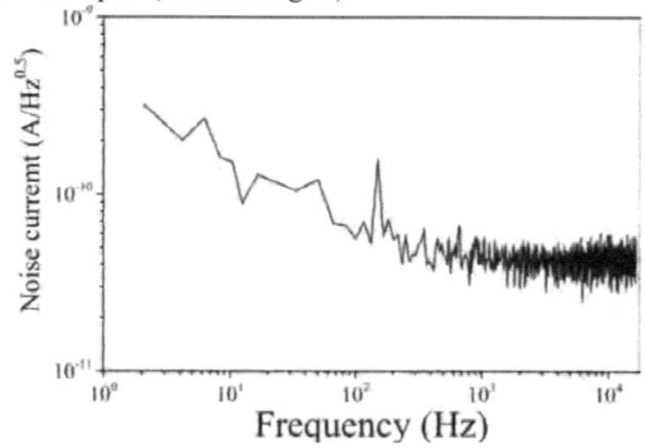

Fig. 3. Measured noise spectrum of the MKING device.

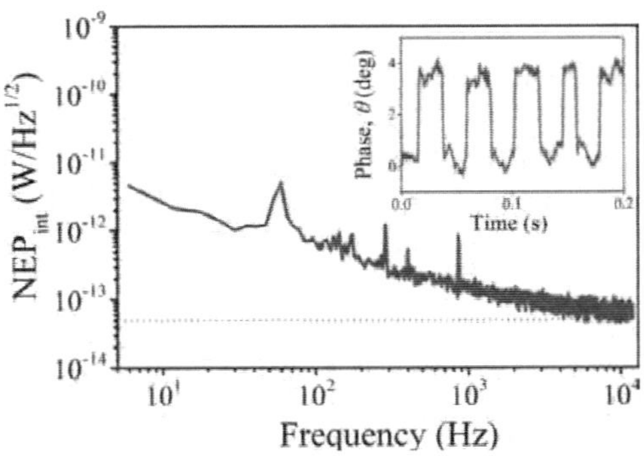

Fig. 4. Measured intrinsic NEP of the detector system. The phase response of the TES-MKING to 160 pW of absorbed THz power (in the inset).

The intrinsic phase-watt responsivity of 2×10^{10} deg/W was obtained in this measurement. To determine intrinsic NEP, we recorded a time trace of the phase without THz signal and plotted spectrum of the phase noise, divided by the measured responsivity (Fig. 4).

C. Discussion

The measured intrinsic NEP reaches 5×10^{-14} W/Hz$^{1/2}$ at white noise level, which is a factor 5 higher than the estimated value. It is probably due to the prevalence of the noise current of the particular MKING over noise of the TES. Indeed, if we take the separately measured TES responsivity and the noise current of the MKING we end up with the same NEP. A reduction of the effective noise temperature of the microwave readout could be achieved using back-end electronics with ADC of higher resolution. This would result in a factor 2 lower noise current of the MKING. Additionally, using the previous version of MKING with lower critical current [11] it would be possible to reach a factor 4 lower NEP.

In order to obtain system NEP ~ 10^{-15} W/Hz$^{1/2}$, which is required for the 300-K background-limited operation, the performance of TES should be improved. A lower system NEP at 4.2 K could be achieved with improved thermal isolation of the bolometer. For a TES on a SiN_x membrane the required thermal isolation has been already demonstrated [12]. A matching of the TES current to the operation range of a MKING with a critical current $I_c \approx 30$ µA seems feasible with the current responsivity $>10^4$ A/W and thus with the noise current $\delta I \approx 10$ pA/Hz$^{1/2}$. Alternatively, antenna-coupled suspended superconducting nano-bolometers can be used, which already demonstrated the required sensitivity [7]. In this case, large membranes are not required and fabrication process is less complex. Both, quasi-optical lens coupling and waveguide coupling, could be used for this type of TES.

Conclusions

The detector system for demonstration of functionality with voltage-biased antenna-coupled TES and kinetic-inductance current amplifier at temperature 4.2 K reached internal NEP ~ 5×10^{-14} W/Hz$^{1/2}$, which is close to the estimated phonon-limited NEP of the particular TES. The sensitivity of the system is limited by the particular current amplifier and the microwave setup. It could be further improved by a factor 2 using baseband ADC with higher resolution. Our analysis shows that NEP ~ 10^{-15} W/Hz$^{1/2}$ is feasible with already existing well-isolated THz TES and MKING with improved design and back-end electronics. The obtained values of Q factor of resonator in current amplifier allow to scale up detector system to >100 pixel per readout channel at 4.2 K.

Acknowledgments

Authors would like to thank M. Schmelz and R. Stolz (IPHT Jena) for the fruitful discussions and help during measurements, A. Stassen and K.-H. Gutbrod for the help with preparation of samples and for fabrication of mechanical parts.

References

[1] J. P. Guillet, B. Recur, L. Frederique, B. Bousquet, L. Canioni, I. Manek-Hönninger, P. Desbarats, and P. Mounaix, "Review of Terahertz Tomography Techniques," *Journal of Infrared, Millimeter, and Terahertz Waves*, vol. 35, no. 4, pp. 382–411, 2014.

[2] E. Bründermann, H.-W. Hübers, M. F. Kimmitt, Terahertz Techniques New York, Springer, p 310, 2012.

[3] T. Robin, C. Bouye, and J. Cochard, "Terahertz applications: trends and challenges," *Terahertz, RF, Millimeter, and Submillimeter-Wave Technology and Applications VII*, Jul. 2014.

[4] P. Dean, A. Valavanis, J. Keeley, K. Bertling, Y. L. Lim, R. Alhathlool, A. D. Burnett, L. H. Li, S. P. Khanna, D. Indjin, T. Taimre, A. D. Rakić, E. H. Linfield, A. G. Davies, "Terahertz imaging using quantum cascade lasers—a review of systems and applications", *J. Phys. D: Appl. Phys.* 47, p. 374008, 2014.

[5] H. Latvakoski, M.G. Mlynczak, D.G. Johnson, R.P. Cageao, D.P. Kratz, K. Johnson, "Far-infrared spectroscopy of the troposphere: instrument description and calibration performance", *Applied optics*, 52(2), pp.264-273, 2013.

[6] E. Heinz, T. May, D. Born, G. Zieger, S. Anders, G. Thorwirth, V. Zakosarenko, M. Schubert, T. Krause, M. Starkloff, A. Krüger, M. Schulz, F. Bauer, and H.-G. Meyer, "Passive Submillimeter-wave Stand-off Video Camera for Security Applications," *Journal of Infrared, Millimeter, and Terahertz Waves*, vol. 31, no. 11, pp. 1355–1369, 2010.

[7] A. Luukanen and J. P. Pekola, "A superconducting antenna-coupled hot-spot microbolometer," *Applied Physics Letters*, vol. 82, no. 22, pp. 3970–3972, Feb. 2003.

[8] P. K. Day, H. G. Leduc, B. A. Mazin, A. Vayonakis, and J. Zmuidzinas, "A broadband superconducting detector suitable for use in large arrays," *Nature*, vol. 425, no. 6960, pp. 817–821, 2003.

[9] J. Luomahaara, V. Vesterinen, L. Grönberg, and J. Hassel, "Kinetic inductance magnetometer," *Nature Communications*, vol. 5, no. 1, Oct. 2014.

[10] A. Kher, P. K. Day, B. H. Eom, J. Zmuidzinas, and H. G. Leduc, "Kinetic Inductance Parametric Up-Converter," *Journal of Low Temperature Physics*, vol. 184, no. 1-2, pp. 480–485, 2015.

[11] S. Doerner, A. Kuzmin, K. Graf, I. Charaev, S. Wuensch, and M. Siegel, "Compact microwave kinetic inductance nanowire galvanometer for cryogenic detectors at 4.2 K," *Journal of Physics Communications*, vol. 2, no. 2, p. 025016, 2018.

[12] M. Arndt, S. Wuensch, C. Groetsch, M. Merker, G. Zieger, K. Peiselt, S. Anders, H.G. Meyer, M. Siegel, "Optimization of the Microwave Properties of the Kinetic-Inductance Bolometer (KIBO)" *IEEE Trans. Appl. Supercond.*, 27(4), pp.1-5, 2017.

Observation of positive gain by a combination of quasiparticle SIS up and down frequency converters

T. Kojima*[1], Y. Uzawa[2,1], and W. Shan[1]

[1] *National Astronomical Observatory of Japan, Mitaka, 181-8588, Japan*
[2] *National Institute of Information and Communications Technology, Koganei, Tokyo 184-0015, Japan*
*Contact: t.kojima@nao.ac.jp

Abstract— Heterodyne instruments have recently attained quantum-limited low-noise performance, particularly in radio astronomy, but it is difficult to develop large heterodyne arrays. In the realization of the heterodyne array, the reduction of power dissipation for semiconductor-based amplifiers remains a major challenge. Alternatively, superconducting parametric amplifiers still seem to have several barriers to application, especially in terms of operating temperature. Here, we show a novel concept of microwave amplification based on up and down frequency-conversion processes using quasiparticle superconductor-insulator-superconductor (SIS) tunnel junctions. The SIS up- and down-converters were connected in cascade and driven by a local oscillator (LO) power supply. In this case, input and output microwave signal frequencies become identical, and the input microwave power can be amplified with bilateral conversion gains.

We demonstrate positive gain using a proof-of-concept test module, which operates with a power dissipation of several μW at a bath temperature of 4 K. The test module used in this experiment is an existing heterodyne mixer block and two SIS converter chips at a radio frequency (RF) band of 100 GHz developed for the four beam receiver system on the 45-m telescope (FOREST). The operation temperature is high compared to those of a superconducting Josephson amplifier and kinetic inductance travelling wave amplifier with a typical operating temperature of several hundred milli-Kelvins. The result indicates that the configuration potentially works as a low-noise and low-power-consumption microwave amplifier at an operating temperature of 4 K with Nb-based SIS junctions.

At the conference, we will present the detailed configuration, and describe measurement and analysis results.

Low Power Cryogenic Rad-hard LNAs for Space

Peter J. Sobis*, Niklas Wadefalk[2], Joel Schleeh[2], Christina Emrich[1], Mats Lindgren[1], and Anders Emrich[1]

[1]*Omnisys Instruments AB, Västra Frölunda, SE-421 32, Sweden*
[2]*Low Noise Factory AB, Göteborg, SE-412 63, Sweden*
*Contact: peter.sobis@omnisys.se

Abstract**—** We present the design and implementation of a packaged cryogenic InP HEMT MMIC LNA for space borne applications. The particular module is part of the submillimetre wave instrument (SWI) band1 and band2 receiver channel development for the ESA Jupiter ICy moons Explorer mission (JUICE) programme. The main objective of the activity has been to develop a general connectorized rad-hard LNA package for space applications, and to qualify the cryogenic InP HEMT MMIC process at Chalmers University of Tehnology which today covers a frequency range up to the W-band with state-of-the-art performance in terms of noise, gain and power dissipation.

The activity has currently moved from a pre-qualification phase to a lot acceptance test (LAT) phase and has up till now included various environmental tests e.g. vibration, total dose radiation tests up to 300kRad on die level, and thermal cycling both to hot and cold temperatures. The InP HEMT MMIC process can be operated at ultra low power with ultra low noise which makes it an excellent candidate for deep space missions and satellite communication links, but also for the next generation of radioastronomy telescopes based on cryogenic receivers e.g. HEB and SIS multipixel cameras.

Beam pattern measurements of a quasi-optical HEB mixer at 2 THz

Yoshihisa Irimajiri[1*], Alvaro Gonzalez[2], Satoshi Ochiai[1], Akira Kawakami[1], and Yoshinori Uzawa[1]

[1] National Institute of Information and Communications Technology (NICT), Koganei, Tokyo, 184-0015 Japan
[2] National Astronomical Observatory of Japan (NAOJ), Mitaka, Tokyo, 181-8588, Japan
*Contact: irimaji@nict.go.jp

***Abstract*— We plan to measure beam patterns of a quasi-optical NbN HEB mixer in the 2 THz-band. A 2 THz AMC (Amplifier Multiplier Chain) is used as the test RF source. A THz-QCL is phase-locked using a SLH mixer (super-lattice harmonic) mixer with 12th harmonic multiplication. A beat note is measured by a heterodyne receiver system with the phase-locked THz-QCL as a local oscillator. The amplitude and the phase of the antenna beam patterns are measured using a lock-in amplifier. If successful, this will be the first time that beam patterns are measured in amplitude and phase at frequencies as high as 2 THz.**

I. Introduction

Low noise heterodyne receivers at terahertz frequency are widely used for radio astronomical observations and atmospheric spectroscopy. Beam pattern of a receiver system is one of the important basic performances for these observations. Optics of an HEB mixer for the ASTE telescope was characterized at 900 GHz [1], and reconfigurable beam pattern measurement system for different frequency band of ALMA was reported, and beam pattern measurements were performed at 1.37 THz, 1.47 THz, and 900 GHz [2]. Experimental characterization of the HIFI FM model was described and measurements were performed at 480 GHz, 802 GHz, 1.1 THz, and 1.6 THz in [3].

We are developing an HEB mixer for the measurement of atomic oxygen line at 2.06 THz for SMILES-2 [4]. We plan to measure beam patterns of the quasi-optical HEB mixer in the 2 THz-band. The measurement set up is based on a near-field beam pattern measurement system at 0.03 to 1.6 THz [2]. If successful, this will be the first time that beam patterns are measured in amplitude and phase at frequencies as high as 2 THz.

II. Experiment setup

A quasi-optical NbN HEB mixer device was fabricated in NICT. Figure 1 shows photographs of the HEB mixer device with a log-spiral antenna, the quasi-optical mixer mount with an AR-coated hyper-hemispherical HRFZ Si lens, and the receiver setup with a parabolic focusing mirror in a 4 K cryostat. (The focusing parabolic mirror may be replaced to an ellipsoidal mirror to narrow the measurement area and shorten the observation time.)

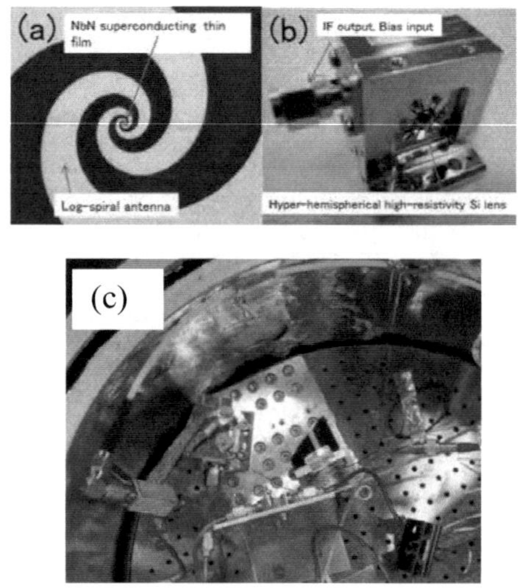

Fig. 1. (a) A quasi-optical NbN HEB mixer device with a log-spiral antenna. The device was fabricated in NICT. (b) A quasi-optical mixer block with an AR-coated hyper-hemispherical HRFZ Si lens. (c) Receiver setup with a parabolic focusing mirror in a 4 K cryostat.

A 2 THz AMC source with a diagonal horn output is used as a test RF source. The AMC source consists of a microwave synthesizer, amplifiers, four doublers, and two triplers. The multiplication factor is 144. The output frequency of the AMC source and the THz-QCL are set to be 2066.4 GHz and 2067.44 GHz, respectively so that the IF frequency goes into the IF band of the HEB mixer. The beat note of 1.04 GHz is detected by the HEB mixer with the THz-QCL as a local oscillator. A 3rd order antenna-coupled DFB QCL array at 2 THz-band provided by MIT group is cooled using a cryotel (CT) cooler. A THz-QCL with lasing frequency of ~2067 GHz is used. The THz-QCL is operated in CW-mode with an output power of ~600 μW at 45 K. The power consumption of THz-QCL is about 1.9 W. The THz-QCL is phase-locked using a SLH mixer (super-lattice harmonic) mixer with 12th harmonic multiplication. A D-band (110-170 GHz) AMC source is used as a LO for the SLH mixer at 172.32 GHz. The multiplication factor of the LO is 12, therefore, the total multiplication factor is 144 which is same as that of the 2 THz AMC source. The amplitude and the phase of the antenna beam patterns are measured using a lock-in amplifier. All oscillators are connected to a same 10 MHz reference. The AMC source is scanned by a XYZ stage. The source is rotated 90 degrees to measure co- and cross-polarization characteristics. Figure 2 shows a block diagram of the measurement setup. The far field beam pattern is calculated from near field measurements. The beam pattern of the diagonal hon of the RF source should be corrected. If possible, a RF source with a waveguide output may be used.

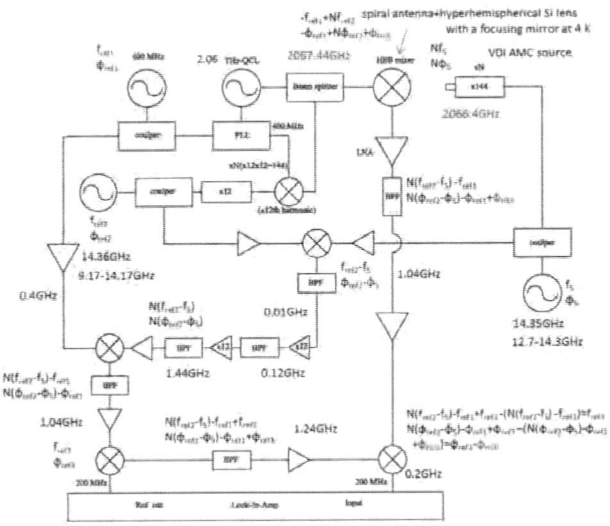

Fig. 2. Setup of beam pattern measurements of a quasi-optical 2-THz HEB mixer. A 2 THz AMC source with a diagonal horn is used for a RF source. The multiplication factor is 144. The output frequency of the AMC source and the THz-QCL are set to be 2066.4 GHz and 2067.44 GHz, respectively so that the IF frequency goes into the IF band of the HEB mixer. The beat note of 1.04 GHz is detected by the HEB mixer with the THz-QCL as a local oscillator. The THz-QCL is phase-locked using a SLH mixer with 12th harmonic multiplication. A D-band AMC source is used as a LO for the SLH mixer at 172.32 GHz. The multiplication factor of the LO is 12, therefore, the total multiplication factor is 144 which is same as that of the 2 THz AMC source. The amplitude and the phase of the antenna beam patterns are measured using a lock-in amplifier. All oscillators are connected to a same 10 MHz reference. The AMC source is scanned by a XYZ stage. The source is rotated 90 degrees to measure co- and cross-polarization characteristics. The far field beam pattern is calculated from near field measurements. The beam pattern of the diagonal hon of the RF source should be corrected.

The setup of the characterization of the amplitude and the phase of the beam pattern is based on the previous work [2, 3]. The measurement set up is designed so that the phases of each microwave oscillators other than that of an HEB mixer (φ_{HEB}) are canceled out. The phase at an IF signal is described as

$$(\varphi_{HEB}+N\varphi_{ref2} - \varphi_{ref1})-N\varphi_S= N(\varphi_{ref2} - \varphi_S) - \varphi_{ref1}+\varphi_{HEB} \quad (1)$$

, where
N: harmonic number
φ_{HEB} : phase of the HEB mixer
φ_S : phase of the RF source
φ_{ref1} : phase of reference (400 MHz) for phase-locking of the THz-QCL
φ_{ref2} : phase of reference (14.36 GHz) for the LO of the SLH mixer.

In order to measure φ_{HEB}, $N(\Phi_{ref2}-\varphi_S)-\varphi_{ref1}$ should be canceled out. The phase differences between the source oscillator of the RF source and that of LO for the SLH mixer ($\Phi_{ref2}-\varphi_S$) is multiplied by 144 using two 12 times frequency multipliers. By mixing this signal with reference source for PLL (400 MHz) and reference of the lock-in amplifier (200 MHz), a phase of $N(\varphi_{ref2}-\varphi_S)-\varphi_{ref1}+\varphi_{ref3}$ is detected. Subtracting equation (1) from this, we can measure φ_{HEB} as follows

$$N(\varphi_{ref2}-\varphi_S)-\varphi_{ref1}+\varphi_{ref3}-(N(\varphi_{ref2}-\varphi_S)-\varphi_{ref1}+\varphi_{HEB})=\varphi_{HEB}+\varphi_{ref3} \quad (2)$$

, where
φ_{ref3} : phase of reference (200 MHz) of the lock-in amplifier.

The important thing is the harmonic number of the RF source (144) is same as that of the LO for the SLH mixer (144: x12x12) for this cancelation.

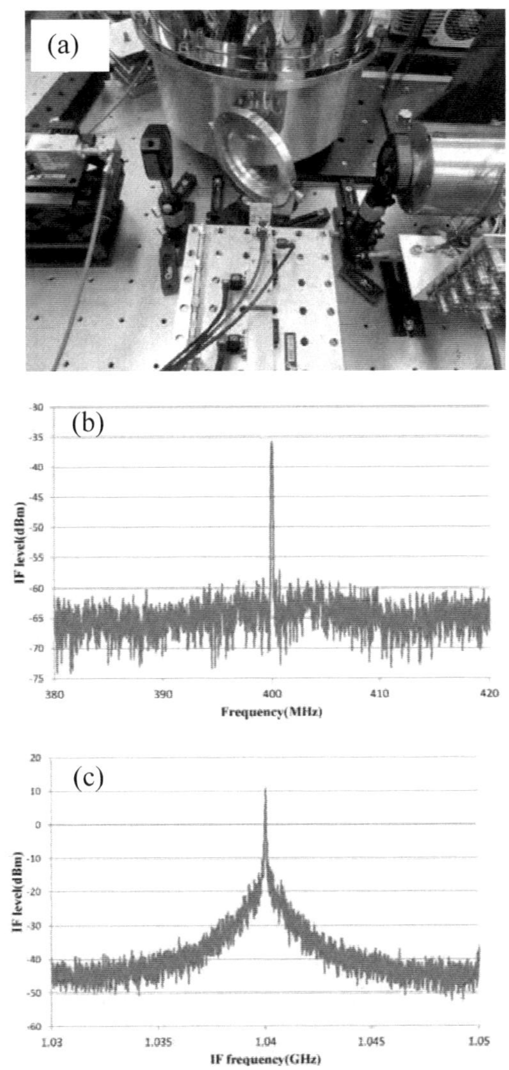

Fig. 3. (a) Photograph of a setup of test measurements. (b) A phase-locked beat note of the 2 THz-QCL detected by a SLH mixer. (c) A beat note between the phase-locked THz-QCL and an AMC source at 1.04 GHz.

III. Preliminary Results

The measurement set up is under construction at the present. In order to confirm the setup can be operated at the designed frequency, we made some experiments (Figure 3 (a)). The THz-QCL is phase-locked at 2067.44 GHz using the SLH mixer. Figure 3 (b) shows the phase-locked beat note of the 2 THz-QCL. A beat note is detected at 1.04 GHz with a SN ratio of more than 50 dB at RBW of 100 kHz as shown in figure 3 (c). We will start the measurements from this spring. In the measurement, we have to pay attention to followings.

- Stability of the signal (We will use a LHe dewar for cooling of the HEB mixer to avoid mechanical vibration of a 4 K cooler. Any other vibrations, e.g. human movement, should be avoided as well. Vibration of a THz-QCL cooler and a vacuum pump would affect to the stability?)
- Calibration measurements are necessary. (Measure the same point at an each scan.)
- Scan range and measurement time should be optimized. (scan step is 75 μm (=λ/2).) (e.g. +/ 7.5 mm scan: 200 x 200(40,000 points)*0.3 sec/point=3.3 hours + calibration + 2 surfaces)
- Measure the beam patterns in as short a time as possible.
- Measure an area as small as possible, ex. at a beam waist using an ellipsoidal mirror.
- Purity of the signal (We need band-pass filters at any points.)
- Linearity of an HEB mixer regarding to an input RF signal.
- Linearity of microwave amplifiers.
- Bias point of an HEB mixer should be stable for RF input.
- Source beam pattern should be corrected. (RF source needs to be modified to a waveguide output?)
- Data processing to calculate a far filed pattern from two near filed patterns.

etc.

Conclusions

We plan to characterize the beam pattern of a quasi-optical HEB mixer at 2 THz. The measurement setup is not completed yet but we confirmed the setup works properly at the designed frequency.

We also plan to develop a waveguide-type HEB mixer with a corrugated feed horn to achieve good beam pattern and polarization characteristics. In order to avoid vibration from a mechanical cooler and a vacuum pump used for cooling a THz-QCL, an another 2 THz AMC source might be tested as a local oscillator of the HEB mixer.

References

[1] A. Gonzalez, T. Soma, T. Shiino, K. Kaneko, Y. Uzawa, and S. Yamamoto, "Optics characterization of a 900-GHz HEB receiver for the ASTE telescope: design, measurement and tolerance analysis", *J Infrared Milli Terahz Waves*, (2014), 35:743–758.

[2] A. Gonzalez, Y. Fujii, T. Kojima, and S. Asayama, "Reconfigurable Near-Field Beam Pattern Measurement System from 0.03 to 1.6 THz", *IEEE Trans. THz Sci. and Technol.*, vol. 6, no. 2, Mar. 2016.

[3] W. Jellema, "Optical Design and Performance Verification of Herschel-HIFI", *ISBN 978-90-367-7443-7*.

[4] S. Ochiai, P. Baron, T. Nishibori, Y. Irimajiri, Y. Uzawa, T. Manabe, H. Maezawa, A. Mizuno, T. Nagahama, H. Sagawa, M. Suzuki, and M. Shiotani, "SMILES-2 Mission for Temperature, Wind, and Composition in the Whole Atmosphere", *SOLA, 2017, Vol. 13A, 13–18, doi:10.2151/sola.13A-003*

Tilted Beam Measurement of VLBI Receiver for the South Pole Telescope

Junhan Kim[1,*] and Daniel P. Marrone[1]
[1]Department of Astronomy and Steward Observatory, University of Arizona, Tucson, AZ 85721, USA
*Contact: junhankim@email.arizona.edu

***Abstract*—We have developed a 230 and 345 GHz very-long-baseline interferometry (VLBI) receiver for the South Pole Telescope (SPT). With the receiver installed, the SPT has joined the global Event Horizon Telescope (EHT) array.**

The receiver optics select the 230 or 345 GHz mixers by rotating the tertiary mirror around the optical axis, directing the chief ray from the secondary mirror to the feed horn of the selected frequency band. The tertiary is installed on top of the receiver cryostat, which contains both mixer assemblies. The feed horns are placed symmetrically across the centerline of the telescope optics and tilted inward by 5.7 degrees from the vertical plane so that their beams intersect at the chief ray intersection on the tertiary mirror.

We have performed vector beam measurements of the SPT VLBI receiver in both frequency bands. The measurements preserved the relative location of the beams, to establish the relative locations of the phase centers of the two horns. Measurements in two parallel reference planes above the cryostat were used to suppress reflected light. To model the beam, we derive a general expression of the electric field vector on the measurement plane for a tilted beam and infer the feed horn position parameters for both frequency bands by fitting models to data with a Markov chain Monte Carlo (MCMC) method. The inferred parameters such as the tilt angle of the feed horn are in good agreement with the design. We present the measurement setup, amplitude and phase pattern of the beam, and the fitting result here.

I. Introduction

The South Pole Telescope (SPT) is a cosmic microwave background (CMB) experiment located at the Amundsen-Scott South Pole station [1,2]. Its 10-meter diameter primary dish was designed with a surface accuracy suitable for submillimeter-wavelength observation. Its geographic location, far from other submillimeter telescopes, provides an opportunity to greatly increase the size millimeter wavelength very-long-baseline interferometry (VLBI) arrays, such as the Event Horizon Telescope (EHT) [3].

We have developed a 230 and 345 GHz VLBI receiver for the SPT. In this paper, we report the vector beam measurement of the receiver using a technique developed for submillimeter receivers (e.g., [4,5]). We use the measurement to characterize the location and tilt of the horn phase centers inside the receiver dewar. In section 2, we describe the receiver and its optical design. In section 3, we show the beam measurement setup and the test result. In section 4, we introduce a general expression of the electric field when the tilt angle of the beam is considered, and then infer the model parameters using Markov chain Monte Carlo (MCMC) method. We have found that the estimated model parameters from the data agree well with the design.

II. SPT VLBI Receiver

The SPT VLBI receiver operates in the 230 and 345 GHz frequency bands. Fig. 1 shows the inside of the dewar. In each band, the RF components – feed horns, ortho-mode transducers (OMT, a polarization separation device), and mixer blocks – are cooled down to 4 K. The 230 GHz receiver uses ALMA band 6 sideband-separating SIS mixers [6]. The 345 GHz mixer is under development and expected to be installed in early 2019.

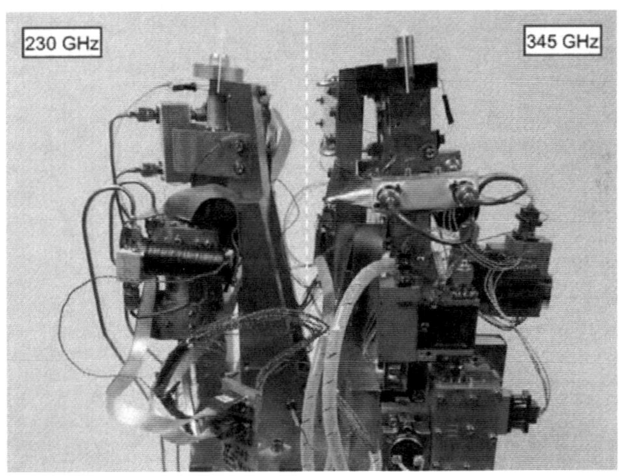

Fig. 1. An interior view of the SPT VLBI receiver. The left and right are 230 GHz and 345 GHz assemblies of the receiver, respectively. Both 230 GHz and 345 GHz feed horns are designed to be 5.74 degrees tilted inward so that the receiver operates at both frequencies using a rotating tertiary mirror above the dewar. The yellow arrows show the beam propagation directions of the feedhorns, and the white dashed line indicates the vertical plane.

The sky signal is delivered to the VLBI receiver with removable secondary and tertiary mirrors that are, independent of the SPT-3G optics (for details, see [7]). The tertiary mirror assembly sits on the receiver dewar and the mirror rotates around the optical axis to switch between 230 and 345 GHz. To couple the chief ray reflected through the tertiary mirror to each feed horn, we designed the feed horns to be tilted toward each other by 5.74 degrees.

III. Vector Beam Measurements

Accurate positioning of the two feedhorns is critical to the coupling of the two frequency bands to the optics in this submillimeter-wavelength receiver. We perform 230 and 345 GHz vector beam measurement of the SPT VLBI receiver to characterize their location. We set up the vector beam measurement system as shown in Fig. 2. The test tone is generated from a Gunn oscillator followed by a frequency tripler (230 GHz: VDI WR-5.1 × 3, 345 GHz: VD1-WR2.8 × 3). We put WR-4 open waveguide probe (0.04 inch × 0.02 inch) whose cutoff frequency is ~140 GHz, on the tripler output, and the transmitter system is mounted on X-Y translation stage composed of two Parker Motion 403XE linear stages. The test tone frequency is chosen such that the intermediate frequency (IF) of the receiver is placed within the IF amplifier operating range. The IF is again mixed down to ~546 MHz to use a K&L Microwave bandpass filter with 10 MHz bandwidth to increase the signal to noise ratio. We can read the amplitude and relative phase of the second IF with the reference input from HP 8648C signal generator. During the beam scan, all the hardware components around the setup are covered with AN-72 broadband microwave absorber to reduce the reflection of the injected tone.

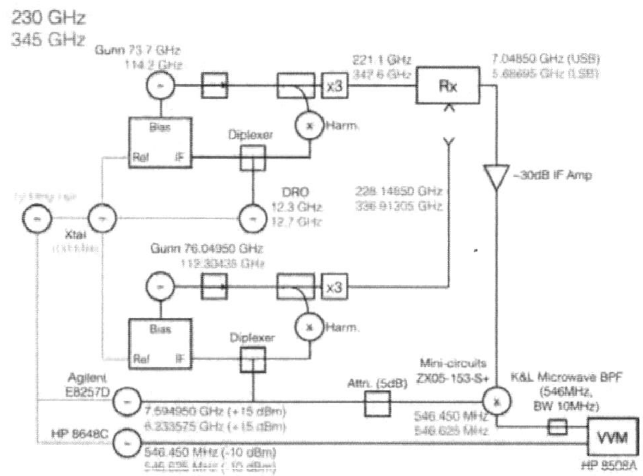

Fig. 2. Schematic of the beam measurement system. Frequencies for the 230 GHz chain are listed in blue, with 345 GHz in red. The signal generators and the phase locked loop for the system are locked to a common 10 MHz signal.

We perform a two-dimensional scan of a region 100 mm × 100 mm in size with a 2.5 mm step, in two parallel planes: The planes are ~200 mm and ~150 mm vertically above the horn

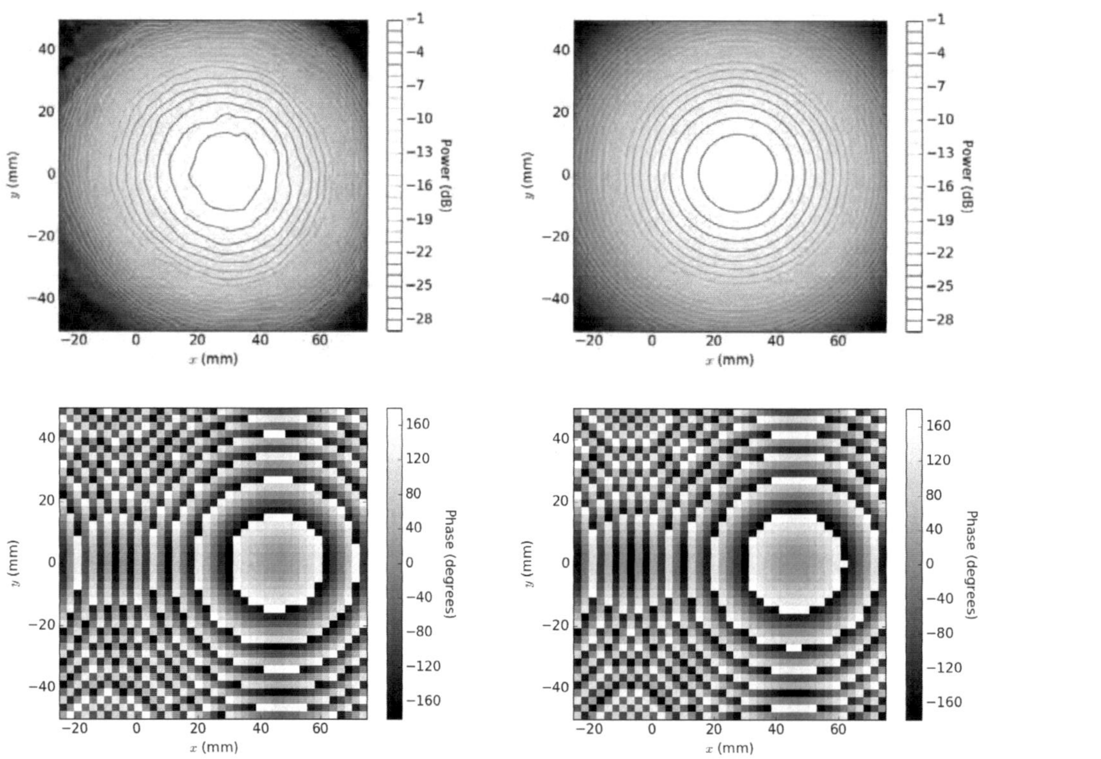

Fig. 3. (*Top left*) The amplitude and (*bottom left*) the phase pattern of 230 GHz beam measurement at $z = 200$ mm. The contours in the amplitude map are in dB units, and the phase map is plotted in degree units. The amplitude map shows more than 30 dB of dynamic range. The phase is not centered due to the tile of the feedhorn. (*Top right*) The amplitude and (*bottom right*) the phase pattern of the beam, using the analytical model with the best-fit parameters.

phase center plane. The 230 GHz scan region is centered on $(x, y) = (25\ \text{mm}, 0\ \text{mm})$ and the 345 GHz region is centered on $(x, y) = (-25\ \text{mm}, 0\ \text{mm})$, to locate the maximum amplitude position near the center of the scanning area. After every column scan, the probe moves to the center position to track amplitude and phase stability. Interpolating reference measurements compensates the time-dependent phase drift during the scan. At each plane we make two maps on planes that are a quarter-wavelength apart in the z-direction. Averaging these two sets of data can further reduce the reflection. Fig. 3 shows the amplitude and phase of 230 GHz at $z = 200$ mm plane. For 230 GHz, we recorded the beams of two different polarizations, by rotating the waveguide probe by 90 degrees.

IV. Analysis

To analyze the data, we derive a general form of the electric field distribution of the tilted Gaussian beam, when the scanning plane is not vertical to the axis of propagation. Then, we fit the model to the data to infer the feed horn position and its tilt angle.

A. Gaussian Beam Propagation of tilted horn

The normalized electric field distribution [8] is

$$E(r, z) = \left(\frac{2}{\pi w^2}\right)^{0.5} \exp\left(-\frac{r^2}{w^2} - jkz - \frac{j\pi r^2}{\lambda R} + j\phi_0\right),$$

where z is the distance along the axis of propagation and r is the perpendicular distance from the axis of propagation. R, w, and ϕ_0 are the radius of curvature, beam radius, and the Gaussian phase shift and expressed as

$$R = z + \frac{1}{z}\left(\frac{\pi w_0^2}{\lambda}\right)^2,$$

$$w = w_0\left[1 + \left(\frac{\lambda z}{\pi w_0^2}\right)^2\right]^{0.5},$$

$$\phi_0 = \arctan\left(\frac{\lambda z}{\pi w_0^2}\right),$$

where w_0 is the beam waist radius.

In Fig. 4, we first define two planes parallel to each other: the horn phase center plane and the scanning plane where the waveguide probe mounted on the translation stage moves. We assume arbitrary horn phase center position $P_h = (x_h, y_h, 0)$ to derive an electric field at the scan position $P_s = (x, y, z_0)$.

The equation of the line l_{beam} is

$$l_{beam}: (x_h, y_h, 0) + t(-\tan\theta_x, \tan\theta_y, 1).$$

The directional vector of axis of propagation on l_{beam} is

$$\vec{b} = (-\tan\theta_x, \tan\theta_y, 1),$$

and the vector between P_h and P_s is

$$\overrightarrow{P_h P_s} = (x - x_h, y - y_h, z_0).$$

The vertical distance from the horn phase center to the wave plane that hits P_s, z_{tilt} is

$$z_{tilt} = \frac{\left|\vec{b} \cdot \overrightarrow{P_h P_s}\right|}{\left|\vec{b}\right|},$$

where

$$\left|\vec{b} \cdot \overrightarrow{P_h P_s}\right| = \left|z_0 - (x - x_h)\tan\theta_x + (y - y_h)\tan\theta_y\right|,$$

and

$$\left|\vec{b}\right| = \sqrt{\tan^2\theta_x + \tan^2\theta_y + 1}.$$

The offset from the axis of propagation on the reference plane is

$$r_{tilt} = \frac{\left|\vec{b} \times \overrightarrow{P_h P_s}\right|}{\left|\vec{b}\right|},$$

where

$$\begin{aligned}\left|\vec{b} \times \overrightarrow{P_h P_s}\right|^2 &= \left[z_0\tan\theta_y - (y - y_h)\right]^2 \\ &\quad + \left[z_0\tan\theta_x + (x - x_h)\right]^2 \\ &\quad + \left[(y - y_h)\tan\theta_x - (x - x_h)\tan\theta_y\right]^2 \\ &= z_0^2\left(\tan^2\theta_x + \tan^2\theta_y\right) \\ &\quad + (x - x_h)^2\left(\tan^2\theta_y + 1\right) \\ &\quad + (y - y_h)^2\left(\tan^2\theta_x + 1\right) \\ &\quad + 2z_0(x - x_h)\tan\theta_x + 2z_0(y - y_h)\tan\theta_y \\ &\quad + 2(x - x_h)(y - y_h)\tan\theta_x\tan\theta_y \\ &= z_0^2\left(\sec^2\theta_x + \sec^2\theta_y\right) + (x - x_h)^2\sec^2\theta_y \\ &\quad + (y - y_h)^2\sec^2\theta_x \\ &\quad - 2z_0\left[z_0 - (x - x_h)\tan\theta_x\right]\left[z_0 \right. \\ &\quad \left. + (y - y_h)\tan\theta_y\right].\end{aligned}$$

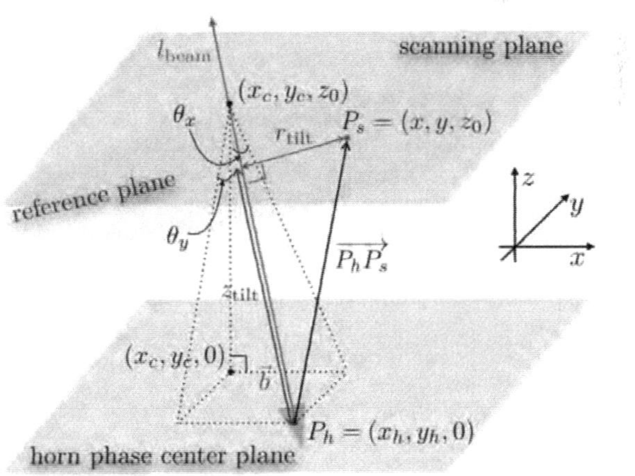

Fig. 4. Schematic of the tilted beam measurement. The feed horn is tilted by θ_x and θ_y on the vertical plane in x and y-direction. The Gaussian beam radiates from the horn phase center $(x_h, y_h, 0)$ along its axis of beam propagation l_{beam} (red arrow). The beam scanning plane and the horn phase center plane are vertically separated by z_0.

We now have the normalized electric field distribution for the tilted beam on the scanning plane

$$E(x,y,z_0) = \left(\frac{2}{\pi w_{tilt}^2}\right)^{0.5} \exp\left(-\frac{r_{tilt}^2}{w_{tilt}^2} - jkz_{tilt} - \frac{j\pi r_{tilt}^2}{\lambda R_{tilt}} + j\phi_{0,tilt}\right),$$

where

$$R_{tilt} = z_{tilt} + \frac{1}{z_{tilt}}\left(\frac{\pi w_0^2}{\lambda}\right)^2,$$

$$w_{tilt} = w_0\left[1 + \left(\frac{\lambda z_{tilt}}{\pi w_0^2}\right)^2\right]^{0.5},$$

$$\phi_{0,tilt} = \arctan\left(\frac{\lambda z_{tilt}}{\pi w_0^2}\right).$$

The center of the beam that hits the scanning plane is where the equation of line l_{beam} intersects with $z = z_0$ plane and

$$x_c = x_h - z_0 \tan\theta_x,$$
$$y_c = y_h + z_0 \tan\theta_x,$$
$$z = z_0.$$

B. Beam Fitting

We fit the beam mapping data to the tilted Gaussian beam profile described in the previous Section IV-A, maximizing the power coupling coefficient between the data and the model electric fields. The power coupling coefficient is an absolute square of the field coupling coefficient

$$K_{ab} = \left|\iint E_a^* E_b dS\right|^2,$$

where E_a and E_b are the electric field distributions of two Gaussian beams. Then, we use Markov chain Monte Carlo (MCMC) sampling with the python package emcee [9].

The corner plot (Fig. 5) is an example of the parameter estimation for one polarization of the 230 GHz beam at $z =$ 200 mm. For all scans, the y- phase center position is less than 0.5 mm and the y-direction tilt angle is less than 0.5 degrees. The best-fit parameters for the receiver tilt angle in the x-direction are listed in Table 1 and an example of the model beam is shown in Fig. 3. The 5.74 degrees x-direction tilt angle from the design is within 1-sigma uncertainty range of all the measurements.

CONCLUSIONS

We have performed the beam measurement of the SPT VLBI receiver at 230 and 345 GHz. We introduced the functional form of the electric field for the tilted beam. The MCMC fitting of the model to the data gives the estimation of the horn parameters, especially the tilt angle of the feed horn that is critical to the beam coupling between the horn and the optical elements.

ACKNOWLEDGMENT

We thank Edward Tong for useful comments on the measurement technique. This work was supported by NSF awards AST-1207752 and AST-1440254.

REFERENCES

[1] J. E. Carlstrom, P. A. R. Ade, K. A. Aird, et al. "The 10 Meter South Pole Telescope," *Publications of the Astronomical Society of the Pacific*, vol. 123, pp. 568-581, May. 2011.
[2] B. A. Benson, P. A. R. Ade, Z. Ahmed, et al., "SPT-3G: a next-generation cosmic microwave background polarization experiment on the South Pole telescope," in *Proc. SPIE, Vol. 9153, Millimeter, Submillimeter, and Far-Infrared Detectors and Instrumentation for Astronomy VII*, 2014, p. 91531P-22.
[3] S. Doeleman, E. Agol, D. Backer, et al. "Imaging an Event Horizon: submm-VLBI of a Super Massive Black Hole," in *Astro 2010: The Astronomy and Astrophysics Decadal Survey*, 2009.
[4] C.-Y. E. Tong, S. Paine, and R. Blundell, "Near-Field Characterization of 2-D Beam Patterns of Submillimeter Superconducting Receivers," in *Proc. Fifth International Symposium on Space Terahertz Technology*, 1994, p. 660.
[5] M.-T. Chen, C.-Y. E. Tong, R. Blundell, D. C. Papa, and S. N. Paine, "Receiver-beam characterization for the SMA," in *Proc. SPIE, Vol. 3357, Advanced Technology MMW, Radio, and Terahertz Telescopes*, 1998, p. 106.
[6] A. R. Kerr, S.-K. Pan, S. M. X. Claude, et al. "Development of the ALMA Band-3 and Band-6 Sideband-Separating SIS Mixers," *IEEE Transactions on Terahertz Science and Technology*, vol. 2, pp. 201-212, Mar. 2014.
[7] J. Kim, D. P. Marrone, C. Beaudoin, et al. "A VLBI receiving system for the South Pole Telescope," in *Proc. SPIE, Vol. 10708, Millimeter, Submillimeter, and Far-Infrared Detectors and Instrumentation for Astronomy IX*, 2018, p. 10708-97.
[8] P. F. Goldsmith, Quasioptical Systems: Gaussian Beam Quasioptical Propogation and Applications, Wiley-IEEE Press, 1998.
[9] D. Foreman-Mackey, "corner.py: Scatterplot matrices in Python", *The Journal of Open Source Software*, vol. 1, p. 24, Jun. 2016.

TABLE I
THE BEST-FIT PARAMETERS FOR THE x-DIRECTION TILT ANGLE

Tilt angle (θ_x)					
230 GHz			345 GHz		
Pol 0	z = 200 mm	5.62 (+0.56 / -0.56)	Pol 0	z = 195 mm	5.80 (+0.64 / -0.64)
	z = 155 mm	5.45 (+0.74 / -0.78)		z = 150 mm	5.69 (+0.88 / -0.89)
Pol 1	z = 200 mm	5.50 (+0.58 / -0.54)			
	z = 155 mm	5.48 (+0.75 / -0.76)			

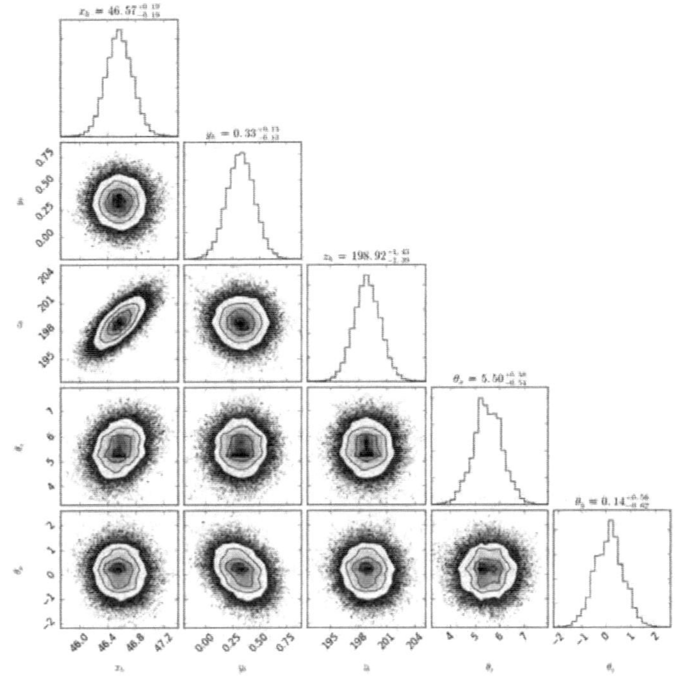

Fig. 5. Parameter estimation result for the 230 GHz measurement at $z = 200$ mm plane using MCMC. The inferred parameters agree well with the design.

A refined method of the data processing for astroclimate measurements in mm-waves

Grigoriy M. Bubnov*[,1,2], Vyacheslav F. Vdovin[1,2], Peter M. Zemlyanukha[1] and Igor I. Zinchenko[1]

[1]Institute of Applied Physics RAS, N.Novgorod, 603950, Russia

[2]Nizhniy Novgorod State Technical University, N.Novgorod, 603950, Russia

*Contact: bubnov@ieee.org

Abstract— **We are presenting an algorithm of post-processing of data obtained through the atmospheric dip method, and the properties of the filters to be used with this algorithm. The algorithm allows to make flagging of erroneous data and thus to decrease the stochastic error of the atmospheric transparency values. The algorithm includes finding and excluding the own noises of the receiver, flagging the data corresponding to broken or stratus clouds, and calculating the tau with different methods. Based on physical parameters of the atmosphere and the hardware, the algorithm helps save from 5 to 20 percent of data which otherwise are not suitable for processing.**

I. Introduction

Atmospheric propagation of terahertz waves strongly depends on the atmospheric conditions, and one of the topical issues of today is to investigate this dependence and develop techniques, instrumentation and mathematical modelling of measurement of the atmospheric transparency. Our improved post-processing method described below in this paper can be applied to data collected with the atmospheric dip method. Our method includes finding the own-noise parameters of the hardware, evaluation of the atmospheric conditions with the measurement data, and, finally, filtration of these data. One of the problems of atmospheric transparency data processing which researchers often have to deal with is gaps in data arrays which occur in partial cloudiness, and we are suggesting a way of dealing with this difficulty.

We have developed instrumentation and methods for investigations of atmospheric propagation of terahertz waves [1] and since 2012 we have gone on 7 astroclimate expeditions and explored atmospheric absorption on some 11 sites [2,3]. Our goal is to find the most appropriate place to build a radiotelescope operating in the millimetre and submillimetre wavebands. For estimating the atmospheric absorption (optical depth, tau, or Nepers) we use an assemblage which includes a radiometric system and a firmware system, the former comprising two self-contained radiometers operating in two different bands of 84-99 GHz (3 mm) and 132-148 GHz (2 mm), a rotary support, and a control. The integral absorption is determined by using the atmospheric dip method, that is, a 5-step approximation of atmospheric brightness temperature as the function of the zenith angle. The zenith angles choice is 0°, 60.48°, 76.32°, 81.36°, 84.24° and 88.56° (the latter corresponds to aiming at the horizon). Each of these angles corresponds to an integer number of steps of the stepper motor by which the mirror is driven, and to more or less equal intervals between the brightness temperatures at each of these directions. The voltage of the receiving detector (we use the negative value of this for the sake of convenience) is in linear dependence on brightness temperature; linearity holds over a wide range of brightness temperature values.

For the case of the clear sky, the experimental points can be fitted well by exponent function, and the result of approximation (the exponent index, Nepers) has a good stability over time. But for the case of partial cloudiness the fitting occurs with wide deviation against the exponent and large scattering of the points in time.

The design of the algorithm was due to the properties of our hardware and the method of the atmospheric depth measurement. The atmospheric dip method is based on measuring the difference of brightness temperature of the atmosphere's own radiation at different elevations. If the gradient of the brightness temperature is broken for some reason (cloudiness or the own noise of the receiver) this results in either fatal calculation errors (the logarithm of a negative number or division by zero) or dramatic increase in inaccuracy.

At the same time, we cannot filter or process experimental data unless such procedure is well-grounded from the physical point of view. Thus, the data processing task is formulated as follows: first, we must take away the error in the measurements as long as it is possible, and then muffle of the remaining noises of the receiver and the atmosphere with the help of mathematical processing.

II. Algorithm

The algorithm includes several steps united by the uniform logic which is being discussed at the end of this chapter. In addition, we have annexed the Python script used at one stage of the algorithm to this paper. The primary data are output by the software controlling the physical interfaces of the MIAP-2 receivers: it gathers the measurements, drives the mirror, reads the ADC readings and records a Raw file indicating voltage of the detectors at each angle. The files are then mathematically processed.

A. Calculation of optimal integration time

Integration time of the atmospheric brightness temperature at each angle can be changed within a wide range of values each time the software controlling the physical interfaces of MIAP-2 is launched. This does not apply to the previously collected data.

As we lower the integration time for a given angle, we increase the stochastic spread of values. If we raise the integration time, the average one-cycle period will also increase, and therefore, short-time fluctuations of the atmospheric conditions will be lost. Figure 1 shows a diagram of finding the optimal integration time for a given angle.

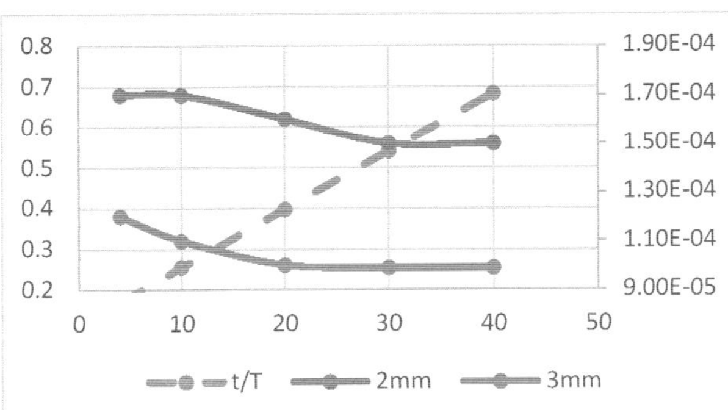

Fig. 1. A diagram showing the dependence of RMS noise of the data (2 mm and 3mm, right-hand axis) and catching of the short-time fluctuations in atmospheric conditions on the integration time at a given angle (left-hand axis).

We grant that the reference time of change of the atmospheric conditions is 7 minutes. This is a tentative parameter defined empirically based on experimental data. The grey dashed line reflects the ratio of the one cycle measurement time to the reference time of 7 minutes. Similarly to the conclusions following from the Kotelnikov theorem, we think it reasonable to grant that the optimal ratio is about 0.5, that is, the time for one-cycle measuring should be twice as little as the time of the change in the atmospheric parameters. The continuous lines show the measured sigma values (RMS) in both channels. These values obviously drop as the integration time grows. Therefore, the optimal integration time for a given angle can be defined as about 20 to 30 seconds.

B. Measuring and filtering the circuit noises

It is evident that the receivers have their own thermal noises. The correction cannot include the atmospheric parameters for the obvious reason that we cannot know them before we measure them. The filtering procedure is based on the parameters of the hardware which are known or can be measured so that we could know that we would filter out of the experimental data only the hardware-generated noises. In order to measure the hardware own noises, the quasi-optical path of the receiver is covered with the black body absorber whose temperature is stabilised with the help of a standard air conditioning device, and then the readings are recorded in the usual manner.

The resulting record of noises is then wavelet-packet decomposed into order-2 coiflets [4] with the help of the PYWT library [5]. Further, the median absolute deviation is calculated in the wavelet space, and thus the low pass filter is defined. We apply this filter to the collected observation data. An example of the filter application is shown in Figure 2.

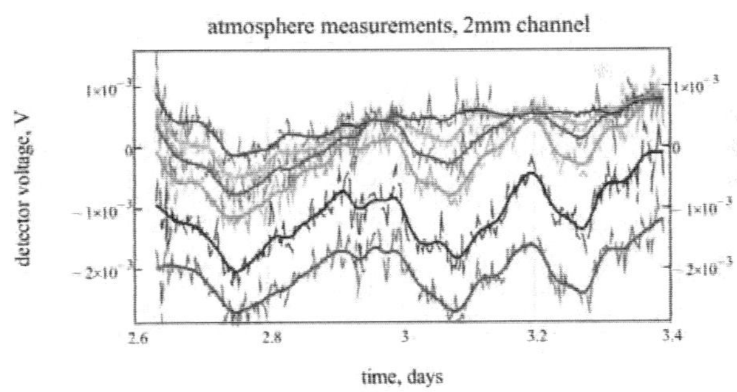

Fig. 2. Filtration of data in the 2-mm channel of the receiver. The topmost plot represents the horizon-aimed measuring, the bottom line the zenith, and the intermediate positions are between them. The thin dashed lines show the measuring, while the thick lines represent the filtered data.

The filtering parameters for the 2-mm and 3-mm channels are calculated independently. For the 3-mm channel the filtration parameters are calculated in a similar manner; we only have chosen the 2-mm channel to be presented here as a more illustrative example. The filtered data are then processed through the further stages of the algorithm.

C. Cloud correlator

When taking measurements in low partial cloudiness there is a high probability that the receiver's beam will cross a cloud and cause the brightness temperature to rise over a relatively short time interval at the given angle. This temporally and spatially local upsurge muddles the evaluation of the optical depth and must be filtered out of the further consideration. In order to pinpoint such upsurges, we calculate the Pearson correlation for each pair of angles and then clear away everything below 0.8 from the data array.

The example in Figure 3 shows the upsurge of the brightness at the beginning of the third day (see the segment between 3.03-3.12 in Fig. 3) as the beam of the receiver was aimed at a cloud at 60.48° thus decreasing the Pearson correlation down to 0.6. In the further processing all data corresponding to the Pearson correlation below 0.8 were excluded.

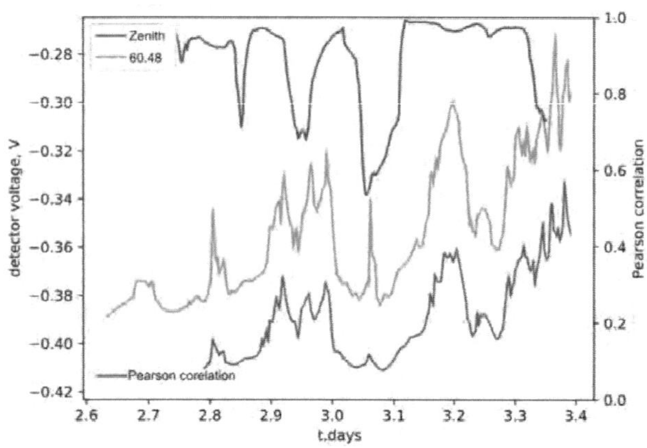

Fig. 3. An example of the Pearson correlation calculation for the zenith and 60.48° pair.

D. *Flagging of overlapping readings*

In high humidity or overcast conditions some values measured at low angles overlap the readings taken at the horizon. This happens mostly due to the dispersion of radiation in the stratus clouds; at the low angles the receiver also senses the thermal radiation of the land surface.

The algorithm tracks down converging of the receiver voltage at juxtaposed angles going over from point to point; if the convergence becomes lower than 2σ, the respective interval will be excluded from the consideration. The horizon and the zenith are reference points and never deleted.

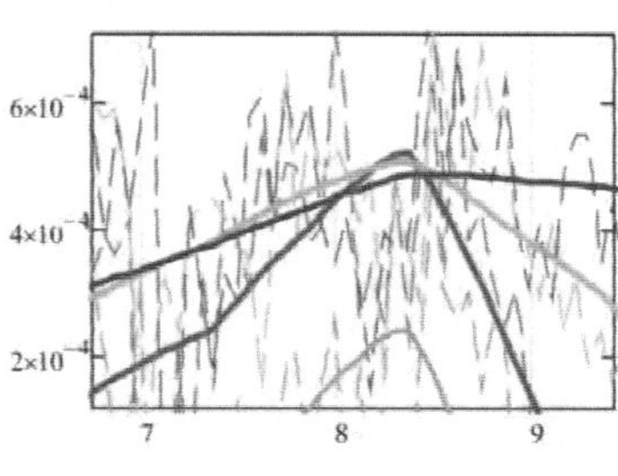

Fig. 4. An example of overlapping of voltages at angles near the horizon: voltages at 84.24° and 81.36° converge and cross the voltage taken at the horizon.

E. *Calculation of tau*

After the voltages of each particular angle are independently processed the data are used for calculating the optical depth (tau or Nepers). At this stage the voltages taken at different angles are fitted to Equation (1) using two methods: the least square fitting and k-Nearest Neighbor.

$$U(\theta) = U_{aver}\left(1 - e^{-\tau \cdot \sec(\theta)}\right) + U_{shift}$$

where τ is optical depth, Nep;

U_{aver} is the mean voltage corresponding to the brightness temperature of the atmosphere at the horizon;

U_{shift} – is the zero-shift voltage conditioned by the hardware properties;

U(θ) – are the measurements at particular angles.

This equation does the fitting of three parameters for each point over the time: U_{aver}, U_{shift} and τ. If for a particular moment 1, 2 or 3 angles were removed from consideration, the fitting is performed for the remaining angles. Exclusion of angles leads to the increase in the theoretical RMS of the evaluation, but not to the data loss as it previously was the case with the older algorithm.

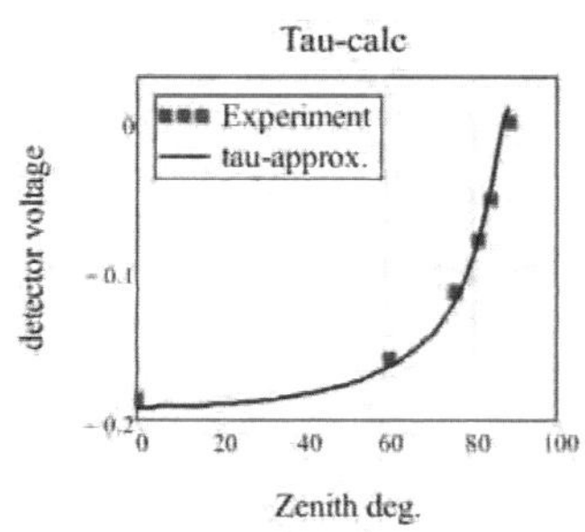

Fig. 5. Fitting of the experimental data with the exponent.

We used two methods for the calculation of tau: the least-squares fitting (LIMFIT [6], LSQ) and the k-Nearest Neighbors (SCIKIT-LEARN [7], kNN). Assuming random noise parameters and the LSQ fitting error we introduce the kNN algorithm for the tau estimation in mostly the same way as discussed in [8]. We use previous measurements and tau estimation as a training set for kNN and LSQ with the m=2 Minkowski metric. As soon as we have enough exposures to fill in neighbouring spaces we obtain RMS for tau $\sqrt{k(k-1)}$, where k is the number of neighbours. In this paper we use 300,000 modelled observations as neighbouring spaces and receive only $\sqrt{k-1}$ RMS reduction. In this case kNN is an addition to LSQ algorithm, not the separate one.

F. Algorithm as a whole

The flowchart in Figure 6 shows data flowing from one stage of the algorithm to the next one indication the data formats. The algorithm is performed in three different software environments. Thee collection of data and physical interface control are done by MIAP-2, the integration time is calculated by MathCAD, and the collected data are processed by Python.

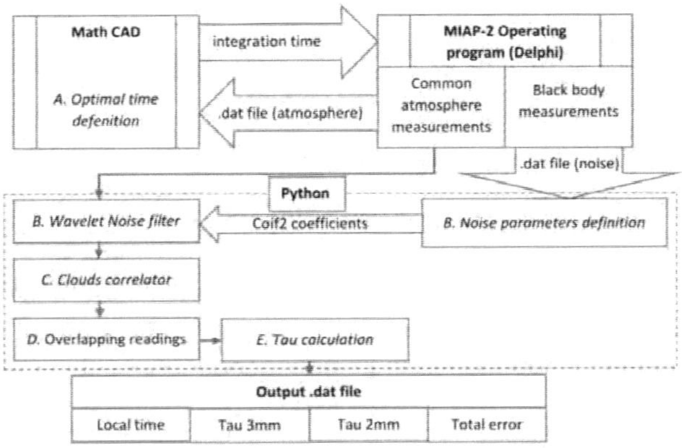

Fig. 6. The algorithm flowchart.

III. Data Processing Results

As an example, we present here data collected from the roof of one of the buildings of the IAP RAS in Nizhny Novgorod in winter time (February to March 2018). All three types of cloudiness are found in the example data: clear sky, low partial cloudiness, and sky overcast with stratum clouds. We chose the optimal integration time to be 20 seconds and recorded the own noises of the receiver applying the same integration time (Fig. 7).

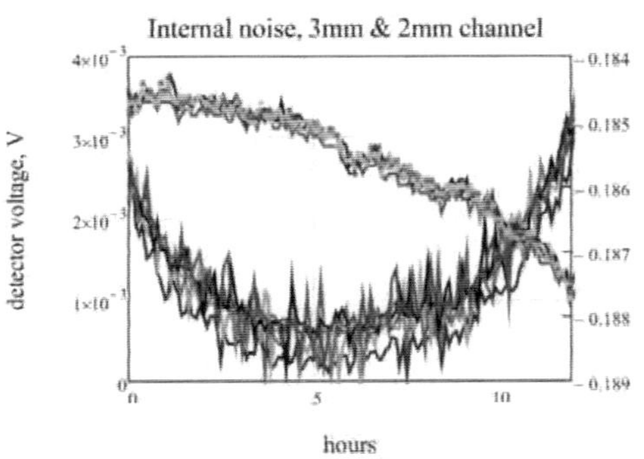

Fig. 7. Record of the hardware own noises over several hours.

The wavelets for the Coif2 filter were calculated as

2mm 0.00070858

3mm 0.00257025

The cloud correlator excluded partial cloudiness from our data arrays, for example, the one that occurred at the beginning of the third day of observations at 60.48° in the 3-mm channel. An example of the overlapping readings flagging is shown for the 2-mm channel at the end of the second day of observations (Fig. 8).

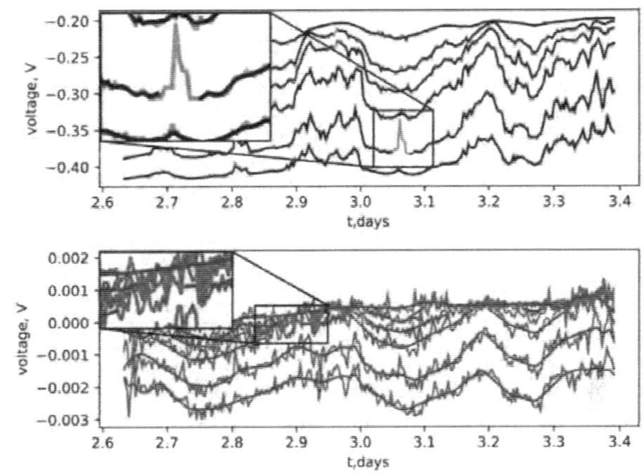

Fig. 8. Examples of algorithmic exclusion of partial cloudiness (top, 3—mm channel) and overlapping readings (bottom, 2-mm channel).

Figure 9 shows the results of the tau calculation for both channels as compared to the old method of direct calculation of an iterative equation by 5 angles. The new algorithm permitted to decrease the RMS deviations and partially to keep the data previously unsuitable for use.

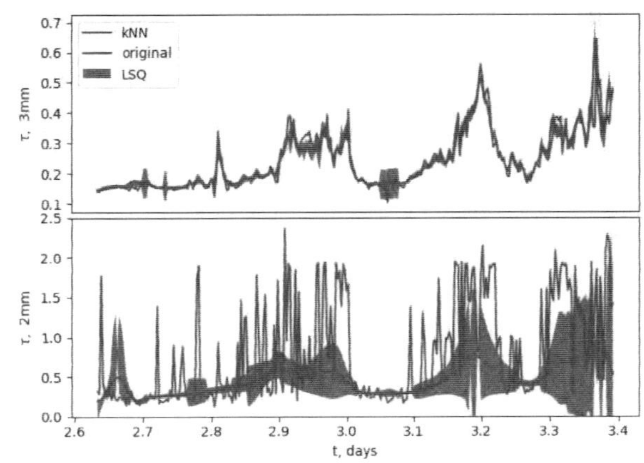

Fig. 9. The results of the algorithm as compared to the standard method.

Conclusions

In this paper we present the algorithm for collection and processing of data obtained with the atmospheric dip method. Its performance is stable and has been broken in for the hardware we dispose of; presently, we are working on a simplified user-friendly interface. The algorithm presented here is optimal for our hardware; however, it is possible that the algorithm will require modifications to be applied to hardware operating in different wavebands.

ACKNOWLEDGMENT

This work was prepared with the financial support of the IEEE Antennas and Propagation Society Eugene F. Knott Memorial Doctoral Research Grant and by the Russian Foundation of Basic Research (RFBR) within Grant 16-32-0038716 mol_a. The authors would like to thank Dmitri Sofronov for the help in editorship.

REFERENCES

[1] V. I. Nosov et al., "A dual-wave atmosphere transparency radiometer of the millimeter wave range", Electronics and Radio Engineering Instruments and Experimental Techniques, Volume 59, Issue 3, pp 374-380, May 2016.

[2] Grigoriy M. Bubnov et al., "Searching for New Sites for THz Observations in Eurasia", IEEE Tr. on TST, Volume: 5, Issue: 1, pp.64-73, 2015.

[3] G. M. Bubnov et al., "The results of the astroclimate investigations in short-millimeter wave on the Suffa plateau", Radiophysics and Quantum Electronics, Vol 59, Issue 8-9, pp.852-861, 2016.

[4] Wavelet browser by PYWAVELETS [Online]. Available: http://wavelets.pybytes.com/wavelet/coif2/

[5] Lee G, Wasilewski F, Gommers R, Wohlfahrt K, O' Leary A, Nahrstaedt H, and Contributors, "PyWavelets - Wavelet Transforms in Python", 2006, [Online], https://github.com/PyWavelets/pywt .

[6] Newville, M., Stensitzki, T., Allen, D. B., & Ingargiola, A. "LMFIT: Non-Linear Least-Square Minimization and Curve-Fitting for Python" 2014, September 21, Zenodo. http://doi.org/10.5281/zenodo.11813

[7] Pedregosa et al., "Scikit-learn: Machine Learning in Python", JMLR 12, pp. 2825-2830, 2011.

[8] Zemlyanukha, P.M., Zinchenko, I.I., Salii, S.V. et al. "The Spatial–Kinematic Structure of the Region of Massive Star Formation S255N on Various Scales" Astron. Rep. (2018) 62: 326. DOI: S1063772918050074.

Development of a cryogenic remote sensing thermometer for CMB polarization experiment

Y. Sakurai [1], T. Matsumura[1], N. Katayama[1], H. Kanai[2], T. Iida[3]
[1] Kavli Institute for the Physics and Mathematics of the Universe, The University of Tokyo, 5-1-5 Kashiwanoha, Kashiwa, Chiba, 277-8583, Japan
[2] Yokohama National University 79-1 Tokiwadai, Hodogaya-ku, Yokohama 240-8501 JAPAN
[3] ispace, inc. Iikurakatamachi Annex Bldg. 6F 3-1-6 Azabudai, Minato-ku, Tokyo, Japan 106-0041
*Contact: yuki.sakurai@ipmu.jp

***Abstract*—We present about the remote sensing thermometer working at cryogenic temperature for a polarization modulator used in cosmic microwave background (CMB) polarization experiment. The polarization modulator employs a superconducting magnetic bearing (SMB) in order to realize a continuous rotation at cryogenic temperature with minimal heat dissipation. The cryogenic remote thermometer is a critical development item in the polarization modulator in order to measure cryogenic, contactless and spinning rotor. We propose a novel idea to take an advantage of a temperature dependent nature in NdFeB permanent magnet. The remnance of the NdFeB peaks about 100 K and decrease as its temperature becomes lower. A simple implementation of a cryogenic Hall sensor at a vicinity of the NdFeB enables to construct a cryogenic remote temperature sensor. We calibrated the thermometer using small NdFeB magnet and applied it to a $\Phi \sim 400$ mm SMB prototype system in a 4 K cryostat with GM cooler. The obtained temperature value is good agreement with the resistive thermometer and the accuracy is $\pm$ 0.4 K in the range from 4 K to 20 K. The results are good enough to use in the polarization modulator.**

I. Introduction

One of the most important research topics in current cosmology and high-energy physics is to study the cosmic inflation theory [1,2]. The theory predicts a rapid expansion of the universe after ~10-38 seconds from the beginning of the universe. The inflation theory predicts that primordial gravitational waves from the cosmic inflation left the characteristic pattern, called B-mode, in a polarization component of a cosmic microwave background (CMB). We can still measure the CMB in microwave range as the oldest light from the big bang. Therefore, the experimental verification of the inflation theory is possible by the precise measurement of the CMB polarization signal. In order to verify the inflation theory, there are many CMB polarization experiments on the ground, balloon, and satellite. The development of related technologies is advancing innovatively.

Since the CMB polarization signal is very weak compared with the intensity signal, we need some dedicated devices or strategies for the precise measurement. A polarization modulator is one of the important instruments to realize the precise measurement [3-7]. It consists of an optical element, a half-wave plate (HWP), and a mechanism that rotates it continuously. The polarization modulator is installed at an aperture of a telescope, and it modulates an incident polarization signal. The modulation is able to reduce the impact of $1/f$ noise to the detectors and to mitigate the systematic uncertainties due to the detector differential characteristics. The HWP has to be maintained at the cryogenic temperature, e.g. below 10 K, in order to reduce noise due to its thermal emission. Therefore, the polarization modulator is required to realize a continuous rotation at cryogenic temperature with minimal heat dissipation.

A general contact type bearing, e.g. a ball bearing, makes a large heat dissipation due to the physical friction. Thus, the polarization modulator often employs a superconducting magnetic bearing (SMB) [8-9] in order to satisfy this stringent requirement. The SMB is contact-less levitation type bearing, which consists of a permanent magnet as a rotor and a superconductor array as a stator. It is possible to realize a contact-less continuous rotation at cryogenic temperature by combining the SMB with a synchronous motor, consisting of rotor permanent magnets and stator coils. The fundamental problem of the SMB is that it is hard to measure the temperature of the levitating and spinning rotor at the cryogenic environment. The temperature is crucial information to estimate the thermal loading from the HWP.

In this paper, we discuss the measurement method of the SMB rotor temperature at cryogenic temperature. We propose a novel idea to take an advantage of a temperature dependent nature in NdFeB permanent magnet. First, we calibrated the thermometer using a calibration setup with small magnets. Then, we evaluated the thermometer performance with a $\phi \sim$ 400 mm SMB prototype system in a 4 K cryostat. Finally, we discuss our results, further challenges, and how to adapt to actual experiments.

II. Cryogenic Remote Sensing Thermometers

The infrared radiation thermometer is widely used in the various application as a remote sensing thermometer. It measures an intensity of an infrared radiation emitted from an object. The intensity is translated to the object temperature. The infrared radiation from the rotor below 10K is significantly weak compared with room temperature. Thus, the detector of the thermometer has to be maintained in sub-Kelvin temperature for this application in order to ensure the sufficient sensitivity. However, installing a dedicated sun-Kelvin cooler for only this thermometer is not realistic.

As another proposal for remote sensing thermometer, we can consider a method of converting a physical wire to inductive contact. In this case, we can use a usual resistive thermometer for the temperature detection. As the readout of the resistive thermometer, an inductor is mounted on the spinning rotor. Another inductor or a SQUID sensor with bias current is mounted on the stator side. We can measure the induced electromotive force by mutual induction of the two inductors. Then, the obtained resistance is translated to the rotor temperature. This method has an advantage that the existing resistive thermometer can be used. However, it is necessary to make the distance of the two inductors as an order of a few μm in order to ensure the sufficient sensitivity. Thus, this method has a significant risk of a physical collision between a rotor and a stator.

The third method is to use a property of a NdFeB permanent magnet. The magnetic field of the NdFeB magnet has strong temperature dependence, especially at cryogenic temperature. Thus, it can be used as cryogenic remote sensing thermometer by installing NdFeB magnet on the rotor and measuring its magnetic field with a Hall sensor. This thermometer is suitable for the polarization modulator because the permanent magnet is already used as a rotor in the SMB.

In this paper, we focus on the third method, i.e. the permanent magnet (PM) thermometer. We describe the calibration method and the verification of the thermometer accuracy.

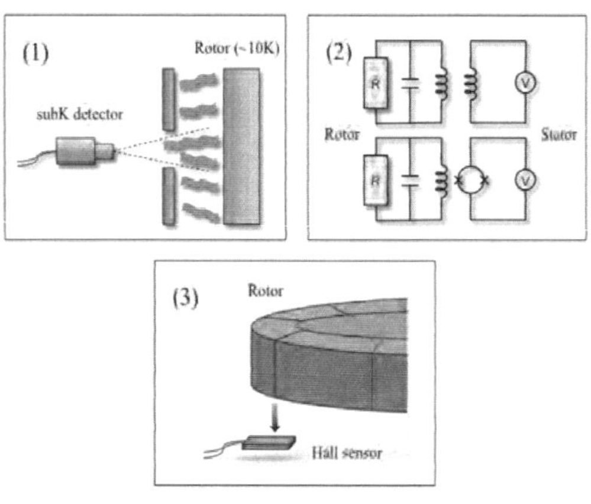

Fig. 1. The conceptual diagrams of (1) the infrared thermometer, (2) the inductive contact thermometer and (3) the permanent magnet (PM) thermometer.

III. Experimental Setup

We constructed two type of experimental setup, i.e. a calibration setup and an application setup using a Φ ~ 400 mm SMB. We install these setups to each dedicated 4 K cryostat with a Gifford-McMahon (GM) cooler in order to maintain an environment below 10 K. The purpose of the calibration setup is to measure the magnetic field of the permanent magnet in the temperature range from room temperature to around 4 K. We prepared two kinds of permanent magnets, NdFeB (N52) and SmCo (R33H), manufactured by Shinetsu Magnet Inc. [10]. They are small pieces of 15 mm × 15 mm × 5 mm without surface treatment. One of the magnets is mounted on an aluminum jig, as shown in Fig. 2. A cryogenic Hall sensor (HGT-3010, Lakeshore [11]) and a resistive thermometer (CX-1010, Lakeshore [12]) are installed neat the magnet. The whole setup is installed in the 4 K cryostat with the most inner shell size of 350 mm × 350 mm × 130 mm. We monitored the

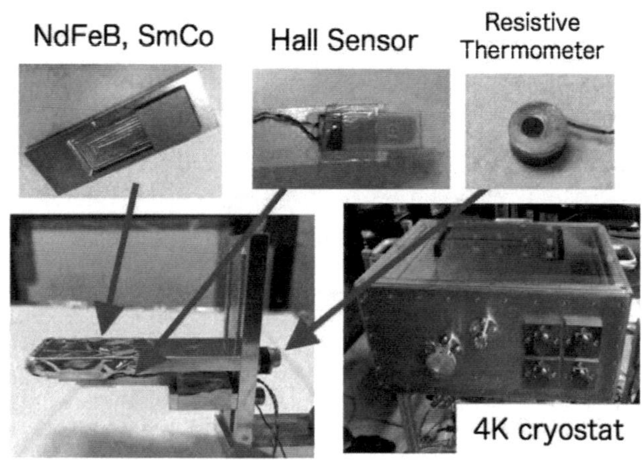

Fig. 2. The experimental setup to measure the temperature dependence of the magnetic field of the permanent magnet. The small piece (15 mm × 15 mm × 5 mm) of the permanent magnet (NdFeB or SmCo) is mounted on the aluminum jig. The Hall sensor and the resistive thermometer are attached near the magnet. The setup is installed in the 4 K cryostat with GM cooler. The 4 K shell size of this cryostat is 350 mm × 350 mm × 130 mm.

magnetic field and the temperature of the permanent magnet with different distances of 1, 2, and 3mm from the magnet surface.

We conducted the Φ ~ 400 mm SMB prototype [13-15] in order to evaluate the developed PM thermometer, as shown in Fig. 3. The SMB consists of a NdFeB permanent magnet as a rotor and a YBCO superconductor array as a stator. The magnet is formed into a ring shape with 16 segmented NdFeB magnets, assembling by a coating glass epoxy. The superconductor array is also formed into a ring shape with 20 three-seeded YBCO tiles, which is covered by a glass epoxy holder. Both the magnet and the superconductor are fabricated and assembled by ATZ [16]. The SMB system is installed in Φ ~ 1 m 4 K cryostat with the GM cooler. The most inner shell diameter is Φ = 850 mm. The cryogenic Hall sensor is installed on top side of the rotor in order to measure the magnetic field of the rotor magnet. The resistive thermometer and heater are mounted on the surface of the rotor magnet. The temperature of the levitating rotor is controlled by the resistive heater. It is monitored by the resistive thermometer. The resistive thermometer and heater are connected with readout wires. Therefore, the rotor is not able to be spinning with this setup. In this setup, the PM thermometer consists of the installed cryogenic Hall sensor and the SMB rotor magnet It is evaluated by comparing with the resistive thermometer. The conceptual design of the experimental configuration is shown in Fig. 4.

Fig. 3. The configuration of the Φ ~ 400 mm SMB system. The top left picture shows the Φ ~ 1 m 4 K cryostat. The top right picture shows after installing the SMB with the Hall sensor, the resistive thermometer and heater. The bottom pictures show the NdFeB rotor magnet and the YBCO superconductor array. The inner diameter of the magnet is ~ 400 mm and it consists of 16 segmented magnets with assembling by coating a glass epoxy. The superconductor array consists of 20 YBCO tiles and it is covered by a glass epoxy.

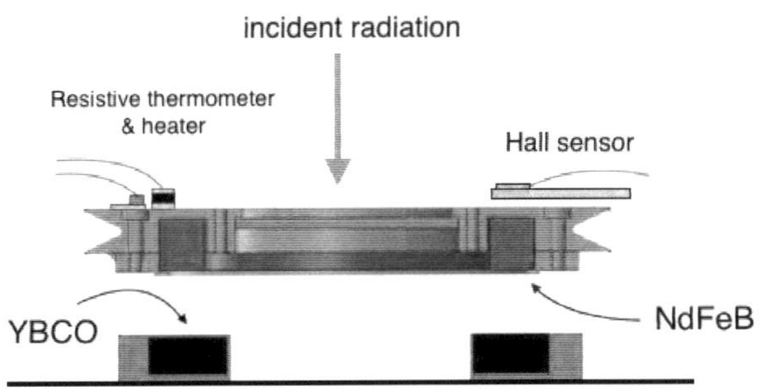

Fig. 4. The experimental configuration to evaluate the PM thermometer. The cryogenic Hall sensor is installed top side of the levitating rotor. The resistive thermometer and heater is mounted on the surface of the levitating rotor. The rotor temperature monitored by the resistive thermometer, and it is controlled by the resistive heater. The PM thermometer consisting of the rotor magnet and the Hall sensor is evaluated by comparing with the resistive thermometer.

IV. Results and Thermometer Performance

A. Calibration

A magnetic field from a permanent magnet has a cryogenic temperature dependence, which differs according to kinds of permanent magnets. We measure the dependence using the calibration setup, as shown in Fig. 2. In the application of the SMB, a rare earth magnet, e.g. NdFeB or SmCo, is often used taking the advantage of their strength of the magnetic field. Other magnets have risks in the cryogenic environment due to their low-temperature demagnetization and brittleness. Thus, the temperature dependences of the NdFeB and the SmCo magnets are measured as the representative permanent magnets.

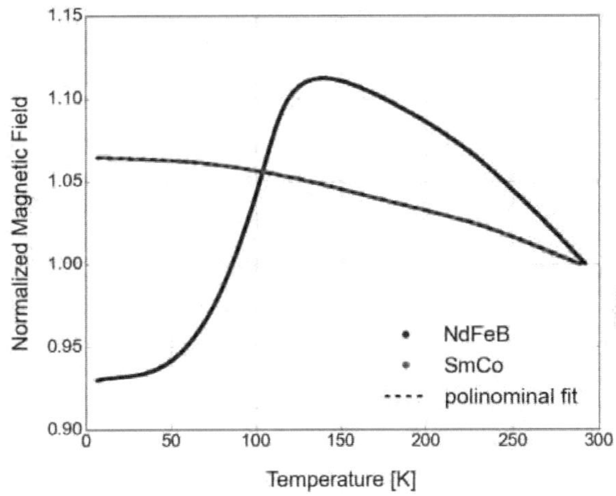

Fig. 5. The temperature dependence of the magnetic field of the permanent magnet. The blue and green lines show NdFeB and SmCo permanent magnet, respectively. The dot lines show the result of the 10th polynomial fit. The vertical axis is normalized to the absolute values of the magnetic field of room temperature (290 K).

Fig 5 shows the measurement result of the magnetic field with

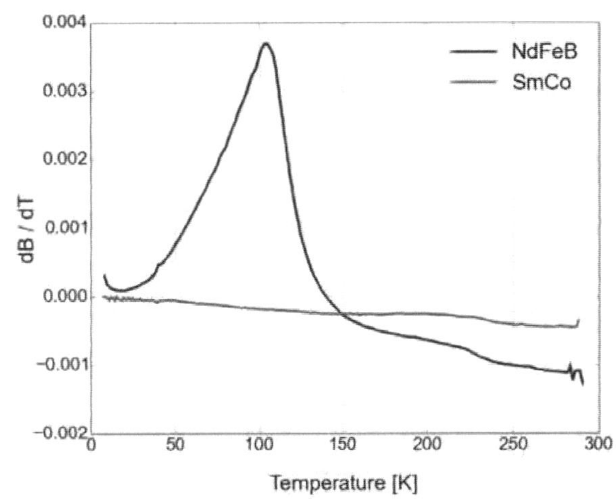

Fig. 6. The gradient of the temperature dependence of the magnetic field from the permanent magnet. The blue and green lines show NdFeB and SmCo permanent magnet, respectively. The vertical axis represents the differential value of Fig. 5.

respect to the rotor temperature for the NdFeB and the SmCo magnets. The gradient of the measured dependence is shown in Fig 6.

From the thermometer point of view, the accuracy depends on the gradient of the variation of the magnetic field. Thus, the NdFeB magnet is suitable to use as PM thermometer compared with the SmCo magnet, as shown in Fig. 6. We use the measured shape of the Fig. 5 as the calibration curve of the PM

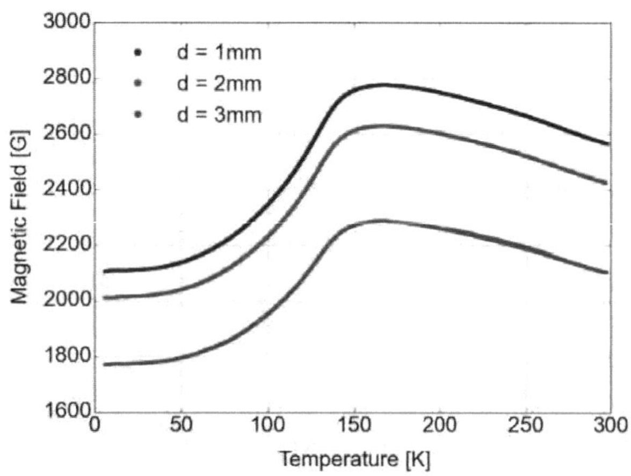

Fig. 7. The temperature dependence of the magnetic field from the permanent magnet with different distances between the Hall sensor and the surface of the magnet. The blue, red and green lines represents the distance is 1, 2 and 3mm, respectively.

thermometer. Once we measure the magnetic field of the NdFeB magnet, we can translate it to the temperature of the magnet. However, the strength of the measured magnetic field follows the distance between the Hall sensor and the permanent magnet. Thus, we also measured the temperature dependence with different distances of 1, 2 and 3 mm, as shown in Fig. 7.

B. Spin reorientation transition of NdFeB

The characteristic shape of the NdFeB magnet in Fig. 5 is known as a spin reorientation transition (SRT) [17-20]. The SRT represents a change of the magnetization direction toward the energy stable state in a crystal. The conceptual diagram of the SRT is shown in Fig. 8. The transition temperature of the NdFeB magnet is known as around 135 K. From the room temperature to the transition temperature, the NdFeB magnetic spin prefers the direction along the c-axis of its tetragonal crystal, i.e. an easy-axis. Below the transition temperature, this spin direction is away from the easy-axis by the canting angle θ, which depends on the magnet temperature. The maximum deviation of the angle is 30 degrees at 4.2 K. The cone defined by θ is referred as an easy-cone. The magnetization direction is tiled by the SRT, and the installed cryogenic Hall sensor is sensitive to the vertical magnetic field. Therefore, the magnetic field rapidly decreases below the transition temperature. The SRT is due to the complex tetragonal crystal structure and the unusual combination of the magnetic anisotropy.

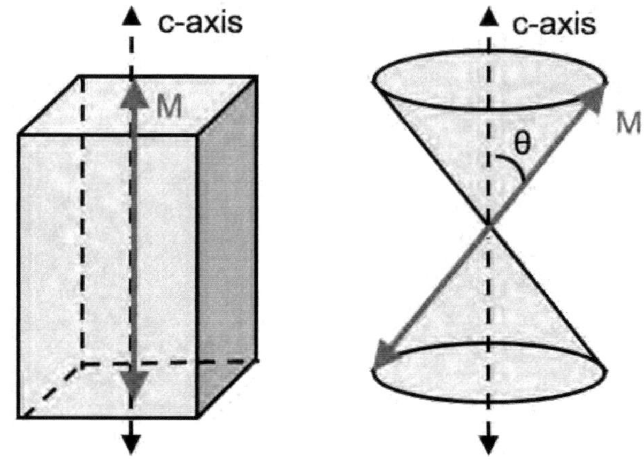

Fig. 8. The conceptual design of the preferred magnetization direction. Above the transition temperature, the magnetization direction, M, is aligned to c-axis in the crystal. Below the transition temperature, M is aligned on the cone surface away from the c-axis at the angle of $\boldsymbol{\theta}$.

C. Performance test using the SMB prototype

We apply the developed PM thermometer to the $\phi \sim 400$ mm SMB prototype. The experimental configuration is shown in Fig. 4. The SMB system is cooled down to around 4 K. We prepare a space of 5 mm between the rotor magnet and the YBCO stator as a levitation height. After the cooling, the rotor is released by the cryogenic actuators. The rotor is levitated by a Meissner effect and the position is fixed by a pinning effect of the type II superconductor. The rotor is slightly dropped from the initial levitation height due to a gravity effect. The dropping distance is estimated by the reduction rate of the magnetic field with respect to the distance between the rotor magnet and the

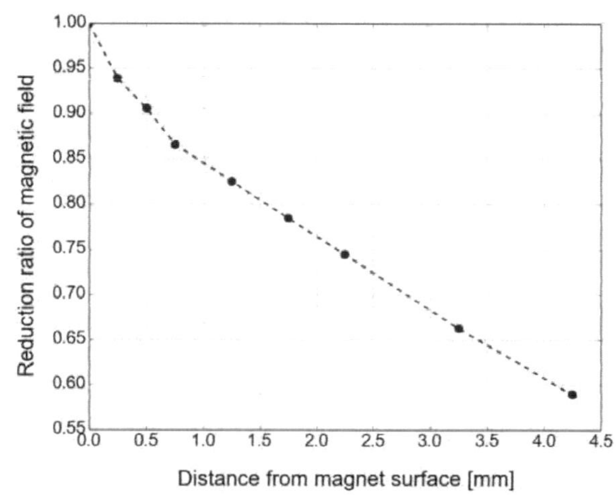

Fig. 9. The reduction ratio of the magnetic field with respect to the distance from the magnetic surface, obtained from the electromagnetic simulation.

cryogenic Hall sensor. Fig. 9 shows the reduction rate obtained by a static electromagnetic simulation [21].

There is no thermal contact with levitating rotor except for the wires of the resistive thermometer and heater. The rotor temperature is increased by the resistive heater by applying a certain voltage and inputting the Joule heat. In order to cool the rotor, the rotor is re-gripped by the cryogenic actuators. During this increasing and cooling cycle of the rotor temperature, we continue to monitor both output signals from the resistive thermometer and from the Hall sensor as the PM thermometer.

The measured magnetic field is translated to the temperature using the calibration curve considering the dropping distance. Then, we evaluate the PM thermometer by comparing with the resistive thermometer. Fig. 10 shows the obtained temperatures

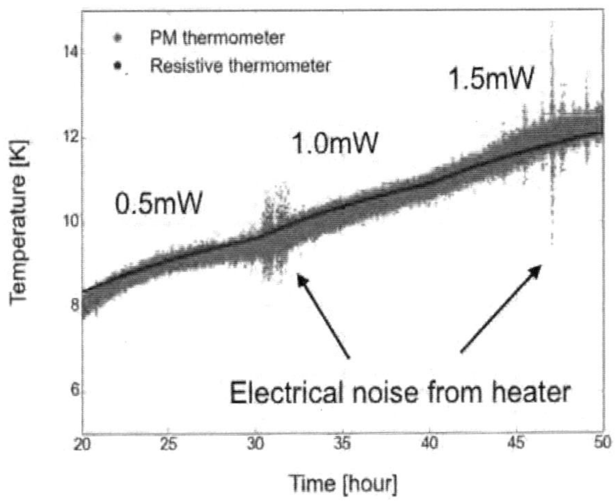

Fig. 9. The comparison of the temperature from the resistive thermometer (blue) and the PM thermometer (red). The levitating rotor is heated by the resistive heater with different input power of 0.5, 1.0 and 1.5 mW.

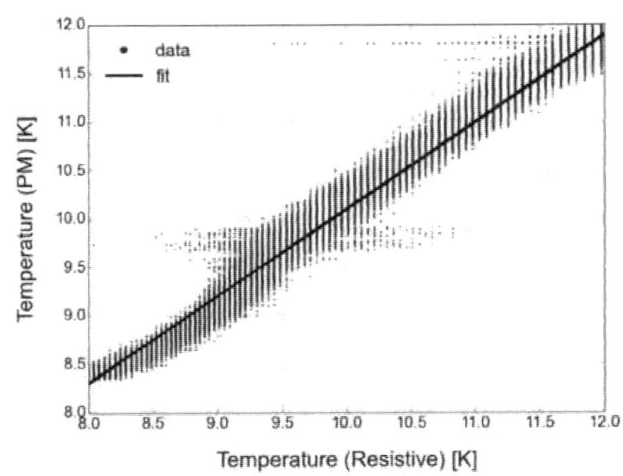

Fig. 10. The linearity between the resistive and the PM thermometer. The blue dot shows the obtained data and the solid line shows a linear fit of the data.

from the resistive thermometer and from the PM thermometer as a function of time. The levitating rotor is gradually heated by biasing the resistive heater with different input powers of 0.5, 1.0 and 1.5 mW. There is a good agreement between two thermometers. The spikes in the PM thermometer output are due to the electrical noise changing the bias voltage. The linearity between the resistive and the PM thermometer is shown in Fig. 11. The temperature accuracy of the PM thermometer in the range from 4 K to 20 K is ± 0.4 K, which is defined by the standard deviation of the temperature variation.

V. DISCUSSIONS

We demonstrated the performance of the PM thermometer using the SMB prototype. The temperatures from resistive thermometer and from the PM thermometer were consistent. The accuracy of the PM thermometer was calculated from the result. In this setup, the wires are connected to the resistive thermometer on the rotor. Thus, it is not possible to measure the rotor temperature during rotation. In order to adapt to the actual modulator, it is necessary to consider the effect of the magnetic field variation during rotation. There is a specific pattern in the magnetic field variation due to the segmented ring magnet in one rotation [18]. Thus, the average value of the variation can be used as the argument of the calibration function.

In addition, the displacement of the rotor magnet in vertical axis due to gravity effect has to be considered. Since the calibration curve can be normalized by the distance between the rotor magnet and the Hall sensor, the displacement can be treated as the vertical geometry factor in the calibration curve.

Therefore, the output of the PM thermometer, T, can be represented as

$$T = F\left(\frac{1}{n}\sum_{0\leq\theta\leq2\pi}^{n} B_\theta, G_Z\right), \tag{1}$$

where F is the calibration function as shown is Fig. 5. B_θ is the measured magnetic field at angle θ, and n is the number of measurement points. θ can be reconstructed by an optical encoder. G_Z is the geometry factor in vertical direction?

From the SMB performance point of view, the SmCo has an advantage to use as the rotor magnet compared to the NdFeB. In case of the NdFeB, the magnetic field pinned to the YBCO at the transition temperature of 94 K and the magnetic field of the rotor magnet below 10 K are about 10 % different. Thus, the SmCo which has less temperature dependence is suitable for the SMB. In this case, the rotor magnet is not able to use as the PM thermometer. However, it is sufficient for the PM thermometer to mount the small piece of the NdFeB magnet on top of the rotor, with installing the Hall sensor close to it. Since the measured magnetic fields are averaged as equation (1), the effect of the rotation is able to be ignored. The demonstration of this method will be followed up by future papers.

CONCLUSIONS

We proposed and developed the PM thermometer as the cryogenic remote sensing thermometer for the polarization modulator used in CMB polarization experiments. We measured the calibration curve between the magnetic field and the temperature using the small piece of the NdFeB magnet and the cryogenic Hall sensor. We confirmed that the characteristic dependence due to the spin reorientation transition of the NdFeB magnet. The calibrated PM thermometer is evaluated using the $\phi \sim 400$ mm SMB prototype comparing with the temperature from the resistive thermometer. We obtained the consistent result between the PM thermometer and the resistive thermometer. The temperature accuracy of the PM thermometer is $\pm$ 0.4 K, which is sufficient for the actual application. We discussed that how to actually adapt the PM thermometer to the SMB of the polarization modulator. The spinning effect and the distance relationship between the magnet and the Hall sensor can be treated by using the average and the geometry factors. Therefore, we successfully developed the cryogenic remote sensing thermometer which is applicable to the SMB of the polarization modulator.

ACKNOWLEDGMENT

This work was supported by MEXT KAKENHI Grant Numbers JP17H01125, JP17K14272, JP15H05441. This work was also supported by JSPS Core-to-Core Program, A. Advanced Research Networks and World Premier International Research Center Initiative (WPI), MEXT, Japan.

REFERENCES

[1] K. Sato, "First-order phase transition of a vacuum and the expansion of the Universe", Monthly Notices of Royal Astronomical Society, 195, 467, (1981).

[2] H. Guth Alan, "The Inflationary Universe: A Possible Solution to the Horizon and Flatness Problems", Phys. Rev. D 23, 347 (1981).

[3] J. H. P. Wua, J. Zuntz, M. E. Abroe, P. A. R. Ade, J. Bock, J. Borrill, J. Collins, S. Hanany, A. H. Jaffe, B. R. Johnson, T. Jones, A. T. Lee, T. Matsumura, B. Rabii, T. Renbarger, P. L. Richards, G. F. Smoot, R. Stompor, H. T. Tran, and C. D. Winant, "MAXIPOL: data analysis and results," The Astrophysical Journal 665, 55 (2007).

[4] Britt Reichborn-Kjennerud, Asad M. Aboobaker, Peter Ade, Franҫcois Aubin, Carlo Baccigalupi, Chaoyun Bao, Julian Borrill, Christopher Cantalupo, Daniel Chapman, Joy Didier, Matt Dobbs, Julien Grain, William Grainger, Shaul Hanany, Seth Hillbrand, Johannes Hubmayr, Andrew Jaffe, Bradley Johnson, Terry Jones, Theodore Kisner, Jeff Klein, Andrei Korotkov, Sam Leach, Adrian Lee, Lorne Levinson, Michele Limon, Kevin MacDermid, Tomotake Matsumura, Xiaofan Meng, Amber Miller, Michael Milligan, Enzo Pascale, Daniel Polsgrove, Nicolas Ponthieu, Kate Raach, Ilan Sagiv, Graeme Smecher, Federico Stivoli, Radek Stompor, Huan Tran, Matthieu Tristram, Gregory S. Tucker, Yury Vinokurov, Amit Yadav, Matias Zaldarriaga, Kyle Zilic, "EBEX: a balloon-borne CMB polarization experiment," Proc. SPIE 7741, 77411C (2010)

[5] T. E. Montroy, P. A. R. Ade, R. Bihary, J. J. Bock, J. R. Bond, J. Brevick, C. R. Contaldi, B. P. Crill, A. Crites, O. Doré, L. Duband, S. R. Golwala, M. Halpern, G. Hilton, W. Holmes, V. V. Hristov, K. \ Irwin, W. C. Jones, C. L. Kuo, A. E. Lange, C. J. MacTavish, P. Mason, J. Mulder, C. B. Netterfield, E. Pascale, J. E. Ruhl, A. Trangsrud, C. Tucker, A. Turner, M. Viero, "SPIDER: a new balloon-borne experiment to measure CMB polarization on large angular scales," Proc. SPIE 6267, 62670R (2006).

[6] A. Kusaka, T. Essinger-Hileman, J. W. Appel, P. Gallardo, K. D. Irwin, N. Jarosik, M. R. Nolta, L. A. Page, L. P. Parker, S. Raghunathan, J. L. Sievers, S. M. Simon, S. T. Staggs, K. Visnjic, "Modulation of cosmic microwave background polarization with a warm rapidly rotating half-wave plate on the Atacama B-Mode Search instrument," Review of Scientific Instruments 85, 024501 (2014)

[7] Charles A. Hill, Shawn Beckman, Yuji Chinone, Neil Goeckner-Wald, Masashi Hazumi, Brian Keating, Akito Kusaka, Adrian T. Lee, Frederick Matsuda, Richard Plambeck, Aritoki Suzuki, Satoru Takakura, "Design and development of an ambient-temperature continuously-rotating achromatic half-wave plate for CMB polarization modulation on the POLARBEAR-2 experiment", Proc.SPIE Int.Soc.Opt.Eng. 9914 (2016) 99142U (2016)

[8] J. Hull, "Topical review: Superconducting bearings, Superconductor Science", 110, 140, Technology 13 (1).

[9] Jeff Klein, Asad Aboobaker, Peter Ade, François Aubin, Carlo Baccigalupi, Chaoyun Bao, Julian Borrill, Daniel Chapman, Joy Didier, Matt Dobbs, Benjamin Gold, Will Grainger, Shaul Hanany, Johannes Hubmayr, Seth Hillbrand, Julien Grain, Andrew Jaffe, Bradley Johnson, Terry Jones, Theodore Kisner, Andrei Korotkov, Sam Leach, Adrian Lee, Lorne Levinson, Michele Limon, Kevin MacDermid, Tomotake Matsumura, Amber Miller, Michael Milligan, Enzo Pascale, Daniel Polsgrove, Nicolas Ponthieu, Kate Raach, Britt Reichborn-Kjennerud, Ilan Sagiv, Radek Stompor, Huan Tran, Matthieu Tristram, Gregory S. Tucker, Amit Yadav, Matias Zaldarriaga, Kyle Zilic, "A cryogenic half-wave plate polarimeter using a superconducting magnetic bearing", Proceedings of the SPIE, Cryogenic Optical Systems and Instruments XIII, Volume 8150, 815004 (2011).

[10] Shinetsu Magnet Inc., Japan. [Online]. Available: www.shinetsu-rare-earth-magnet.jp/e/ . Accessed on: Jun. 1, 2018.

[11] Lakeshore, USA. [Online] Available: www.lakeshore.com/products/Hall-Magnetic-Sensors/Pages/Specifications.aspx . Accessed on: Jun. 1, 2018

[12] Lakeshore,USA.[Online]Available: www.lakeshore.com/products/Cryogenic-Temperature-Sensors/Cernox/Models/Pages/Specifications.aspx. Accessed on: Jun. 1, 2018.

[13] T. Matsumura, H. Kataza, S. Utsunomiya, R. Yamamoto, M. Hazumi, N . Katayama, "Design and Performance of a Prototype Polarization Modul ator Rotational System for Use in Space Using a Superconducting Magn etic Bearing", IEEE Transactions on Applied Superconductivity, Volum e: 26, Issue 3, (2016).

[14] Y. Sakurai, T. Matsumura, H. Kataza, S. Utsunomiya, R. Yamamoto, "E stimation of the heat dissipation and the rotor temperature of supercondu cting magnetic bearing below 10 K", IEEE Transactions on Applied Sup erconductivity, Vol. 27, Iss. 4 (2017).

[15] Y. Sakurai, T. Matsumura, H. Sugai, N. Katayama, H. Ohsaki, Y. Terao, Y. Terachi, H. Kataza, S. Utsunomiya, and R. Yamamoto, "Vibrational c haracteristics of a superconducting magnetic bearing employed for a prot otype polarization modulator", J. Phys.: Conf. Ser. 871, 012091 (2017).

[16] Adelwitz Technologiezentrum GmbH (ATZ), Germany. [Online]. Avail able: www.atz-gmbh.com. Accessed on: Jun. 1, 2018.

[17] J. F. Herbst. R2Fe14B materials: Intrinsic properties and technological a spects. Reviews of Modern Physics, 63(4):819–898, 1991.

[18] J. F. Herbst, J. J. Croat, F. E. Pinkerton, and W. B. Yelon. Relationships between crystal structure and magnetic properties in Nd2Fe14B. Physica l Review B, 29(7):4176–4178, 1984.

[19] D. Givord, H.S. Li, and J.M. Moreau. Magnetic properties and crystal str ucture of Nd2Fe14B. Solid State Communications, 50(6):497–499, 1984 .

[20] Stanley R. Trout, Using Permanent Magnets at Low Temperatures. TN 0 302 p.1 June 2003.

[21] Y. Terachi, H. Ohsaki, Y. Terao, Y. Sakurai, T. Matsumura, H. Sugai, S. Utsunomiya, H. Kataza and R. Yamamoto: "Numerical Analysis of Fun damental Characteristics of Superconducting Magnetic Bearings for a Po larization Modulator," IOP Publishing, Journal of Physics Conference Se ries, Vol. 871. No. 1, p.012094, 2017.

A Programmable Waveguide Calibration Load

Jacob W. Kooi[1*], Rodrigo A. Reeves[2], Arthur W. Lichtenberger[3], Theodore J. Reck[1], Andy K. Fung[1], Sander Weinreb[4], James. W. Lamb[4], Rohit S. Gawande[1], Kieran. A. Cleary[4], and Goutam Chattopadhyay[1].

[1]*NASA Jet Propulsion Laboratory, Pasadena, CA 91109, USA*
[2]*CePIA, Astronomy Department, Universidad de Concepcion, Chile.*
[3]*University of Virginia, Charlottesville, VA 22904-4743, USA*
[4]*California Institute of Technology, Pasadena, CA 91125, USA*
[*]Contact: Jacob.W.Kooi@jpl.nasa.gov

Abstract—

We have developed a programmable, accurate, and miniaturized calibration load for use in millimeter and submillimeter low-noise amplifier characterization. The proposed solution uses a thermally conductive vane attenuator with a small thermal mass, integrated heater, and silicon thermometer. In the present design we utilize a 125 μm thick z-cut crystal quartz vane due to its relatively low dielectric constant, high cryogenic thermal conductivity, chemical robustness, and small thermal contraction. To provide adequate attenuation, the bottom side of the quartz fabrication wafer has an nm thick resistive Ti metal layer deposited.

On the top of the quartz wafer a pattern of Au is deposited to allow adhesion of the heater resistor, thermometer, and internal heat strap. The z-cut quartz vane is mounted on three low thermally conductive Torlon posts, centered on the maximum E-field, and positioned across the waveguide.

With this approach the quartz vane, protruding all the way into the waveguide, approximates a blackbody with a physical temperature T. When operated in a PID loop, these properties combine to facilitate a programmable calibration load with a switching speed of < 10 s between 25 - 75 K.

Measurement verification, including linearity, will be provided. In addition, it will be shown that the W-band design operates overmoded to ~230 GHz at which point the TE30 mode begins to effectively bypass the vane attenuator. The concept is in principle scalable to terahertz frequencies.

Characterization of low-loss reflectors for spaceborne microwave radiometers

Axel Murk[*1], Karl Jacob[1] , David Marote[2]

[1]University of Bern, Institute of Applied Physics, Sidlerstr. 5 Bern, 3012, Switzerland
[1]Airbus Defence and Space SAU (ASE), Avenida de Aragon 404, 28022 Madrid, Spain
[*]Contact: axel.murk@iap.unibe.ch

Abstract— Every reflector of a microwave radiometer will introduce a certain loss which affects the sensitivity and radiometric accuracy of the instrument. These losses depend on material properties of the reflector, the frequency, as well as on the incidence angle and polarization. An accurate measurement of the losses is required in order to select the optimum reflector coating and to correct the losses in the calibration process.

We characterized different reflector samples of the Ice Cloud Imager (ICI, 183-670 GHz) and the Microwave Sounder (MWS, 23-230 GHz) instruments of the Meteorological Operational Satellite - Second Generation (MetOP-SG). The samples were provided by Airbus Defence and Space Madrid and Friedrichshafen. One purpose of these tests was to determine the influence of the coating materials and processes, of the surface roughness, and of aging due to long term storage which was simulated by accelerated lifetime tests in high humidity and high temperature. The second purpose was to provide a lookup table of the resulting reflector losses depending on the frequency bands, incidence angles and variable polarizations during the ICI and MWS operation. These values, and their associated uncertainty, are needed for the radiometric error model of the two instruments.

It is not trivial to determine the reflector losses with the required measurement accuracy of about 0.1%. For the lowest frequency of the MWS instrument at 23 GHz we used a custom built circular waveguide cavity. For all higher frequency bands up to 640 GHz we used a free space cavity in which the sample was inserted either at normal incidence or at an incidence angle of 45° with either TE or TM polarization. The resulting resonance lines of the cavity were measured with a submm vector network analyzer. The quality factor Q of these resonances is a measured of the total cavity losses. It was determined by a simultaneous least squares fit of the amplitude and phase of the complex valued measurements. In the data analysis all additional loss factors need to be considered, including atmospheric losses, coupling losses and truncation losses. In the end we compare the measurement results with the different frequencies and incidence cases with an analytical model based on ideal reflectors with finite a conductivity.

The talk will give an introduction to the theory of reflector losses and the used resonators. After an overview of the test results it will discuss the resulting calibration bias of the MetOP-SG instruments.

Flexible, Thermal Isolating and Low-loss Rectangular Dielectric Waveguide for THz Superconducting Receivers

Hao-Tian Zhu[1,2,4,5*], Dong Liu[2], Jie Hu[2], Sheng Li[2], Sheng-Cai Shi[2] & Quan Xue[1,3,4]

[1]*State Key Laboratory of Millimeter Waves, City University of Hong Kong, Hong Kong SAR, China.*
[2]*Purple Mountain Observatory, Key Laboratory of Radio Astronomy, Chinese Academy of Sciences, Nanjing, 210008, China.*
[3]*School of Electronic and Information Engineering, South China University of Technology, Guangzhou, 510640, China.*
[4]*Shenzhen Key Lab of Millimeter-Wave and Wideband Wireless Communications, Shenzhen Research Institute, City University of Hong Kong, Shenzhen, 518057, China.*
[5]*Poly-Grames, Ecole Polytechnique de Montreal, Montreal, QC, H3T 1J4, Canada.*
[*]Contact: zhuhaotian_2007@126.com

Abstract—The dielectric waveguide is a potential transmission line in Terahertz (THz) applications for its low-loss. A new application of the rectangular dielectric waveguide (RDW) for a THz superconducting receiver system for radio astronomy is investigated and demonstrated for the first time. The RDW can propagate the solid THz source, functioning as pumping signal (local oscillator), to superconductor–insulator–superconductor (SIS) mixer and prevent heat into the cryostat. A series of the RDWs with different lengths were fabricated and measured in cryostat and room temperature, respectively. According to the measured results at 3.4 K environment, the attenuation constant of the RDW is 0.034±0.012 dB/mm, operating in 240-300 GHz. For a comparison, the measured attenuation constant of the RDW in room temperature is 0.069±0.020 dB/mm. An E-bend and a twist based on a 118-mm RDW were also measured in both 3.4 K and 300 K environments. The measured DSB system noise temperature is 99.9±19.1 K over 240-300 GHz. Comparing to traditional pumping signal feeding schemes, this approach well outperforms the traditional ones in feeding flexibility, insertion loss, and thermal isolation for cryostat.

The flexible, thermal isolated and low-loss RDW can be widely applied for the single beam SIS mixer superconducting receivers and THz vector filed measurement instrumentations. Moreover, the RDW can be widely used in the multi-beam superconducting receivers for radio astronomy.

THz Suborbital Payload Optical System Pointing Stability for Target Studies of Compact Sources

Kristina Davis*[1], Chris Groppi[1], Imran Mehdi[2], Jonathan Kawamura[2], José Siles-Perez[2], Paul Scowen[1], Phil Mauskopf[1]

[1]Arizona State University, Tempe, AZ 85287, USA
[2]Jet Propulsion Laboratory, California Institute of Technology, Pasadena, CA 91109, USA
*Contact: kdavis32@asu.edu

Abstract—Previous suborbital missions to observe line emission surveys in the terahertz (THz) frequency spectrum have focused on large survey mapping observing modes or all-sky imaging. These observing modes require highly accurate pointing knowledge, but require only a modest degree of pointing stability. In contrast, target surveys of compact sources, for instance for future missions to observe planetary atmospheres, proto-planetary disks, etc. require both pointing accuracy and stability. Even for balloon-borne survey missions, pointing stability is a concern especially during the commissioning phase of a mission, where problems with pointing stability can stall the process and result in a loss of valuable observing time. Sub-arcsecond pointing stability through motion control of optical subsystem components has been developed outside the THz frequency regime, as in [1] and [2]. However, requirements at those frequencies make the pointing control system cost prohibitive due to the requirement of either cryogenic actuators or large mass and power consuming actuators to independently move the instrument cryostat. For line emission surveys using hot-electron bolometer (HEB) receiver instruments, detector noise in the 1-5 Thz regime is higher than thermal emission at these frequencies, and so the mission requirements do not necessitate cold optics. It is therefore possible to design a low cost and complexity pointing stability system using a simple tip/tilt mirror in a control loop with commercial off-the-shelf gyroscopes, such as those demonstrated on STO-2 or BRRISON/SSIPP [3]. We present the design constraints for a pointing system for the case of a 2.5 m class observatory set to observe the 2.7 THz emission from hydrogen-deuteride (HD) in nearby protoplanetary disks.

[1] L. Jones-Wilson et al., "A sub-arcsecond pointing stability fine stage for a high-altitude balloon platform," 2017 IEEE Aerospace Conference, Big Sky, MT, 2017, pp. 1-15. doi: 10.1109/AERO.2017.7943590
[2] Miller, P. Scowen, T. Veach, "Focal plane actuation by hexapod for the development of a high resolution suborbital telescope," Proc. SPIE, vol. 9912, no. 99126B, July 22, 2016. doi:10.1117/12.2230834
[3] J. Diller, C. DeForest, G. Laurant, J. Brownsberger, "Solar Science from Manned Suborbital Vehicles-the SwRI Solar Instrument Pointing Platform", IEEE Aerospace Conference, 2015. doi: 10.1109/AERO.2015.7119126

Tunable and Reconfigurable Terahertz Devices Based On Photo-Induced Electromagnetic Band Gap Structures

Jun Ren[1]*, Yijing Deng[1], Zhenguo Jiang[1], Md. Itrat Bin Shams[1], Patrick Fay[1], and Lei Liu[1]
[1]University of Notre Dame, Notre Dame, 46556, USA
*Contact: jren1@nd.edu

***Abstract*—In this paper, we report the design and simulation of two optically controlled tunable and reconfigurable G-band (140-220 GHz) waveguide devices based on electromagnetic band gap (EBG) structures. The first device based on a pre-patterned EBG structure can be reconfigured between a band-stop filter (BSF) and a transmission line. The second device based on photo-induced EBG structures using a mesa-array shows increased level of tunability and reconfigurability with a BSF center frequency tunable from 175-200 GHz. The proposed tunable and reconfigurable THz devices are promising for a wide range of applications including multiple-frequency-band THz wireless communications.**

I. Introduction

The realization of tunable and reconfigurable terahertz (THz) devices is essential to the implementation of advanced THz imaging and adaptive THz wireless communication systems, but is challenging using conventional approaches such as employing metamaterials, liquid-crystal-based devices, graphene modulator arrays, thermal sensitive elements, etc. [1, 2]. Although electromagnetic band gap (EBG) structures have been regarded as one promising solution for achieving dynamical control of broadband electromagnetic waves for a wide range of applications, existing EBG-based devices suffer from several key drawbacks including the requirement of complex fabrication processes [3] and limited tunability/reconfigurability [4], and therefore their usage as building blocks for high-performance tunable and reconfigurable THz devices is limited at present.

Recently, our group proposed and developed a novel optical control methodology for realizing tunable and reconfigurable THz devices [5-8]. The photogenerated free carriers in semiconductors were used to spatially modulate incident THz waves for implementing various functionalities. A number of demonstrations, including reconfigurable coded-aperture imaging masks [5], dynamically beam steering and forming antennas [6], and optically controlled variable waveguide attenuators/modulators [7, 8], have been reported. Besides, such an optical tuning methodology has been applied into the design of photo-induced EBG component at microwave frequencies [9]. An optically controlled tunable and reconfigurable microwave frequency filter based on the photo-induced EBG components has been studied and demonstrated, whose functionality can be reconfigured between a band-stop filter (BSF) and a transmission line, and its center frequency can be tuned from 8-12 GHz by changing period of the virtual patterns illuminated onto the device [9].

In this paper, we further extend our previous work and apply the optical control methodology into the design of photo-induced EBG components for operation at THz frequencies. We will present two optically controlled G-band (140-220 GHz) waveguide components based on EBG structures for implementing tunable and reconfigurable THz filters. The first design features a pre-patterned EBG ground, and its functionality can be reconfigured between a filter with a center frequency at around 180 GHz and a transmission line by changing the illumination condition of the device. The second design employs a photo-induced EBG ground implemented on a Ge mesa-array structure for achieving a higher level of tunability and reconfigurability. Simulations show that the center frequency of the filter can be tuned from 175-200 GHz by projecting virtual EBG patterns with different periods onto the Ge mesa-array structure. The demonstrated tunable and reconfigurable THz devices are promising for applications in a range of THz systems including high-speed adaptive wireless THz communication systems.

II. Reconfigurable G-Band Waveguide Filter Based On Pre-Patterned EBG

A. Design

Figure 1 shows the structure of the optically controlled reconfigurable G-band waveguide filter based on pre-patterned EBG. It was implemented in a back-to-back waveguide configuration, and consists of two offset input and output G-band waveguide channels connected by an E-plane microstrip chip. The microstrip chip is placed in the microstrip channel, whose dimension has been optimized to achieve single mode propagation over the full waveguide frequency band. The microstrip chip consists of a 5-um-thick Ge thin film on the backside of a 120-um-thick quartz substrate. A ~6 mm long section of transmission line with waveguide probes at both ends was patterned on the topside of the quartz. The waveguide probes adopted a rectangular microstrip design, similar to the one presented in [10]. A uniplanar EBG ground plane with a one-dimensional circular hole array was patterned on the backside of the Ge. The period of the hole array and the diameter of each hole are 620 μm and 100 μm, respectively, to generate a distinctive stop-band response at around 180 GHz. For both the top conductor and the ground plane, the thickness of the gold is 2 μm. An opening was designed at the bottom of the waveguide block so that the backside of the microstrip chip can be illuminated for realizing tunability and reconfigurability. In experiment, 808-nm infrared laser diodes are employed for illumination.

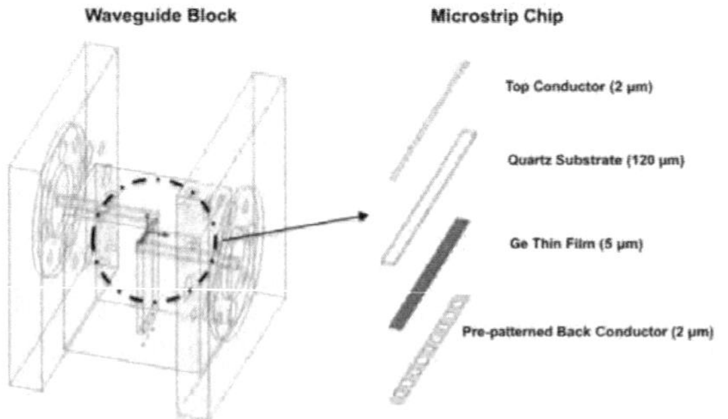

Fig. 1. Schematic drawing of the optically controlled reconfigurable G-band waveguide filter based on pre-patterned EBG. Detailed structure of the microstrip chip is shown.

B. Simulation

Full-wave HFSS simulation has been performed on the proposed waveguide filter above in the frequency range from 140-220 GHz. With no illumination, the simulation shows that the device exhibits a band-stop filter (BSF) behavior with a center frequency at around 180 GHz (see Fig. 2 (a)) as expected. The insertion loss in the pass-band is below 2 dB, and the stop-band rejection is larger than 12 dB. Fig. 2 (b) shows the simulated device response when the backside of the microstrip chip is illuminated with a light intensity of 34 W/cm^2. On the basis of the physics-based model we developed [11], such a level of incident light intensity results in a spatially averaged photoconductivity of 3×10^5 S/m in the illuminated Ge thin film with 5 μm thickness (assuming a conservative value of 40 μs for the effective carrier lifetime [8]). As a result, the BSF response disappears, and the device shows a transmission line response with an average of 1.5 dB insertion loss and better than 15 dB return loss over the full waveguide frequency band. Such a level of light intensity can be easily provided in experiment by illuminating the ~6 mm long EBG pattern area using our high power infrared laser diodes. These results indicate that the proposed G-band waveguide filter with pre-patterned EBG structure can be used to implement reconfigurable functionalities with high performance.

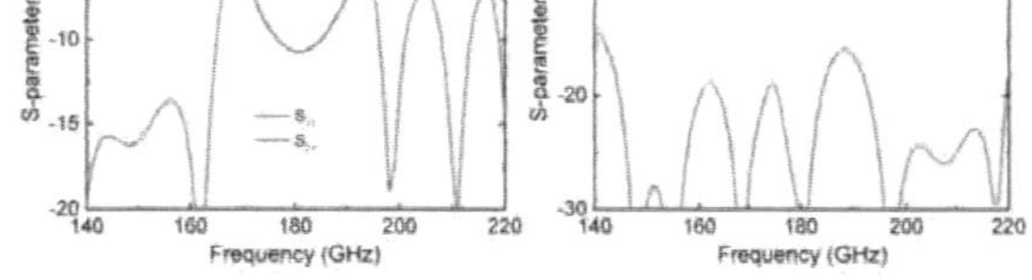

Fig. 2. Simulated frequency response of the optically controlled reconfigurable G-band waveguide filter based on pre-patterned EBG (a) without illumination, and (b) under 34 W/cm^2 illumination.

III. Tunable And Reconfigurable G-Band Waveguide Filter Based On Photo-Induced EBG

A. Design

Although the functionality of the waveguide device described in Section II can be reconfigured between a BSF and a transmission line, no tunability can be realized due to the employment of the pre-patterned gold layer. In order to demonstrate the full potential of the proposed optical control methodology in implementing tunable and reconfigurable THz devices, another G-band waveguide component based on photo-induced EBG structures was designed and simulated. Figure 3 illustrates the waveguide structure of the component. As shown in Fig. 3 (a), the component is implemented in the same back-to-back waveguide block as the first component described in Section II. Different from the first design, a 5-μm-thick Ge thin film without any pre-patterned gold layer was employed as the ground plane of the microstrip chip. By projecting required photo-patterns onto the Ge ground plane using a DMD chip unit in conjunction with laser diodes, EBG structures were virtually generated. For achieving circuit operation at THz frequencies, the required feature size is even smaller than the diffusion length of the Ge (~0.5 mm), which prevents reliable transfer of the photo-patterns from the DMD chip unit to the Ge. To overcome the limitation of achievable spatial resolution, a mesa-array structure was created on the Ge thin film. Detailed study on the mesa-array structures has been reported previously [9][11]. In this work, a simplified version of the mesa-array structure with trenches only cut along the transverse direction was employed. The trench size and mesa size were designed to be 0.1 μm and 25 μm, respectively, to ensure that the mesa-array structure will not degrade the circuit performance while providing sufficiently high resolution. Rectangular (rather than circular) EBG patterns were projected onto the Ge mesa-array to create photo-induced EBG structures. The dimension of the photo-patterns was chosen to be integer multiples of the mesa size so that optical alignment can be easily achieved in experiment.

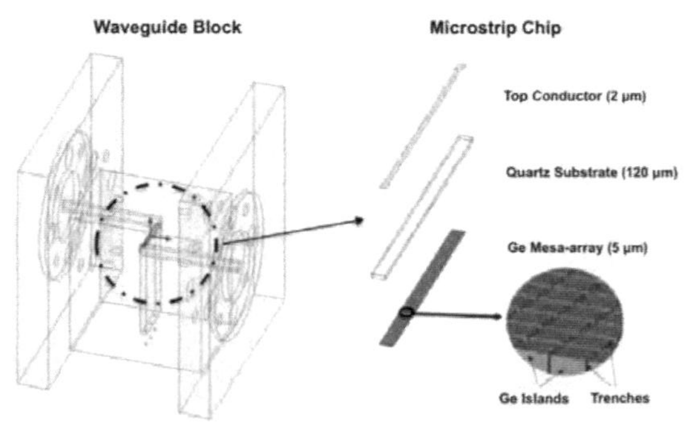

Fig. 3. Schematic drawing of the optically controlled reconfigurable G-band waveguide filter based on photo-induced EBG. Detailed structure of the microstrip chip with a mesa-array is shown.

B. Simulation

Figure 4 (a) shows the simulated S-parameters of the device from 140-220 GHz when a 34 W/cm^2 incident light is uniformly illuminating the mesa-array structure (as shown in Fig. 4 (b)). A spatially averaged photoconductivity of 3×10^5 S/m was used in the simulation for the illuminated Ge mesa-array with 5 μm thickness. It is seen that the device shows a transmission line response with ~2 dB insertion loss and better than 10 dB return loss over the full waveguide frequency band.

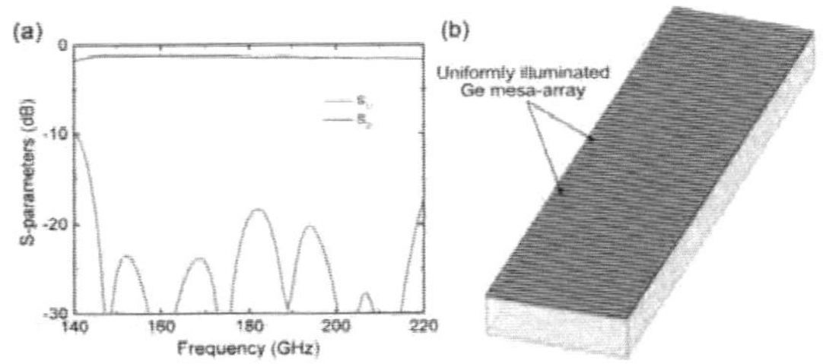

Fig. 4. (a) Simulated frequency response of the optically controlled reconfigurable G-band waveguide device (b) when the mesa-array structure is illuminated with 34 W/cm^2 uniform light.

When the mesa-array structure is illuminated with virtual EBG patterns, the device will be reconfigured to BSFs. This is demonstrated in Fig. 5. It is seen that when a virtual EBG pattern with a period of 625 μm and an un-illuminated area length of 275 μm is imposed onto the mesa-array, a BSF response with a center frequency at 175 GHz was shown. Since the center frequency of the BSF is inversely related to the period of the EBG pattern, it can also be tuned by imposing virtual EBG patterns with different periods. When the period of the photo-induced EBG patterns was varied from 625 μm to 525 μm, and the length of the un-illuminated pattern was changed from 275 μm to 250 μm, another BSF response with a center frequency at 200 GHz was observed. The stop-band rejection exceeds 10 dB in both cases, and the return loss in the stop-band is less than 3 dB. These results indicate that the waveguide component can not only achieve reconfigurable performance between a BSF and a transmission line, but also realize broadband tuning of the BSF center frequency in the THz domain with the much-improved spatial resolution enabled by the mesa-array structure. In addition, it should be noted that by employing a mesa-array with even smaller mesa size, even higher spatial resolution can be achieved, enabling continuous tuning of the center frequency.

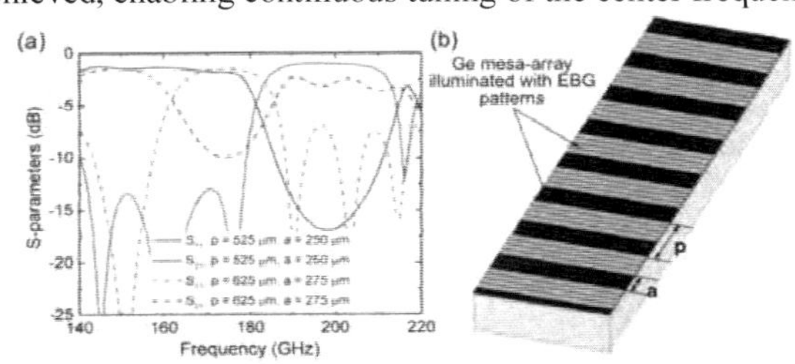

Fig. 5. (a) Simulated frequency response of the optically controlled reconfigurable G-band waveguide filter based on photo-induced EBG in response to (b) EBG patterns with various periods.

On the basis of the above study, it has shown that the proposed optically controlled waveguide component based on photo-induced EBG can achieve high tunability and reconfigurability, and is promising for the implementation of high-performance dynamically tunable and reconfigurable filters operating at THz frequencies.

CONCLUSIONS

Two optically controlled tunable and reconfigurable G-band waveguide devices based on EBG structures have been designed and simulated. The first device based on a pre-patterned EBG structure shows reconfigurable performance between a BSF and a transmission line. The second device based on photo-induced EBG structures presents increased level of tunability and reconfigurability by employing a mesa-array structured ground. A broad tuning range of BSF center frequency from 175-200 GHz has been demonstrated in simulations. The proposed tunable and reconfigurable THz devices based on photo-induced EBG are promising for a wide range of applications including multiple-frequency-band THz wireless communications.

ACKNOWLEDGMENT

This work is partially supported by the National Science Foundation (NSF) under the grant of ECCS-1711631 and ECCS-1711052, and a subcontract from the Harvard-Smithsonian Center for Astrophysics under grant PTX-Smithsonian 17-SUBC-400SV787007. The authors would like to thank the support from Center for Nano Science and Technology (NDnano) and Advanced Diagnostics & Therapeutics (AD&T) at the University of Notre Dame.

REFERENCES

[1] M. Tonouchi, "Cutting-edge terahertz technology", Nature photonics, 1(2), pp. 97-105, 2007.

[2] K. C. Huang, and Z. Wang, "Terahertz terabit wireless communication", IEEE Microwave Magazine, 12(4), pp. 108-116, 2011.

[3] A. Delustrac, F. Gadot, E. Akmansoy, and T. Brillat, "High-directivity planar antenna using controllable photonic bandgap material at microwave frequencies," APhys. Lett., Vol. 78, 4196, 2002.

[4] L. Mercier, M. Thevenot, P. Blonby, and B. Jecko, "Design and characterization of a smart periodic material including MEMS," Proc. 27th ESA Antenna Technology Workshop on Innovative Periodic Antennas: Elecromagnetic Bandgap, Left-Handed Materials, Fractal and Frequency Selective Surfaces, Santigao de Compostela, Spain, March 2004.

[5] M. I. B. Shams, Z. Jiang, S. Rahman, J. Qayyum, L.-J. Cheng, H. G. Xing, P. Fay, L. Liu, "Approaching real-time terahertz imaging with photo-induced coded apertures and compressed sensing," *IET Electron. Lett.*, vol. 50, no. 11, pp. 801-803, 2014.

[6] M. I. B. Shams, Z. Jiang, J. Qayyum, S. Rahman, P. Fay, and L. Liu, "A Terahertz reconfigurable photo-induced fresnel-zone-plate antenna for dynamic two-dimensional beam steering and forming," International Microwave Symposium, 1–4, Phoenix, Arizona, 2015.

[7] Z. Jiang et al., "Investigation and demonstration of a WR-4.3 optically controlled waveguide attenuator," *IEEE Trans. THz Sci.& Tech.*, vol. 7, no. 1, pp. 20-26, 2017.

[8] J. Ren et al., "High-performance WR-4.3 optically controlled variable attenuator with 60-dB range," *IEEE Microw. Wire. Comp. Lett.*, vol. 28, no. 6, pp. 512-514, 2018.

[9] J. Ren et al., "Photo-induced electromagnetic band gap structures for optically tunable microwave filters", *Progress in Electromagnetics Research*, vol. 161, no. 1, pp. 101-111, 2018.

[10] J. L. Hesler, W. R. Hall, T. W. Crowe, R. M. Weikle, B. S. Deaver, R. F. Bradley, Djerafi, and S. K. Pan, "Fixed-tuned submillimeter wavelength

waveguide mixers using planar schottky-barrier diodes", IEEE MTT, vol. 45, pp. 653–658, 1997.

[11] A. Kannegulla, M. I. B. Shams, L. Liu, and L. -J. Cheng, "Photo-induced spatial modulation of THz waves: Opportunities and limitations," *Opt. Exp.*, vol. 23, no. 25, pp. 32098– 32112, 2015.

High Performance Photo-Induced Substrate-Integrated Waveguide for Tunable and Reconfigurable THz Circuits

Y. J. Deng*, M. I. B. Shams, J. Ren, P. Fay, and L. Liu
Department of Electrical Engineering, University of Notre Dame, Notre Dame, IN 46556, USA
*Contact: ydeng2@nd.edu

***Abstract*— We present a novel approach for realizing tunable/reconfigurable THz circuits using photo-induced substrate-integrated waveguide (PI-SIW) architectures. In this approach, fixed metallic vias in conventional SIW are replaced by photo-induced conductive plasma sidewalls. Full-wave HFSS simulation has shown an insertion loss of 4.15 dB/mm at 280 GHz with photo-induced sidewalls formed by 80 W/cm^2 light intensity (550 nm wavelength) on high resistivity silicon (HRS) wafer. To further improve the PI-SIW performance, a pillar-array structure is proposed to suppress carrier diffusion while same time increasing the achievable photoconductivity. With such pillar-array structures, the insertion loss at 280 GHz has been reduced to as low as 0.96 dB/mm, and a value of 0.81 dB/mm can be potentially achieved for a 300 W/cm^2 light intensity.**

I. Introduction

Recently, electromagnetic waves in the THz region have attracted increased interests owing to the promising applications in radio astronomy, medical imaging, security screening, high-speed communication and defense. Tunable and reconfigurable THz circuits have become highly desired in advanced THz sensing, imaging and adaptive wireless communication systems. A variety of approaches such as mechanical tuning [1], thermal agitation [2], electrical tuning using Schottky or PIN diodes [3-4], and metamaterial approaches [5] have been reported to demonstrate THz tunable/reconfigurable components. However, most of those approaches rely on prepatterned circuits, resulting in limited tunability and versatility. In addition, more advanced THz tunable circuits such as phase shifters, delay lines, high-level switches (e.g., SPDT, DPDT) are still quite challenging to realize.

In this paper, we report an alternative approach for realizing the above tunable/reconfigurable THz circuits using photo-induced substrate-integrated waveguide (PI-SIW) architectures. On the basis of the optical THz spatial modulation (OTSM) technology reported in our previous work [6, 7], fixed metallic vias in conventional SIW structures can be replaced with conductive plasma SIW sidewalls by projecting programmable light patterns on semiconductors. For a prototype demonstration, a PI-SIW structure with a height of 100 um and width of 285 um was designed using HRS to achieve the fundamental mode cut-off frequency of 150 GHz. Commercial FEM solver ANSYS-HFSS was utilized for circuit simulation. Simulated results showed an insertion loss of 4.15 dB/mm at 280 GHz with photo-induced sidewalls formed by 80 W/cm^2 light intensity (550 nm wavelength). This high insertion loss is primarily attributed to relatively low photoconductivity achieved and inclined sidewalls due to lateral carrier diffusion. To further improve the PI-SIW performance, a pillar-array structure was applied to suppress carrier lateral diffusion (for straight sidewalls) while same time increasing the achievable photoconductivity. With the proposed pillar-array structure, the insertion loss of a PI-SIW at 280 GHz can be reduced to as low as 0.96 dB/mm, and a value of 0.81 dB/mm can be potentially achieved for a 300 W/cm^2 light intensity. The PI-SIW structures proposed and reported in this paper have low insertion loss and strong reconfigurability, making them promising for achieving tunable/reconfigurable circuits for advanced THz sensing, imaging and communication.

II. PI-SIW Architectures

A Demonstration of PI-SIW Structures

In the PI-SIW structures, fixed metallic vias in conventional SIWs are replaced with conductive plasma SIW sidewalls by projecting programmable light patterns on semiconductors, as is illustrated in Fig. 1. The vertical carrier concentration profile and photoconductivity of silicon wafer illuminated by continuous waves can be calculated from the well-known expressions in [7]. A thin layer of indium tin oxide (ITO) was used to replace the top conductor in a conventional SIW for fulfilling such regional illumination using a combination of a DMD chip and a laser source.

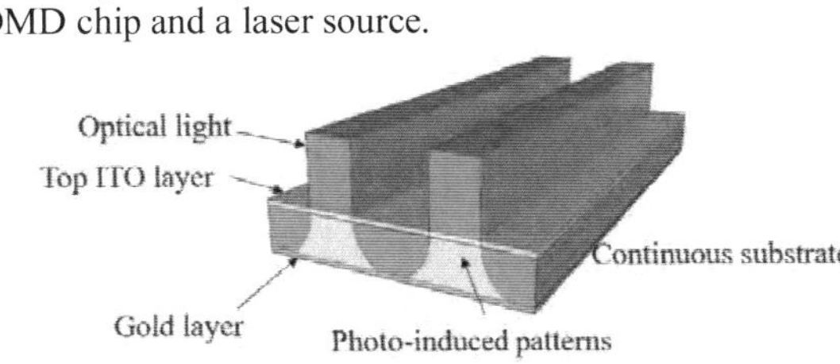

Fig. 1. PI-SIW on continuous HRS substrate

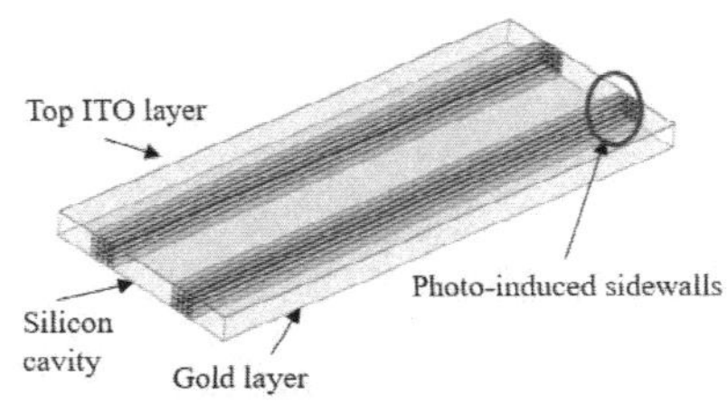

Fig. 2. PI-SIW structure on continuous HRS substrate in HFSS

For a prototype demonstration, 80 W/cm^2 light intensity is used to illuminate on continuous HRS wafer [8]. As shown in Fig. 2, PI-SIW with a height of 100 μm, width of 285 μm and length of 2 mm was designed using full-wave HFSS.

Nevertheless, optical light illuminating on continuous HRS substrate will cause inclined sidewalls due to lateral carrier diffusion as illustrated in Fig. 3. Consequently, such PI region displays relatively low and inconsistent carrier concentration. Being able to inhibit lateral diffusion of carriers will improve the spatial resolution of the photopatterns, while at the same time increase the photo-induced carrier concentration and maintain high modulation speed. Based on this purpose, pillar-array structure, a matrix of isolated islands each with lateral dimension well below the THz wavelength (at the operating frequency), is introduced. Photo-induced substrate with pillar-array structure in Fig. 3 constrains photogenerated carriers within every single pillar and reduces lateral carrier diffusion. A much higher free carrier concentration as well as photoconductivity can thus be achieved in PI regions with pillar arrays.

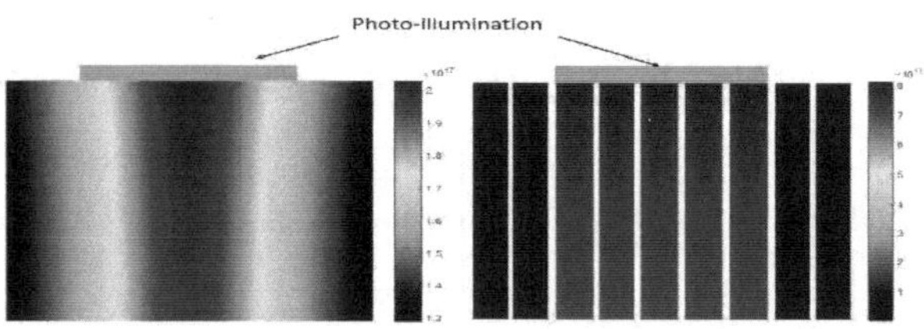

(a) (b)

Fig. 3. Photo-induced substrate (a) without and (b) with pillar-array structure

For the verification of PI-SIW structure demonstrated in Fig. 4 (a), pillar arrays were devised with unit size of 10 um x 10 um, trench width of 0.5 um and SIW cavity length of 986.5 um. SIW structure was redesigned in ANSYS-HFSS with the aforementioned pillar-array design.

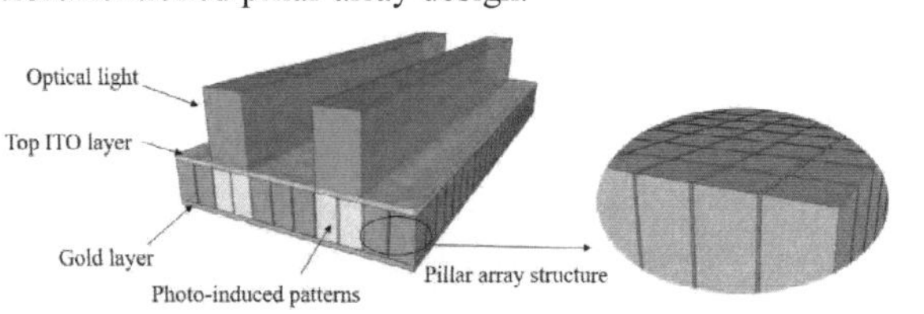

(a) (b)

Fig. 4. (a) PI-SIW structure with (b) pillar-array structure

B PI-SIW Simulated Results and Discussion

First, PI-SIW on continuous HRS substrate is simulated in ANSYS-HFSS. The continuous PI sidewalls are formed with a light intensity of 80 W/cm^2. An insertion loss of 4.15 dB/mm is observed at 280 GHz in Fig. 5. Such high insertion loss is primarily attributed to relatively low photoconductivity achieved and inclined sidewalls due to lateral carrier diffusion.

Subsequently, for the performance prediction of pillar-array structures, PI-SIW on continuous HRS substrate with pillar-array sidewalls are simulated. This is an extreme and ideal circumstance when the width of pillar-array trenches goes to zero while remains the abovementioned properties of pillar-array construction. As shown in Fig. 5, the corresponding insertion loss under light intensity of 80 W/cm^2, 100 W/cm^2, 200 W/cm^2, 300 W/cm^2 are 1.50 dB/mm, 1.38 dB/mm, 0.94 dB/mm, 0.81 dB/mm respectively. Under the light intensity of 80 W/cm^2, the insertion loss of PI-SIW with pillar-array sidewalls at 280 GHz is ~2.6 dB/mm superior than that of PI-SIW with continuous sidewalls. By increasing optical light intensity, photoconductivity of pillar-array sidewalls is increased and the insertion loss of PI-SIW structure is consequently diminished. An insertion loss as low as 0.81 dB/mm can be achieved by PI-SIW with pillar-array sidewalls at a light intensity of 300 W/cm^2.

Finally, a practicable structure for fabrication and measurements, PI-SIW with pillar-array construction, is simulated in ANSYS-HFSS. As displayed in Fig. 5, the insertion loss is 0.96 dB/mm at 280 GHz with a light intensity of 300 W/cm^2, which is only ~0.15 dB/mm inferior than its extreme case where trench width is zero and the substrate becomes continuous. The shifting of cutoff frequency (~4 GHz) is caused by an equivalent permittivity lower than silicon due to air-filling trenches. The E-field distribution of proposed PI-SIW with pillar-array configuration is given in Fig. 6.

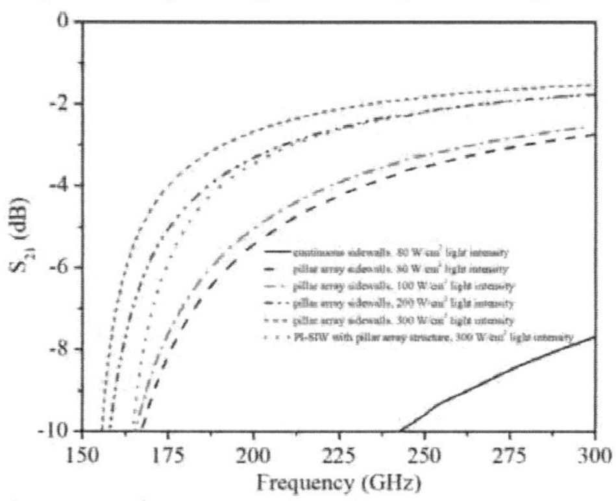

Fig. 5. S-parameter frequency responses for PI-SIW structures under light intensity of 80 W/cm^2, 100 W/cm^2, 200 W/cm^2, 300 W/cm^2 (transmission line length of 2 mm)

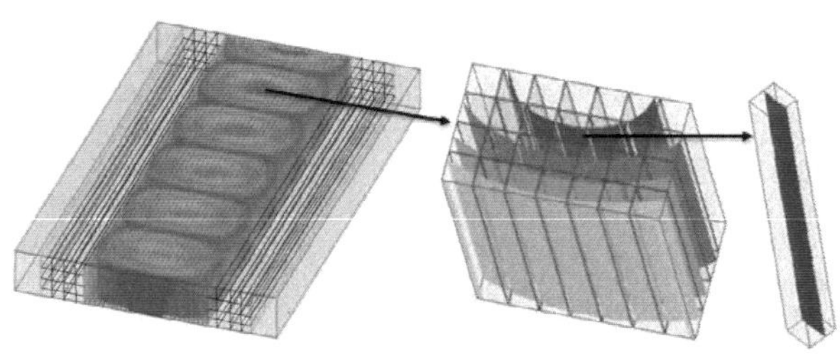

Fig. 6. Electric field distribution of PI-SIW with pillar-array structure and its zoom-in figures

CONCLUSIONS AND FUTURE WORKS

A novel approach for realizing tunable/reconfigurable THz circuits using photo-induced substrate-integrated waveguide (PI-SIW) architectures is reported in this work. Full-wave HFSS simulation has shown an insertion loss of 4.15 dB/mm at 280 GHz with PI sidewalls formed by 80 W/cm^2 light intensity. Such high insertion loss is attributed to relatively low photoconductivity achieved and inclined sidewalls due to lateral carrier diffusion. Pillar-array construction is therefore introduced to constrain photogenerated carriers within every isolated pillar and suppress lateral carrier diffusion. Simulated results indicate that the insertion loss of PI-SIW with pillar-array structure can be reduced to as low as 0.96 dB/mm, and a value of 0.81 dB/mm can be achieved with 300 W/cm^2 light intensity. The proposed high-performance PI-SIW has huge potential in more advanced THz tunable/ reconfigurable circuits such as phase shifters, delay lines and high-level switches (e.g. SPDT, DPDT).

On the basis of the above work: THz tunable circuit designs based on the proposed high-performance PI-SIW structures with pillar arrays will soon be designed and simulated. Meanwhile, pillar-array technique for realizing high resolution circuits and components will be fabricated. For SIW measurements, SIW-to-CPW transitions will be devised and fabricated. Furthermore, on-wafer measurements using THz VNAs and probe station systems for the performance of the developed circuits based on PI-SIWs will be conducted.

Acknowledgment

This work is partially supported by the National Science Foundation (NSF) under the grant of ECCS-1711631 and ECCS-1711052, and a subcontract from the Harvard-Smithsonian Center for Astrophysics under grant PTX-Smithsonian 17-SUBC-400SV787007. The authors would like to thank the support from Center for Nano Science and Technology (NDnano) and Advanced Diagnostics & Therapeutics (AD&T) at the University of Notre Dame.

References

[1] J. Li et al, "Mechanically tunable terahertz metamaterials," Appl. Phys. Lett., vol. 102, no.12, 2013, Art. ID 121101.

[2] L. Liu et al, "Temperature control of terahertz metamaterials with liquid crystals," IEEE Trans. THz Sci. Technol., vol. 3, no. 6, pp. 827-831, 2013.

[3] Z. Jiang, S. M. Rahman, P. Fay, J. L. Hesler and L. Liu, "Tunable 200 GHz lens-coupled annular-slot antennas using Schottky varactor diodes for all-electronic reconfigurable terahertz circuits," Electronics Letters, vol. 49, no. 23, p. 1428 – 1430, 2013.

[4] S. Preu, G. H. Döhler, S. Malzer, L. J. Wang, and A. C. Gossard, "Tunable, continuous-wave Terahertz photomixer sources and applications," J. Appl. Phys., vol. 109, no.6, 2011, Art. ID 061301.

[5] J. Wu et al., "Tuning of superconducting niobium nitride terahertz metamaterials," Opt. Express, vol. 19, no. 13, p. 11922 (1-6), 2011.

[6] L. Cheng and L. Liu, "Optical modulation of continuous terahertz waves towards cost-effective reconfigurable quasi-optical terahertz components," Opt. Exp., vol. 21, no. 23, pp. 28657-28667, 2013.

[7] A. Kannegulla, M. I. B. Shams, L. Liu, and L.-J. Cheng, "Photo-induced spatial modulation of THz waves: opportunities and limitations," Opt. Exp., vol. 23, no. 25, pp.32098-32112, 2015.

[8] Y. Zhou and S. Lucyszyn, "Modelling of reconfigurable terahertz integrated architecture SIW structures," Prog. Electromagn. Res., vol. 105, pp. 71–92, 2010

Membrane Integrated Asymmetric Dual E-plane Probe Ortho Mode Transducer at 424 GHz

Peter J. Sobis[*], Vladimir Drakinskiy[2], Arvid Hammar[1], Johanna Hanning[2], Anders Emrich[1], Elena Saenz[3], and Jan Stake[2]

[1]*Omnisys Instruments AB, Västra Frölunda, SE-421 32, Sweden*
[2]*Chalmers University of Technology, Göteborg, SE-412 96, Sweden*
[3]*European Space Agency – ESA/ESTEC, Noordwijk, PO Box 299*
NL-2200 AG, The Netherlands
[*]Contact: peter.sobis@omnisys.se

Abstract—We present a practical low loss GaAs membrane integrated ortho mode transducer (OMT) at 424 GHz composed of two orthogonal open ended E-plane probes in a circular waveguide. The proposed structure is suitable for building compact dual polarisation receivers in the lower THz range (~100 GHz – 1 THz) and is readily integrated directly to the first receiver element, in this case a subharmonic GaAs Schottky diode mixer, without any additional penalty in terms of loss or manufacturability.

A 3-port OMT split block aluminium module, with rectangular WR-2.2 waveguide interfaces for the separated V/H polarisations and 375 μm diameter circular waveguide for the common V/H port has been designed and fabricated. The simulated nominal response of the membrane OMT reaches X-pol levels below 20 dB over 10% bandwidth with a worst case insertion loss of 0.3 dB. The simulated inter probe leakage (isolation) and circular waveguide reflected X-pol levels where around 20 dB. For the particular OMT test additional E-plane probe transitions were integrated on-chip to be able to couple the separated V/H polarisations to the rectangular waveguide test ports.

Measurements on two circular waveguide interconnected back to back OMT modules using VDI WM-570 (WR-2.2) VNAX frequency extenders, show X-pol and isolation levels below 10 dB and return loss levels below 15 dB including effects of standing waves. Translated to a single OMT transition this corresponds to a X-pol and isolation of about 15 dB. The loss of a single OMT probe transition including the membrane circuit and circular waveguide loss was estimated to about 0.2 dB. The results look promising and the next step will be to test the integrated OMT concept in a 424.7 GHz dual polarization receiver currently being pre-developed by Omnisys Instruments AB for the ISMAR instrument.

Development of a transmit-array for heterodyne receiver

Duccio Delfini[1*], Martina Wiedner[1], Massimiliano Casaletti[2], Julien Serrazin[2], François Joint[1], Thibaut Vacelet[1] and Yan Delorme[1]

[1]LERMA, Observatoire de Paris, PSL Research University, CNRS, Sorbonne Universités, UPMC Univ. Paris 06, F-75014, Paris, France

[2]Sorbonne Universités, UPMC Univ Paris 06, UR2, L2E, F-75005, Paris, France

*Contact: duccio.delfini@obspm.fr

***Abstract*—The development of large multipixel heterodyne imaging arrays is a challenge in modern THz spectroscopy. For large arrays (>100 pixels) the ease of fabrication and the reliability become an important factor. Stacking individual pixels next to each other, might not be the optimal solution for large arrays. There has already been lots of good progress by designing blocks for multiple mixers (e.g. for Supercam, JPL R&D work), which simplifies the task but still requires a lot of machining.**

Here we are exploring possibilities of further simplifying the design. One idea is to replace the horn or lens – that requires individual mechanical fabrication – with an array of planar lenses followed by or integrated in an array of mixers fabricated on a single silicon wafer. Here we are looking at transmit – arrays to take the role of focusing element. Each pixel will require one transmit-array, but an array of ~ 10X10 transmit-arrays for 10X10 pixels could be fabricated on a single wafer in a single process in the clean room. The idea is that this array of planar lenses as well as an array of mixers can be fabricated and aligned in one single process for large (~100) pixel arrays.

We have designed, fabricated and tested a prototype transmit-array for 600 GHz. First we calculated the phase shifts required of each cell of the transmit-array to focus the incoming planar wave front. In analogy to the classical bulky lens, the larger phase shift/delays are required in the center of each pixel to delay the incoming waves and allow constructive interference at the focal point. In order to obtain the necessary phase shift, we required that each transmit-array is composed of three layers of 2D metamaterials separated by two sapphire layers. We modelled the cell as a transmission line where each metamaterial layer is represented by a parallel impedance. Finally, the behavior of the cell is described by its reflection and transmission coefficients, which depends algebraically on the three parallel impedances. The phase of the transmission coefficient must be equal to the phase shift required by the cell. We performed numerical simulations in order to obtain the required phase shift, while keeping the reflection coefficient to -25 dB to maximize the efficiency. The prototype we designed so far allows the focusing of the light in one dimension.

We fabricated this prototype for a single pixel: this planar lens consists of a structure of metal cells we fabricate by depositing gold on three layers of 100 microns sapphire wafers, two layers of the golden cells will be recto/verso on the same wafer, the third layer on a second wafer. The dielectric plates will be aligned with a microscope to micron precision. We also measured the focusing capabilities of the planar lens.

I. Introduction

The heterodyne receiving pixels are made of very directive antennas. Normally the choice of the receiver is based on the beamwidth specification, which is approximatively 2,2°. Nowadays two solutions exist: the horn antenna, or the double slot integrated in a silicon lens. These solutions are good for single receiver, or in general for an array of few elements. When an attempt is made to increase the number of pixels it becomes difficult to realize them in a compact configuration. In this case a special attention must be given to flat elements, like the array of phased microstrip antennas. It consists of an array of superconducting patch elements drawn on a substrate which separates them by a ground plane. This configuration has the drawback of a too thin substrate which is difficult to realize. At this point the best solution seems to use the double slot antenna, where the bulk lens is replaced by a planar thin lens. A widespread literature exists about planar lens, also called transmit-arrays. The transmit-array consists in an array of single cells. Each cell provides the incoming wave with a phase shift to generate a constructive interference at the output, in the focal point. Many examples exist in literature of cells whose reflection coefficient and phase shifts are calculated numerically. Each cell is optimized to furnish a certain phase shift and a low reflection coefficient. They can be made of square rectangles, crosses or other metal motifs designed on a dielectric substrate, or dielectric cylinders on a different dielectric substrate, slots etc… In the totality of the cases, such shapes are too complicated to be rescaled for THz applications. In this work we are going to provide a transmit-array whose cells are made of simple structure which can be fabricated. Moreover we are going to provide a mathematical model which allows to calculate the phase shift and reflection coefficient of every cell, without resorting to numerical optimization in EM simulation softwares.

II. Objective

- The first objective of this work is to substitute the bulk lens of the double slot antenna, with a flat lens, which is easier to fabricate for large compact configurations. Such flat lens has the function of reducing the beamwidth of the double slot antenna, which must be less than 2,2°. The beamwidth is directly linked to the beam waist of the quasi-optical equivalent Gaussian beam. The size of the beam waist decreases with the beamwidth. Since the beamwidth of the double slot is very large, the objective of the flat lens is to reduce the beam waist of the incoming light, to the beam waist of the double slot. Since the dimension of the transmit-array is quite small compared to the beam waist of the incoming light, and since the beam waist of the double slot is very small, we can assume that the transmit-array converts an incoming planar wave into a cylindrical wave, with the center on the double slot receiver.

- The radiation which is incoming on the heterodyne is generated very far away from the observer. The resulting signal is very weak, therefore we don't want that it would be reflected away and lost. For this reason the transmit-array must be designed so that a very low reflection coefficient is achieved. In the analytical design the value of the reflection coefficient will be set to -25 dB.

III. METHOD

The transmit-array is composed by several cells stacked together into an array (fig 1)

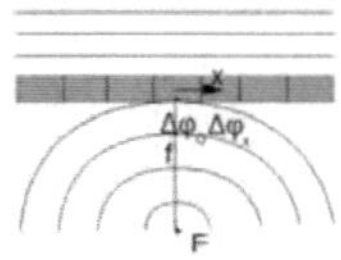

Fig. 1. Transmit-array scheme

In order to transform an incoming planar wave into a cylindrical wave, the single cell of the transmit-array, placed at a distance x from its center, must provide a phase shift given by eq. 1:

$$\Delta\varphi_x = \Delta\varphi_0 + \beta\left(\sqrt{f^2 + x^2} - f\right) \quad \text{(eq.1)}$$

Where $\Delta\varphi_0$ is the phase shift of the central cell, $\Delta\varphi_x$ is the phase shift of the cell a the distance x from the center, β is the wave number and f is the distance between the focal plane and the array. Eq. 1 represents the first condition that the cell must satisfy. The second condition consists in the minimization of the radiation reflected by the cell. Our objective is to determine the shape of the cell to achieve such requirements. Let's consider the generic cell made of three metal motifs separated by two dielectric substrates of the same material (fig. 2)

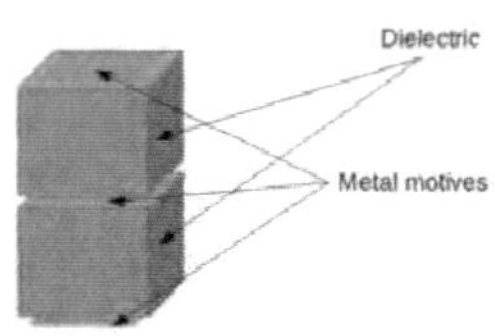

Fig. 2. Scheme of the three layers generic cell

Each layer (i.e. the metal motif) can be seen as an impedance in the equivalent circuit of the cell (fig.3):

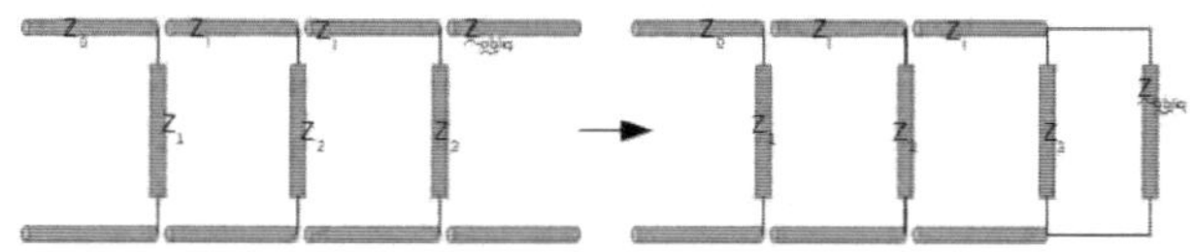

Fig. 3. Equivalent circuit of the cell

where Z_1 , Z_2 and Z_3 represent the three values of the impedances; Z_l represents the impedance of the substrate, Z_0 is the impedance of the free space and $Z_{obliq} = Z_0/\cos(\theta_{TE})$ is the impedance of the free space which is seen by the outcoming beam, that is inclined by the TE angle θ_{TE} towards the focal line. The calculation of the equivalent impedance of the circuit allows to determine the reflection coefficient as:

$$\Gamma = \frac{Z_{eq} - Z_0}{Z_{eq} + Z_0} \quad \text{(eq.2)}$$

where Z_{eq} is the equivalent impedance. The forward part of the input and output signal with respect to the circuit is given by:

$$V_i^+ = \frac{V_i + Z_0 I_i}{2} \quad \text{(eq.3)}$$

and

$$V_o^+ = \frac{V_o + Z_{obliq} I_o}{2} \quad \text{(eq.4)}$$

respectively. So we have:

$$\Delta\varphi_x = \sphericalangle V_o^+ - \sphericalangle V_i^+ \quad \text{(eq.5)}$$

$$\Gamma = -25\, dB \quad \text{(eq.6)}$$

The value of the three impedance is varied until eq. 5 and eq. 6 are satisfied. It is possible to determine the shape of the motifs as a function of the impedances by means of the paper of O. Luukkonen (IEEE June 2008). They generally consist of an array of metal patches (or strips if the imaginary value of the calculated impedance is positive). Since only the dimension of the patch which is parallel to the electric field is important, the patches can be stacked together to form strips, as shown in figure 4 (the metal path is in blue);

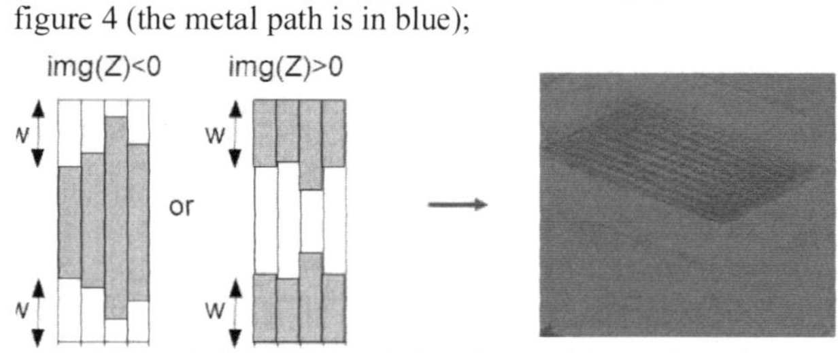

Fig. 4. Relation between the impedance and the metal motif

An array of metal strips orthogonal to the electric field will be created, so that the linearly polarized electric field will focus on a line which contains the HEB.

IV.NUMERICAL SIMULATIONS

Figure 5 shows a qualitative study of the transmit-array in the transmitting mode. It consists of a FEKO simulation where a dipole is placed on the focal line of the transmit-array. It is clearly evident that the transmit-array transforms the cylindrical wave into a planar wave.

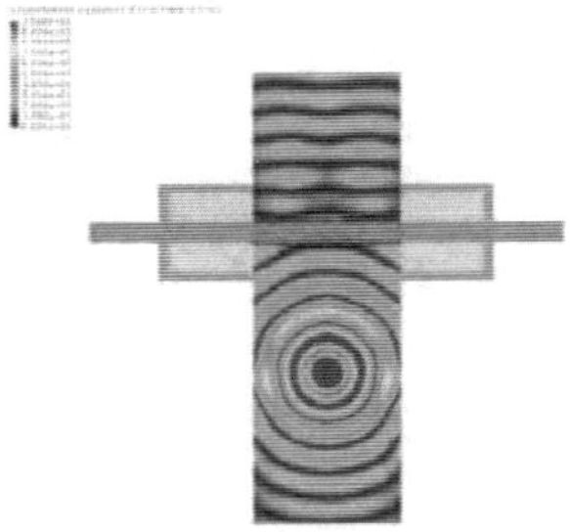

Fig. 5: Near field from a dipole through the transmit-array in the transmitting mode

A more quantitative study can be done in the receiving mode by assuming an incoming planar wave of 600 GHz. The transmit-array focal length is assumed to be 1,5 mm. A frame is placed around the transmit-array to block the interaction between the field downstream the array, and the external field. In this case an overall efficiency of 45% is achieved with a beam waist radius of 21% of the edge of the transmit-array. Such reduced efficiency is due to the blocking effect of the frame. Without the frame the transmitted power would be equal to the incoming power, but the focusing effect would be weakened. The 1,5mm focal length transmit-array is of great interest for a possible application for heterodyne receivers, however it is too difficult to test. For this reason a 20mm focal length transmit-array has been designed and simulated. In this case simulations show an efficiency of 53% within a beam radius of 42% of the transmit-array edge.

V. TEST

For the experimental setup we need a high frequency signal generator, which is the RPG chain, and a detector, which is the terahertz pyroelectric detector or QCM, from the QCM instrument Ltd. The experimental setup is shown in fig. 6.

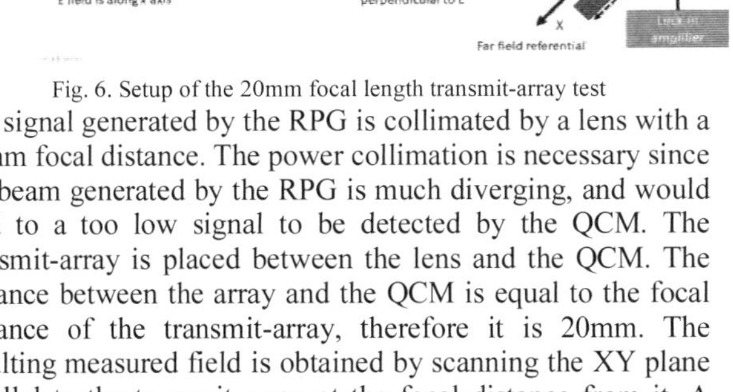

Fig. 6. Setup of the 20mm focal length transmit-array test

the signal generated by the RPG is collimated by a lens with a 50mm focal distance. The power collimation is necessary since the beam generated by the RPG is much diverging, and would lead to a too low signal to be detected by the QCM. The transmit-array is placed between the lens and the QCM. The distance between the array and the QCM is equal to the focal distance of the transmit-array, therefore it is 20mm. The resulting measured field is obtained by scanning the XY plane parallel to the transmit-array at the focal distance from it. A metallic disk with a hole of 1,5mm in the center is applied to the QCM to increase its resolution. Smaller the hole, higher the resolution, however the hole cannot be too small because the signal measured by the QCM would be too small.

We performed two main measurements: the first measurement is made with an empty frame on the sapphire dielectric substrate, while the second is made with the transmit-array and the frame together. We should see an increment of the collimation between the first and the second measurement, which can be quantified by the directivity. Figures 7 and 8 show the field measurement on the XY plane for the frame only and the frame with the transmit-array respectively:

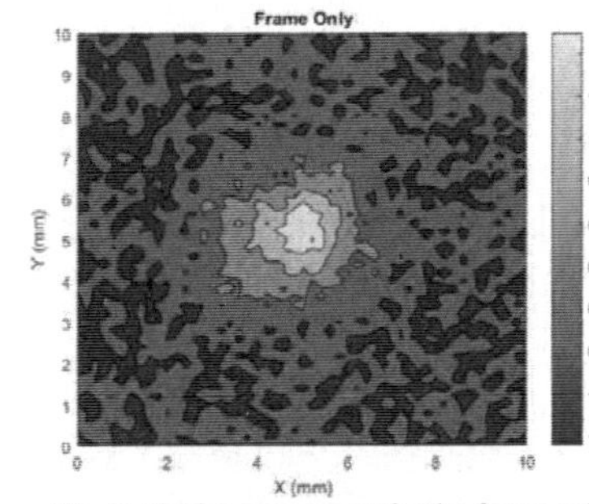

Fig. 7. Field measurement for the frame only

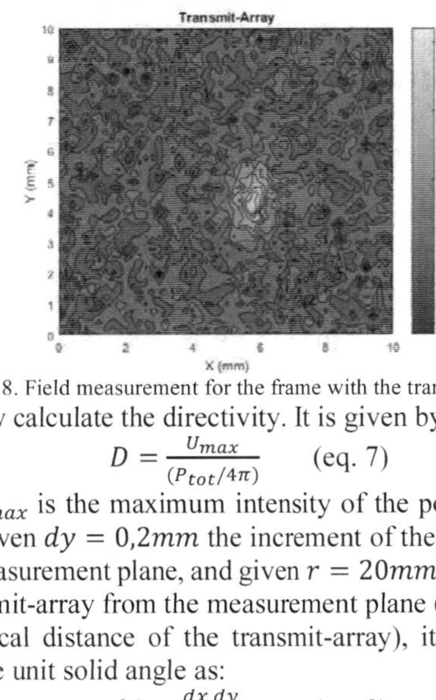

Fig. 8. Field measurement for the frame with the transmit-array

Let's now calculate the directivity. It is given by:

$$D = \frac{U_{max}}{(P_{tot}/4\pi)} \qquad \text{(eq. 7)}$$

where U_{max} is the maximum intensity of the power unit solid angle. Given $dy = 0{,}2mm$ the increment of the single element in the measurement plane, and given $r = 20mm$ the distance of the transmit-array from the measurement plane (which is equal to the focal distance of the transmit-array), it is possible to define the unit solid angle as:

$$d\Omega = \frac{dx\,dy}{r^2} \qquad \text{(eq. 8)}$$

Then the directivity can be expressed as:

$$D = 4\pi \frac{U_{max}}{\iint_{\Omega} U(x,y) d\Omega} \approx \frac{4\pi}{d\Omega} \frac{U_{max}}{\sum_{n,m} U(n,m)} \qquad \text{(eq.9)}$$

where $x = (n-1)dx$ and $y = (m-1)dy$. The results are shown in tab. 1

	linear	dB
Frame only	718	28,6
Transmit-array with frame	1258	31,0

Tab. 1. Directivity of the frame only and transmit-array with frame

The transmit-array shows a small increase in the directivity. This is the best result we can achieve with our instrumentation, where the standard deviation of the signal is 12% of the peak value.

Conclusions

A transmit-array design has been introduced which transforms an incoming planar wave into a cylindrical wave. Thus the signal is focused on a line parallel to the transmit-array plane and in the same direction of the electric field. Simulations show a good focusing effect for the 1,5mm focal length transmit-array, but the efficiency is reduced by the presence of the frame, which is necessary to avoid the interaction of the field downstream the transmit-array with the external field. Even if the converging of the 1,5mm transmit-array towards the focal line is good, it is difficult to test with the available instrumentation. For this reason a 20mm focal length prototype has been designed, simulated and tested. Simulations show a reduced focusing effect, due to the reduced curvature of the wave downstream the array, so that the beam waist is greater with respect to the 1,5mm case. The tests are performed with the QCM by measuring the far field in the focal plane of the transmit-array. Results show a very small increase of the directivity of the transmit-array, with respect to the case where the frame only is considered. This effect shows that the transmit-array is working, even if a more precise measurement is not possible with our instrumentation.

References

[1] L. Dussopt, H. Kaouach, J. Lanteri and R. Sauleau, "Circularly-Polarized Discrete Lens Antennas in the 60-GHz Band" *Radioengineering,* vol.20 no.4,.pp 733-738, December 2011

[2] H. A. Malhat, N. A. Eltresy, S. H. Zainud-Deel and K. H. Awadalla "Nano-Dielectric Resonator Antenna Reflectarray/Transmitarray for Terahertz Applications" *Advanced Electromagnetics,* vol.4 no.1,.pp 36-44, June 2015

[3] C. G. M. Ryan, J. R. Bray and Y. M. M. Antar "A Broadband Transmitarray Using Double Square Ring Elements" *13th International Symposium on Antenna Technology and Applied Electromagnetics and the Canadian Radio Sciences Meeting,* 2009

[4] C. G. M. Ryan, M. R. Caharmir, J. Shaker, J. R. Bray and Y. M. M. Antar "A Wideband Transmitarray Using Dual-Resonant Double Square Rings" *IEEE Transactions on antennas and propagation,* vol.58 no.5,.pp 1486-1493, May 2010

[5] A. Clemente, D. Laurent, R. Sauleau, P. Potier and P. Pouliguen "1-bit Reconfigurable Unit-Cell for Transmit-Array Applications in X-Band" *IEEE International Symposium on Antennas and Propagation (APSURSI)*,.pp 684-687, 3-8 July 2011

[6] J. Y. Lau and S. V. Hum "Reconfigurable Transmitarray Design Approaches for Beamforming Applications" *IEEE Transactions on Antennas and Propagation*,. vol 60, no. 12, pp 5679-5689, December 2012

[7] O. Luukkonen, C. Simovski, G. Granet, G. Goussetis, D. Lioubtchenko,A.V.Raisanen and S.A. Tretyakov "Simple and Accurate Analytical Model of Planar Grids and High-Impedance Surfaces Comprising Metal Strips or Patches" *IEEE Transactions on Antennas and Propagation*,. vol 56, no. 6, pp 1624-1632, June 2008

WM380 (675-700GHz) Band-Pass Filters in Milled, Split-Block Construction

Daniel Koller*, Eric W. Bryerton, and Jeffrey Hesler
Virginia Diodes, Inc., Charlottesville, VA 22903, U.S.A.
*Contact: koller@vadiodes.com

***Abstract*—Filters play an invaluable role in RF analysis and communications hardware, blocking unwanted signals, limiting bandwidth, attenuating harmonic components, etc. In waveguide construction for mm-wave frequencies, where the powers are generally low, the iris-coupled-resonator band-pass filter has proven to be very practical and amenable to easy machining. Described here is a WM380 (WR1.5) filter for the 500-750 GHz waveguide band having a pass-band of 675-700 GHz.**

Measured results of a batch of filters are shown and compared to simulations to illustrate just how well these filters can be made. Furthermore, the filters can be tuned by simple mechanical means and data are presented to illustrate how easily the filters can be adjusted. Additional modifications to the filter topology to simplify machining, and allow other construction techniques to be utilized are also demonstrated. Ultimately it should be possible to push the split block technology to manufacture iris coupled resonator filters for use at frequencies of well over 1 THz.

I. INTRODUCTION

The specific motivation for the work described here was the development of a 640 GHz integrated-block heterodyne polarimeter, intended to serve as a proof-of-concept for future instruments in which many complex mm-wave components are integrated into a single machined part[1]. As shown in the block diagram in Figure 1 below, a pair of identical filters is required to pass only the upper sideband of an input signal, from 675-692 GHz. The two filters must be closely matched, and more importantly need to meet a tight, and absolute frequency specification as determined by the science for which the instrument is designed.

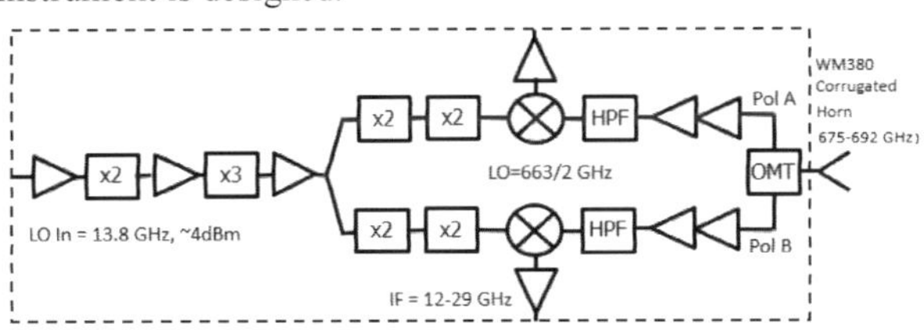

Figure 1 Block diagram of an Integrated 640 GHz Heterodyne Polarimeter

The filters chosen for the project are of the iris-coupled resonator design, adapted to a split-block construction, which lends itself readily to CNC machining[2]. In such a device, thin inductive irises milled into the block divide the waveguide into a series of coupled resonators. A typical resonator length for the WM380 band in question is on the order of 250 microns, yet state-of-the-art CNC milling allows a machining tolerance of +/-2.5 microns. In a worst-case scenario, the errors can be independent, yielding a length variation of the resonators of +/-2%. The corresponding variation in frequency for a 675 GHz filter is approximately +/-14 GHz, which is more than half of the target pass-band and will clearly not work for the instrument pictured above. A possible solution is to design the filters to operate at frequencies nominally higher than the desired pass band and to tune them down mechanically to the desired values. Several techniques for doing so are described below.

II. FILTER DESIGN I

A filter, hereafter referred to as X1, was designed with a 3-dB pass-band of 670-710 GHz to allow for some initial compensation for machining tolerances. X1 is of a "conventional" design in which machined irises, each 1/10 the waveguide width in thickness, stand out from the waveguide sides. The irises are of different heights. Construction is of the split-block design in which half of the structure is located in part of a metal block with the other half in an identically machined block, and the structure is split down the middle along the E-symmetry plane of the waveguide. By pushing state-of-the-art milling machines to their limits, five blocks were delivered with mechanical tolerances better than +/-2.5 microns in all dimensions. A micrograph of one of the filters is shown in Figure 2.

Figure 2 Micrograph of the X1 filter looking in from the split-plane, showing the thin irises.

RF performance of all five filters was measured on a Keysight Vector Network Analyzer with a pair of Virginia

Diodes WM380 frequency extension heads, using an SOLT calibration. Results for all five units are shown in Figure 3 below. Note that two of the filters (#1-01 and #1-05) are shifted down in frequency relative to the others, presumably due to tolerance variation. Otherwise, the filters are very tightly grouped, indicating good control over the machining.

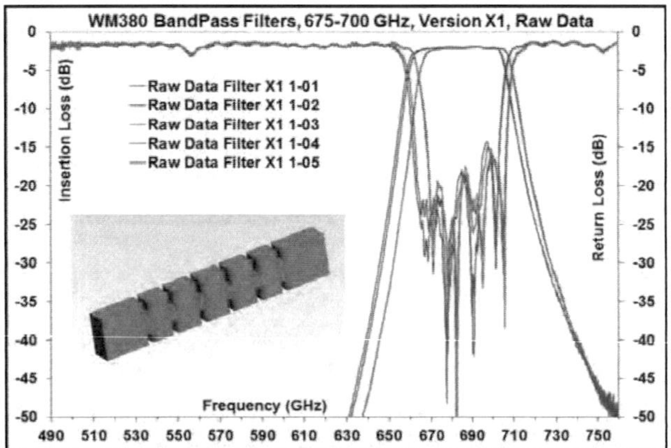

Figure 3 Measured performance of the X1 filter. Inset shows the as-assembled filter geometry.

The expected behaviour of the X1 filter upon intentionally introducing various gaps between the block halves was simulated in HFSS and the results are shown in Figure 4. The lower edge of the pass band shifts linearly at the rate of -1.9 GHz per micron of introduced gap. The upper band edge shifts slightly less, at a rate of -0.9 GHz per micron so the overall pass-band of the filter increases as the filter response is shifted down. The return loss of the filter also worsens as the gap is increased, but the decay is well behaved. The simulations hint at an elegant way to "tune" the filter by introducing a gap between the block halves in a controlled manner. Because the filter is E-plane split, RF leakage from the gap is small and does not appreciably deteriorate the insertion loss of the device.

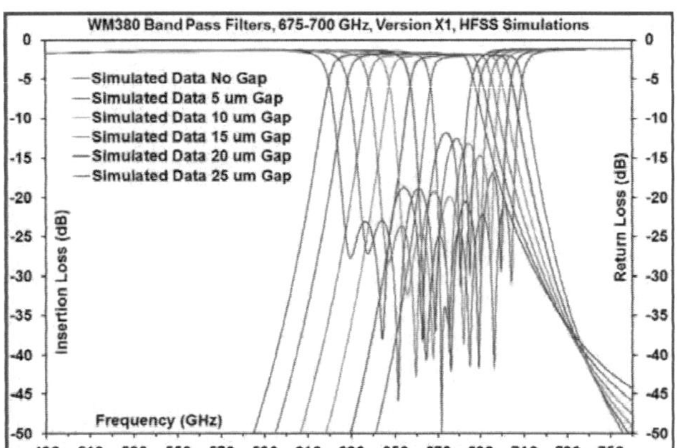

Figure 4 HFSS simulations of X1 filter, "tuned" by separating the halves at the split-plane.

III. FILTER DESIGN II

A second series of filters was also designed and built. The X2 structure is intended to be more amenable to scaling to higher frequencies, where a two-level electroforming technique might provide higher dimensional accuracy. All the irises are of the same height, but the widths vary. The design was carried out by optimizing in FEST-3D but forcing the iris heights to be constant, while maintaining the same target performance as the X1 design[3]. A micrograph of half of an X2 filter is shown in Figure 5.

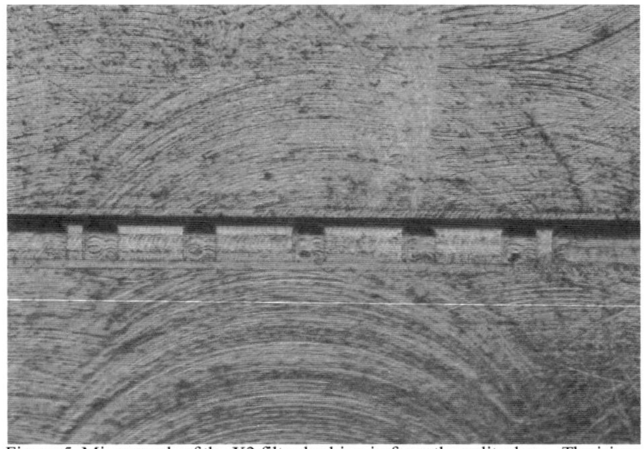
Figure 5 Micrograph of the X2 filter looking in from the split-plane. The irises are all the same height, but the "widths" vary.

All five units were measured on a VNA and the results are presented in Figure 6, which also includes an image of the unusual filter structure in the inset. Note that all five filter responses are relatively closely matched, and as with the X1 design, there is a net shift down in frequency from the design target.

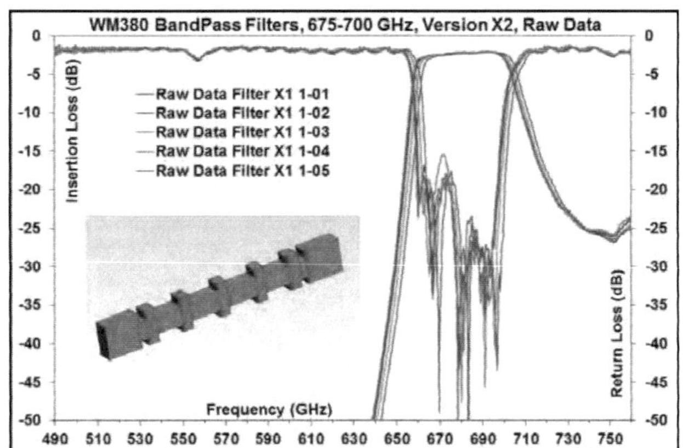

Figure 6 Measured performance of the X2 filter. Inset shows the as-assembled filter geometry.

The simulated effect of tuning the filter by introducing a gap between the block halves is again explored in Figure 7 and some notable differences are evident. The lower and upper edges of the band shift downward at the rate of 2.6 and 2 GHz per micron respectively, a rate much greater than that of the X1 design. In addition the bandwidth increases by only 0.6 GHz per micron, vs. 0.95 GHz per micron for the X1 design. Overall, the X2 design is much more sensitive to tuning by introduction of a gap between the block halves and the bandwidth changes more slowly.

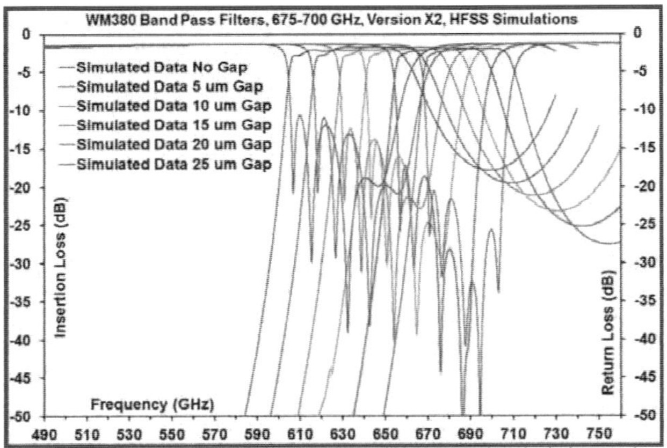

Figure 7 HFSS simulations of X2 filter, "tuned" by separating the halves at the split-plane.

IV. Fixed Tuning with Mylar and Aluminum Shims

For the initial series of experiments, small Mylar squares, each 5 microns thick, were placed *between* the clamp screws of filter X1 #1-02[4]. Because the spacers were not placed directly under the clamp screws, the block deformed when clamped and the desired gaps were less than expected. The blue curves in the shift summaries below were obtained.

In the next series of tests, larger sheets of Mylar were used that cover the full length of the block and the results for multiple filters with differing gaps are presented in Figure 8. Note that the filters were not identical so the resulting frequency shifts appear to be non-linear and curves for filters with the same gap are not coincident. When the frequency shifts relative to the un-gapped filter are determined, the response is indeed linear as will be further summarized below.

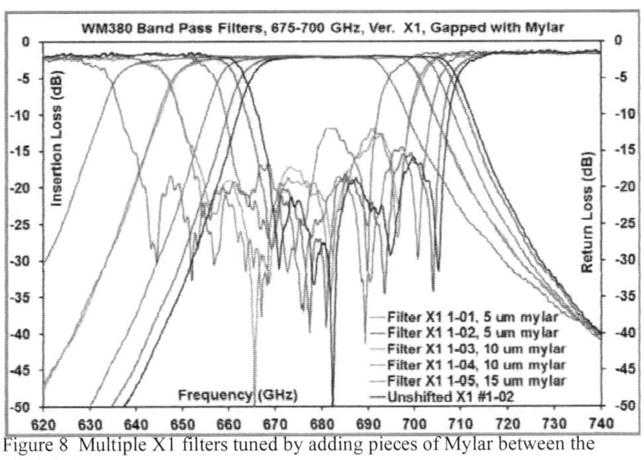

Figure 8 Multiple X1 filters tuned by adding pieces of Mylar between the block halves.

Finally, a 14 micron thick sheet of aluminum foil was used as shim material. One, and later two pieces of foil were laid down across the entire central section of the block straddling the position of the filter to ensure a uniform gap. The results are shown in Figure 9 along with the positioning of the foil as illustrated in the inset. The resulting filter tuning was linear and as predicted by the HFSS simulations.

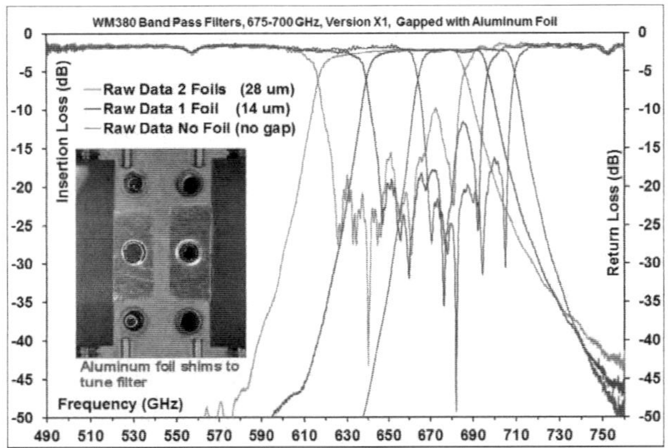

Figure 9 Filter X1 #1-02 tuned by adding layers of rolled aluminum foil.

V. Shift Summary

The performance of the shimmed X1 filters is summarized in Figures 10 and 11 below, with linear fits to the data for comparison. The non-linearity of the tuning with small Mylar squares is evident, and it also apparent that the larger Mylar sheets (not shown, but roughly the size and shape of the aluminum foil pieces shown above) are indeed effective.

Aluminum shims appear to be the most robust, and shift the frequency response of the filter by nearly the same amount as expected from the simulations. Corresponding plots for Filter X2 are provided in Figures 12 and 13.

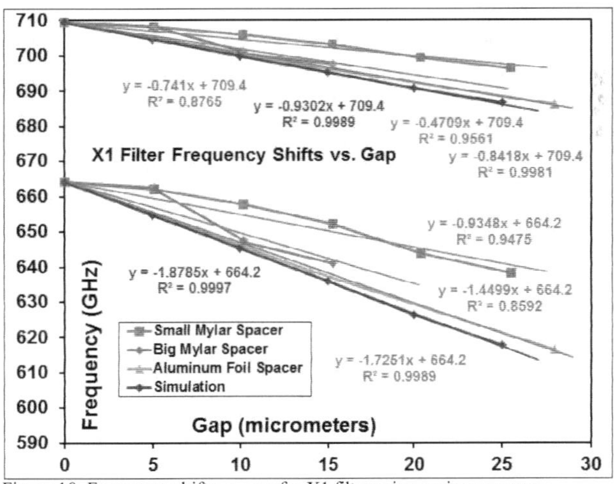

Figure 10 Frequency shifts vs. gap for X1 filter using various spacers.

Stability

Mylar is an established material for space-qualification, but stability of the filters remains a concern. Two X1 filters, each with 2 layers of 5 micron thick Mylar, were baked in an oven at 110C for 50 hours. Additionally two X2 filters, one with only one 5 micron layer, and one with 15 microns of Mylar, were similarly baked. There was no measurable change in the filter response of any of the filters.

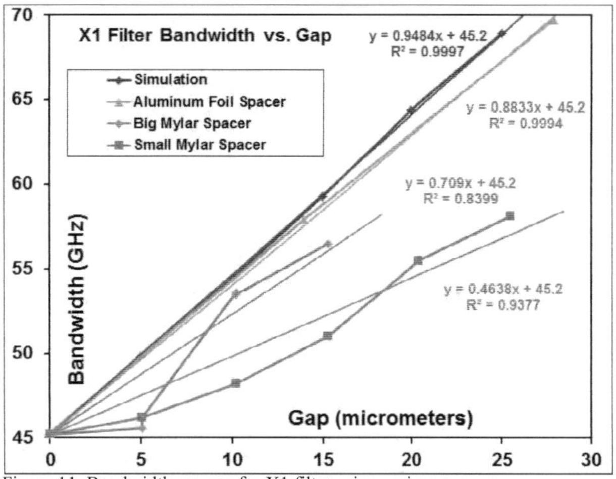

Figure 11 Bandwidth vs. gap for X1 filter using various spacers.

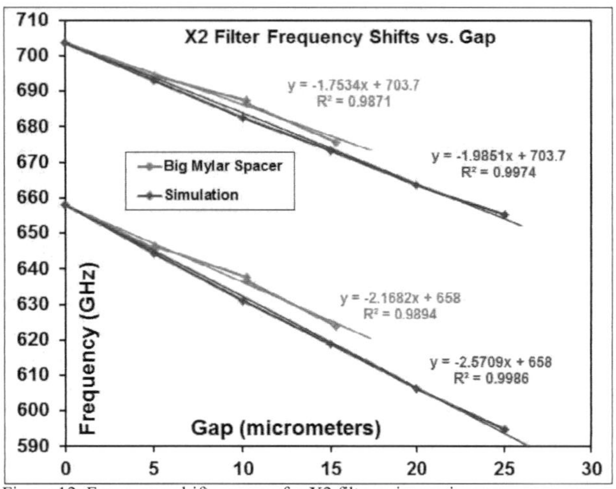

Figure 12 Frequency shifts vs. gap for X2 filter using various spacers.

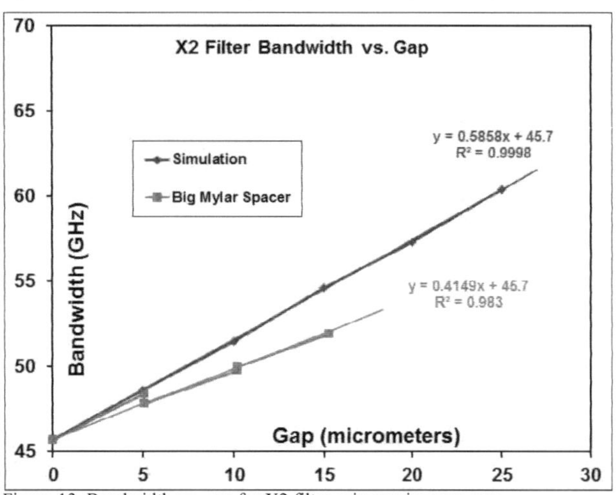

Figure 13 Bandwidth vs. gap for X2 filter using various spacers.

A more stringent test is the so-called "Bellcore" cycle, in which parts are subjected to a 70C bake followed by a -40C soak in a 1 hour period, for 1000 hours[5]. As before, two X1 filters, each with 10 micron Mylar spacers, and two X2 filters, one with a 5 micron spacer, the other with 15 microns, were put through Bellcore cycling. As before, no filter experienced any measurable change in performance, indicating that the Mylar and the clamping are stable under the tested conditions.

Conclusions

A WM380 675-700 GHz band-pass filter, with an inductive shunt, iris-coupled-resonator structure has been made in split-block construction. The filter design is "routine", well understood and has been in the literature for decades. What is unique is its frequency of operation. At 675 GHz, construction required pushing to the absolute limits of machining tolerances to make a repeatable filter capable of hitting absolute frequency targets. Despite the demonstrated reproducibility of the pass-band, the push towards even higher frequency, and smaller dimensions will require some sort of tuning of the filters. A predictable, stable means of shifting the pass band of the as-designed filter down by discrete steps with Mylar or aluminum shims has been demonstrated.

A second design, with a much simpler mechanical structure has been similarly proven. Rather than utilizing thin irises milled into the waveguide, the simplified design has features milled to only two depths. The intention is that the filter could be made more easily with CNC milling, but also by electroforming or etching techniques where an "etch to a depth" or "deposit to a height" fabrication method would benefit from having only two steps in the structure. While there is no theoretical design paradigm for developing the filter, a structure is easily obtained by optimizing a model, with constraints, using commercially available simulation software.

Acknowledgment

This research was carried out during the development of a 640 GHz Heterodyne Polarimeter, built in support of a NASA Goddard Space Flight Center SBIR Phase II Contract, #NNX17CG06C.

References

[1] Eric W. Bryerton, Jeffrey L. Hesler, Daniel Koller, Yiwei N. Duan, and Thomas W. Crowe, "A Compact Integrated 675-693 GHz Polarimeter", accepted for publication in the Proceedings of the 43st International Conference on Infrared, Millimeter, and Terahertz waves (IRMMW-THz), 2018.
[2] G. Matthaei, L. Young, and E.M.T. Jones, *Microwave Filters, Impedance Matching Networks and Coupling Structures,* New York: McGraw Hill, 1964, first ed.,pp. 450-464.
[3] FEST3D, www.fest3d.com, Aurora Software and Testing S.L.U. (CST/3DS).
[4] E.g. 5 micrometer thick clear Mylar, 7.0g /m^2, available from http://www.homefly.com/products.asp?id=31 [as of 12 June 2018].
[5] Bellcore (Bell Communications Research, Morristown, N.J.) Document TA-NWT-000765 and GR-765.

Wideband Passive Circuits for Sideband Separating Receivers

Hawal Rashid[1*], Denis Meledin[1], Vincent Desmaris[1], and Victor Belisky[1]
[1]Group for Advanced Receiver Development (GARD), Chalmers University, Gothenburg, 41296, Sweden
*Contact: hawal.rashid@gmail.com

Abstract— **In this work, we present the design and characterization a new generation of compact wideband quadrature hybrids operating both at room and cryogenic temperatures for prospective use in the IF chain of 2SB THz receivers.**

The compact wideband design employs a multi-section topology (coupled line coupler – Lange coupler– coupled line coupler) and thin-film technology with gold plated transmission lines and air bridges to connect the fingers of the Lange coupler (middle section).

The hybrids were designed to have the amplitude and phase imbalance better than 0.6 dB and ±3° respectively over a 3.5-12.5 GHz and 4-16 GHz frequency bands. The experimental verification of the assembly at 293 K and 4 K shows very good agreement between the measurements and simulations with amplitude imbalance better 0.5 dB and maximum phase imbalance of ±1.5°.

I. Introduction

The demand for wideband millimeter-submillimeter wave multi-pixel receivers is continuously increasing. So far, only DSB multi-pixel heterodyne systems have been demonstrated [1]. Implementing of wideband sideband separating mixers into multi-pixel design is hindered however, by the fact that as the number of pixels increases, also the number and the size of passive and active components (LNA) increases, which leads to unpractical receiver pixel footprint. A typical sideband separating receiver consists of among others, RF and IF quadrature hybrids, bias-T, isolator and low noise amplifier, see Fig. 1. One of the key figures of merit in sideband separating receivers is the image rejection ratio, which is mainly dependent on the overall amplitude and phase impedance of the receiver chain [2]. This sets tough requirements on both the RF and IF quadrature hybrids.

In order to shrink the wideband receiver pixel footprint, compact and wideband passive and active device are required. One way to reduce the size is to use substrates with high dielectric constant and to integrate the bias-T with the wideband IF hybrid [2]. Furthermore, by using a balanced LNA topology [3] (employing the wideband IF hybrids) would yield a both compact and wideband LNA with sufficient return loss, which practically removes the need for having an isolator in the receiver chain [2]. Hence, the importance of wideband quadrature hybrid with excellent amplitude and phase imbalance performance [4-6]. This way, only two devices are required instead of four, which saves space considerably and improve the performance. Furthermore, it is essential for integration and fabrication purpose that the circuits are planar.

To meet these goals, both compact wideband IF (one octave) and RF circuits has been developed and demonstrated [4-6]. Some of these wideband circuits have successfully been integrated into ALMA Band 5 receivers [6] or in the SEPIA receiver [7], which demonstrates image rejection ratios typically better than 15 dB which is significant improvement [6]. However, to meet the future demands of wideband (IF bandwidth of two octaves or more) radio astronomy receivers, new generation of circuit with bandwidth and enhanced performance have been designed.

In the next section, two compact wideband hybrids are presented, one operating at 3.5-12.5 GHz, and one at 4-16 GHz. Measurements of the manufactured circuits show exceptional performance at both room and cryogenic temperatures.

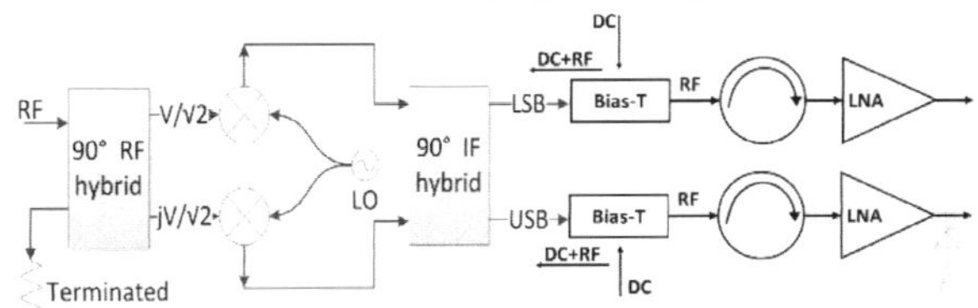

Fig. 1. Schematic of a sideband separating receiver

II. High performance wideband hybrids

A. Page Layout

The requirement on the amplitude and phase imbalance was discussed in [6], were the maximum allowed imbalance was 0.6 dB and ±3° respectively over one octave fractional bandwidth. However, future receivers require at least the same imbalance performance over two octaves or more. Hybrids employing single-section topologies cannot provide the required amplitude imbalance for bandwidths of greater than one octave at best [8, 9]. Therefore, multi-section layout must be used to meet the requirements upcoming wideband/multipixel receivers. A multi-section layout of three sections have been used here, since three sections is enough to meet the requirements, and the same time keep the circuit compact.

B. Substrate selection

The possibility to use a substrate with reasonably high dielectric permittivity allows reaching compact design with minimum insertion loss; the latter depending on dielectric loss and conducting loss in the transmission lines. The miniature hybrid chip allows it to be integrated into virtually any sideband separating mixer, which is especially advantageous for multi-pixel receivers or low noise balanced amplifier layouts. Moreover, due to the multi-section design, it would be advantageous to use a substrate with high dielectric constant in order to minimize the overall chip dimensions. Therefore, Alumina substrate was selected as it exhibits excellent mechanical properties, low loss tangent, good thermal conductivity, and a dielectric constant of 9.6 at the design frequency.

C. Three section hybrids

The proposed design employs a symmetrical three-section coupler. The even mode characteristic impedance and for each stage of the multi-section coupler can be found in tables for a given equi-ripple across the band [10]. Two hybrids with different bandwidths were designed. The design procedure in both cases are the same. First step is to calculate the coupling coefficients of each section of the hybrid is calculated. The center section has the highest coupling coefficient, while the adjacent sections are loosely coupled. Therefore, the first and the third sections consist of coupled line coupler structure whereas the middle section coupler consists of a Lange coupler. Through these choices of couplers, the final structure becomes planar and the hybrid dimensions are completely determined by photolithography process. The miniaturized 3-section hybrid chip are made using thin-film technology, which utilizes gold plated transmission lines and air bridges to connect the fingers of the Lange coupler (middle section). A schematic of the three-section hybrid is illustrated in Fig. 2.

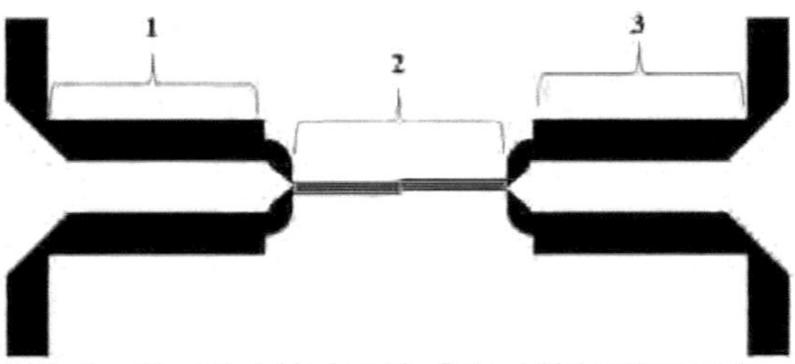

Fig. 2: The 3-section IF hybrid schematic, first and third sections are the coupled line couplers whereas the second section is the Lange coupler.

The initial dimensions for the multi-section coupler were calculated with Keysight ADS [10] LineCalc, using the coupling coefficients described above. The complete hybrid structure was optimized with Keysight Momentum [11]. The total length and width of the hybrid are 18.3 mm and 5 mm respectively. Fig.3 and Fig.4 shows the simulated performance of the 3.5-12.5 GHz and 4-16 GHz hybrid with excellent amplitude imbalance performance. Moreover, by employing superconducting Nb lines as described in [12] the performance can be further improved.

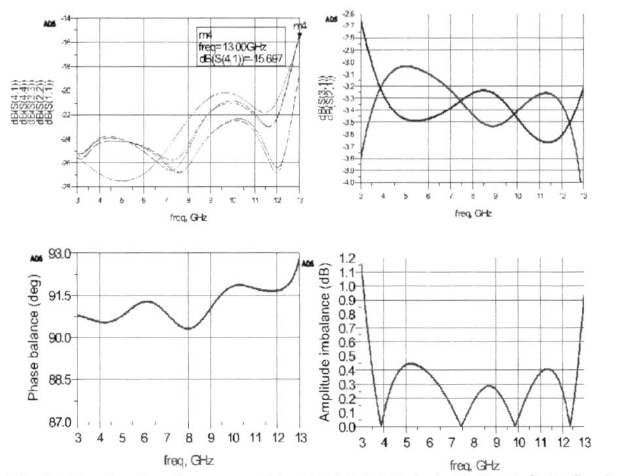

Fig. 3: Simulated performance of the 3.5-12.5 GHz hybrid. Top left: Reflection and isolation plot (dB). Top right: Though and coupled plot (dB). Bottom left: Phase balance (deg). Bottom right: Amplitude imbalance (dB).

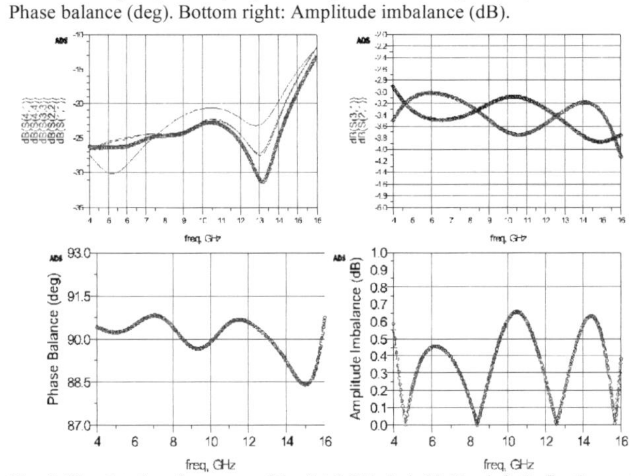

Fig. 4: Simulated performance of the 4-16 GHz hybrid. Top left: Reflection and isolation plot (dB). Top right: Though and coupled plot (dB). Bottom left: Phase balance (deg). Bottom right: Amplitude imbalance (dB).

III. Measurement Results

A. Characterization setup

The fabricated chips were mounted in a housing featuring 4 K-type connectors, which was then inserted in a cryostat for cooling down the samples down to 4K, using a close cycle machine. 4-port S-parameters measurements were carried out at room temperature (293 K) and at cryogenic temperatures (~4 K) using a Rohde&Schwarz ZVA40 vector network analyser (VNA). Owing to the lack of calibration standards at cryogenic temperatures, the S-parameters measurements at 4K were conducted with using a special calibration procedure for cryogenic temperature, as described in details in [6]. The measured performance of the hybrids is shown in Figs. 5-9.

B. 3.5-12.5 GHz hybrid measured at 293 K

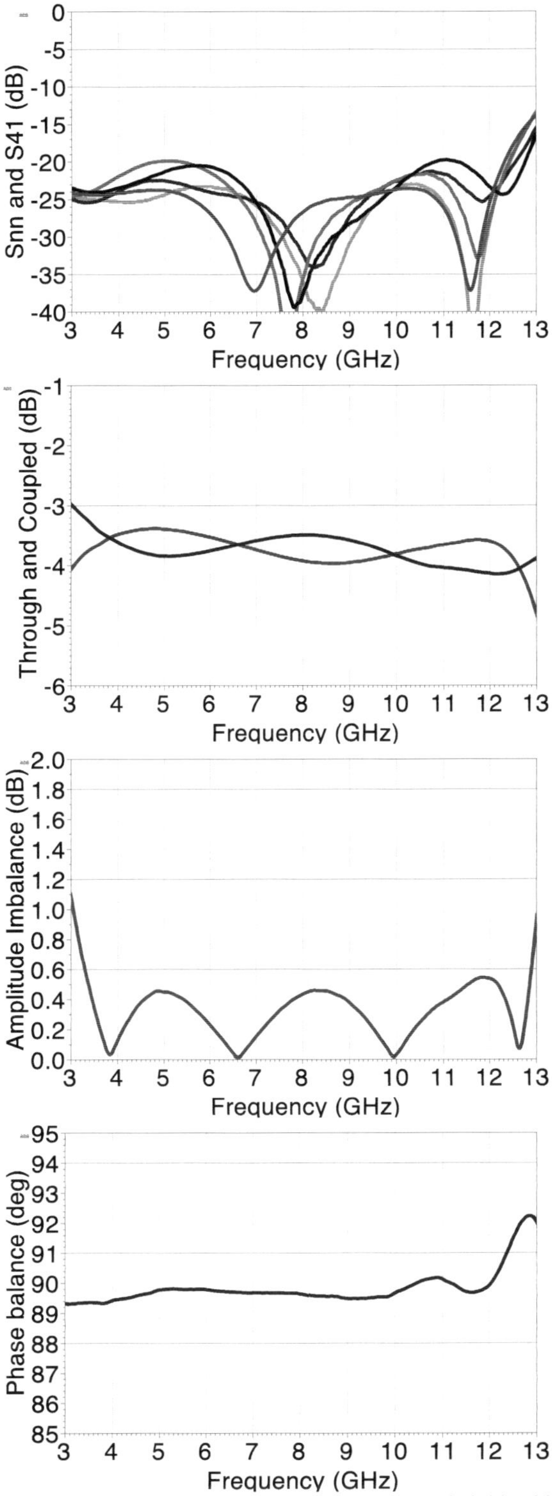

Fig. 5: Measured performance of the 3.5-12.5 GHz hybrid at 293 K. Figures from top to bottom represent: Reflection and isolation plot (dB); Though and coupled plot (dB); Amplitude imbalance (dB); Phase balance (deg.)

C. 3.5-12.5 GHz hybrid measured at 4 K

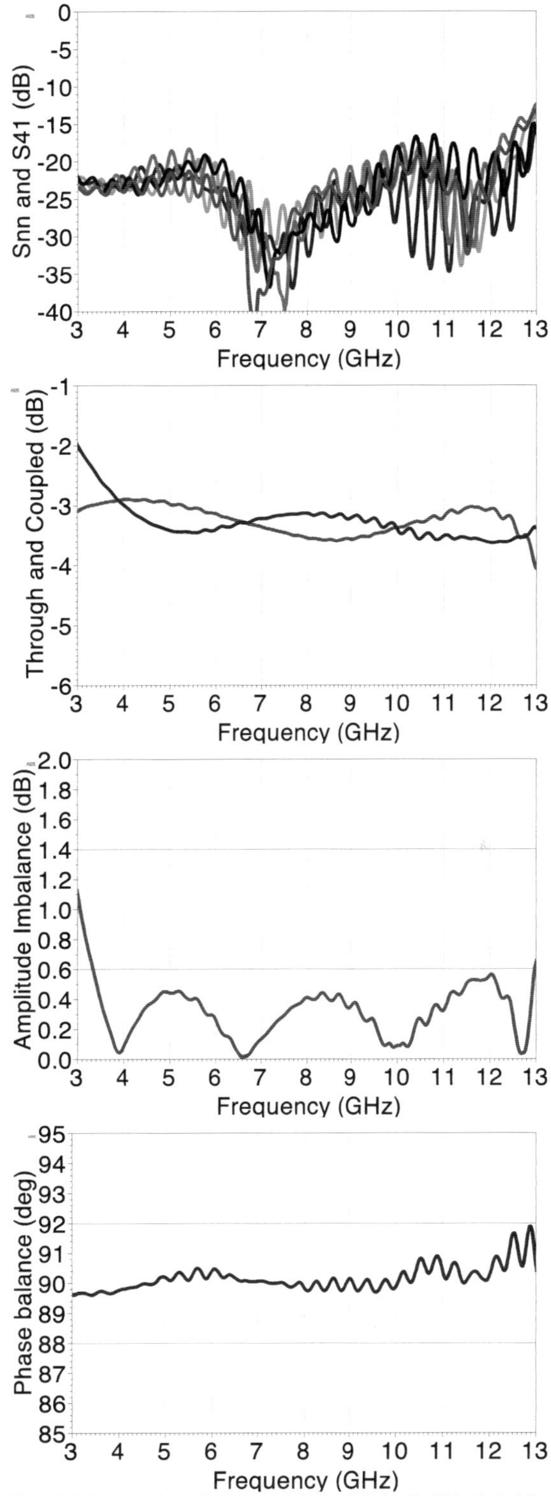

Fig. 6: Measured performance of the 3.5-12.5 GHz hybrid at 4 K. Figures from top to bottom represent: Reflection and isolation plot (dB); Though and coupled plot (dB); Amplitude imbalance (dB); Phase balance (deg.).

D. 4-16 GHz hybrid Measured at 293 K

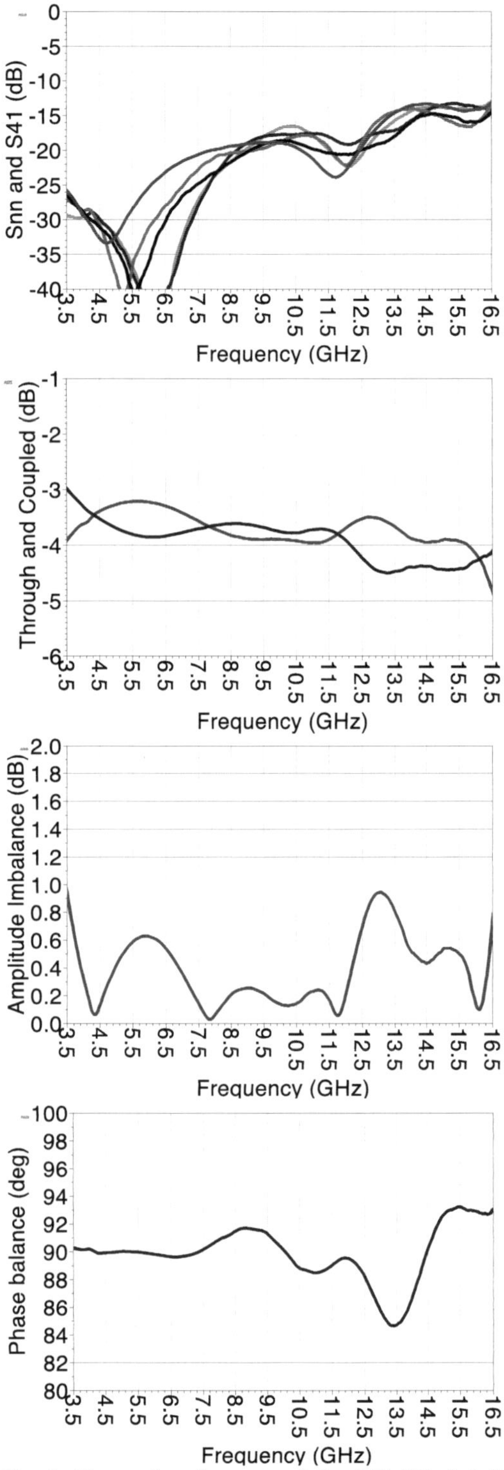

Fig. 7: Measured performance of the 4-16 GHz hybrid at 293 K. Figures from top to bottom represent: Reflection and isolation plot (dB); Though and coupled plot (dB); Amplitude imbalance (dB); Phase balance (deg.)

E. 4-16 GHz hybrid Measured at 293 K

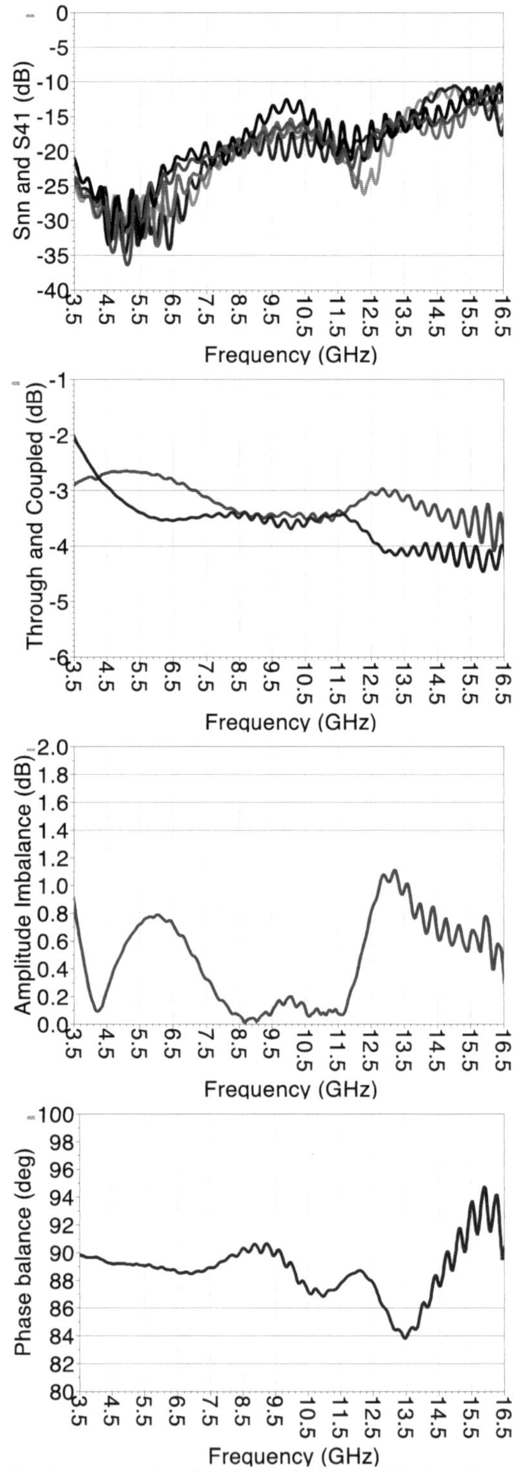

Fig. 8: Measured performance of the 4-16 GHz hybrid at 4 K. Figures from top to bottom represent: Reflection and isolation plot (dB);

Though and coupled plot (dB); Amplitude imbalance (dB); Phase balance (deg.)

F. 4-12 GHz superconducting hybrid measured at 4K

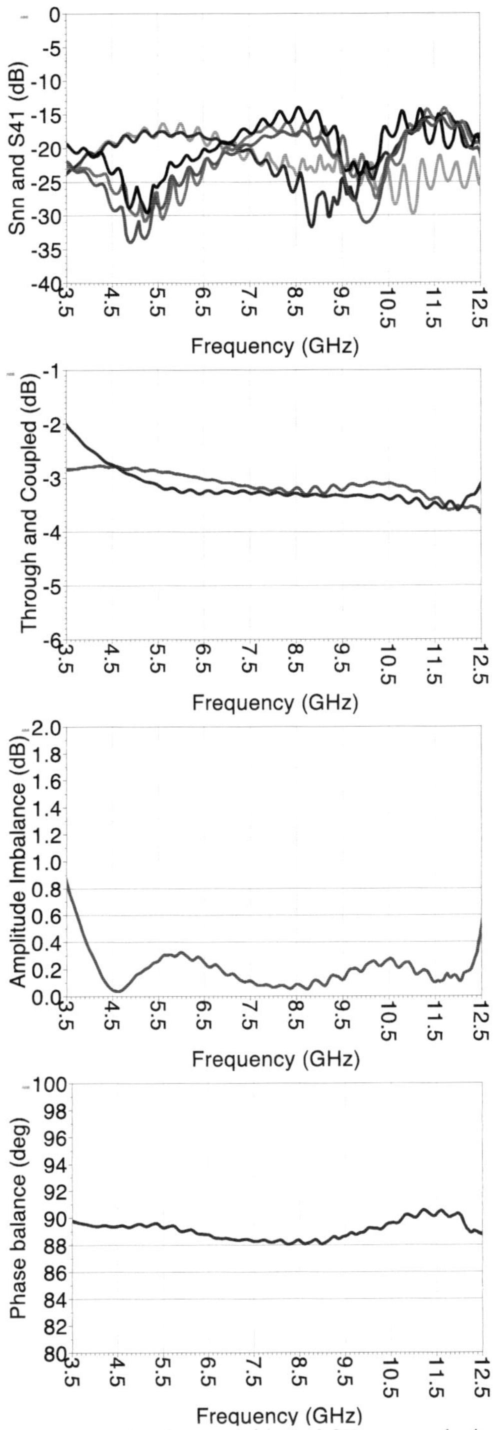

Fig. 9: Measured performance of the 4-12 GHz superconducting hybrid at 4 K. Figures from top to bottom represent: Reflection and isolation plot (dB); Though and coupled plot (dB); Amplitude imbalance (dB); Phase balance (deg.).

IV. Discussion

Comparing fig. 5 to 6, it can be seen that the amplitude and phase imbalance 3.5-12.5 GHz hybrid is not affected by the change in temperature, even though the insertion loss is clearly improved when the devices are operated at 4K, most likely due to the increase of conductivity of the metallic microstrip lines when cooling down the device.

Despite a slight overcoupling, we observe a similar relative independence of the amplitude and phase imbalance as well as an improvement of the insertion loss for the 4-16 GHz hybrids, as show on as shown on fig 7 and fig 8. However for this particular hybrid, we observe a discrepancy in return loss between the measurement results Fig.7 and the simulation in Fig. 3 is probably due to due mounting or/and fabrication

The 4-12 GHz hybrid made of superconducting transmission lines shows measured 4-12 GHz hybrid with almost ideal performance and is very suitable for integration into 2SB THz receivers based on superconducting mixers.

V. Conclusions

In this work, we report on the design and characterization of a 3.5-12.5 GHz and 4-16 GHz quadrature 3 dB directional couplers. The compact size of the hybrid chip allows it to be integrated into virtually any sideband separating (2SB) mixer operating at room temperature or cryogenic temperatures and is furthermore especially advantageous for multi-pixel 2SB receivers or balanced amplifier layouts.

The hybrid was fabricated using in-house thin-film technology on Alumina substrate and uses air bridges to inter-connect the fingers of the Lange coupler. The four-port S-parameters measurement showed very good agreement with the simulation results

Acknowledgment

The authors would like to acknowledge Sven-Erik Ferm for his help with the fabrication of the mechanical fixtures to for the measurement tests.

References

1] C. E. Groppi and J. H. Kawamura, "Coherent Detector Arrays for Terahertz Astrophysics Applications," *Terahertz Science and Technology, IEEE Transactions on,* vol. 1, pp. 85-96, 2011.

[2] I. Malo-Gomez, J. D. Gallego-Puyol, C. Diez-Gonzalez, I. Lopez-Fernandez, and C. Briso-Rodriguez, "Cryogenic Hybrid Coupler for Ultra-Low-Noise Radio Astronomy Balanced Amplifiers," *Microwave Theory and Techniques, IEEE Transactions on,* vol. 57, pp. 3239-3245, 2009

[4] E. Sundin, A. Ermakov, V. Desmaris, H. Rashid, D. Meledin, M. Bylund and V. Belitsky, "Cryogenic IF Balanced LNAs Based on Superconducting Hybrids for Wideband 2SB THz Receivers", *Proceedings of the ISSTT2017*, Cologne, Germany, 2017.

[5] H. Rashid, D. Meledin, V. Desmaris, and V. Belitsky, "Novel Waveguide 3 dB Hybrid With Improved Amplitude Imbalance," *Microwave and Wireless Components Letters, IEEE,* vol. 24, pp. 212-214, 2014.

[6] H. Rashid, D. Meledin, V. Desmaris, A. Pavolotsky, and V. Belitsky, "Superconducting 4-8 GHz Hybrid Assembly for 2SB Cryogenic THz Receivers," *Terahertz Science and Technology, IEEE Transactions on,* vol. 4, pp. 193-200, 2014.

[7] H. Rashid, V. Desmaris, V. Belitsky, M. Ruf, T. Bednorz, and A. Henkel, "Design of Wideband Waveguide Hybrid With Ultra-Low Amplitude Imbalance," *Terahertz Science and Technology, IEEE Transactions on,* vol. PP, pp. 1-8, 2015.

[6] V. Belitsky *et al.*, "ALMA Band 5 receiver cartridge: Design, performance, and commissioning", *Astronomy and Astrophysics*, vol. 611, A68, 2018.

[7] V. Belitsky *et al.*, "SEPIA – a new single pixel receiver at the APEX telescope", *Astronomy and Astrophysics*, vol. 612, A23, 2018.

[8] B. Billade, H. Rashid, V. Desmaris, and V. Belitsky, "Superconducting 4 - 8 GHz IF Hybrid for Low Noise mm-Wave Sideband Separation SIS Receiver," *Microwave and Wireless Components Letters, IEEE,* vol. 22, pp. 589-591, 2012.

[9] H. Rashid, P. Y. Aghdam, D. Meledin, V. Desmaris, and V. Belitsky, "Wideband Planar Hybrid With Ultralow Amplitude Imbalance," *IEEE Microwave and Wireless Components Letters,* vol. 27, pp. 230-232, 2017.

[10] E. G. Cristal and L. Young, "Theory and Tables of Optimum Symmetrical TEM-Mode Coupled-Transmission-Line Directional Couplers," *Microwave Theory and Techniques, IEEE Transactions on,* vol. 13, pp. 544-558, 1965.

[11] ADS. *www.keysight.com.*

[12] V. Belitsky *et al.*, "Superconducting microstrip line model studies at millimetre and sub-millimetre waves". *International Journal of Infrared and Millimeter Waves*. 27. Pp. 809-834, 2006.

Prototype Design of a Dielectrically Embedded Mesh Lens

Jeremy D. Whitton[1,*], Philip D. Mauskopf[1], Paul F. Goldsmith[2], Kristina K. Davis[1], Christopher E. Groppi[1]
[1]School of Earth and Space Exploration, Arizona State University, Tempe, AZ 85283, USA
[2]Jet Propulsion Laboratory, California Institute of Technology, Pasadena, CA 91109, USA
*Contact: jeremy.d.whitton@gmail.com

Abstract—**Here we present a prototype design for a dielectrically embedded mesh lens consisting of stacked layers of printed circuit board (PCB) material and embedded copper elements. The dielectrically embedded mesh lens consists of layers of dielectric which contain sub-wavelength-dimension metal elements laid out in a grid fashion, and is both flat and lightweight. It has been demonstrated that the sizes of these metal elements can be varied according to their position in the apparatus, using models based on transmission line theory, to create a lens which focuses a plane wave at millimeter wavelength to a Gaussian beam with very low transmission loss, even without the use of antireflective coating. We present the phase design for our lens which was designed, using transmission line theory and electromagnetic modelling software, to operate at 20GHz. We further present an analysis of the transmission line components which will make up the lens.**

I. Introduction

There is currently a burgeoning field of research, both in industry and academia [2], into the use of CubeSats for remote sensing purposes, both active, for observing the Earth and other nearby systems from space, and passive, which can be used in addition for astronomical observations [1]. For remote sensing in astrophysical applications particularly, signal-to-noise ratios are very low, and so long integration times are often required. This problem is accentuated by the fact that CubeSats are small: 10cm x 20cm x 30cm, with a weight budget of 8kg, is a common CubeSat form factor [1], which puts tight constraints on the sizes of CubeSat systems, and especially lens or antenna apertures. This requirement can be ameliorated somewhat by the use of low-loss systems and antennas or lenses that have as large an aperture size as is feasible. One suggested way to maximize the signal obtained by a CubeSat is the use of inflatable antennas, deployable after the CubeSat is launched [3]. This technology is still, to some degree, in its infancy.

Another possible approach is to use a metamaterial lens as an aperture. Metamaterials, which involve the structure embedding of metal meshes within dielectric substrates, have recently seen much advancement and development [6]. Metamaterial lens technology is also quite recent, though there have been advancements made which allows them to be designed and fabricated to operate at millimeter wavelengths [4]. A metamaterial lens offers a number of attractive qualities for use in CubeSats: they can be made flat and thin (in contrast to a conventional geometrically curved lens), and therefore do not count heavily against the CubeSat's small weight budget. It also makes it easier to place and stow the lens, if a deployable design is necessary or desirable. Furthermore, these design techniques by [4], ensure that no anti-reflection coating is necessary to minimize reflection losses. Finally, such lenses have been found to, at least theoretically, have less than half a dB of loss, which is significantly better than that of a Fresnel zone plate lens, which can exhibit on the order of 3 to 4dB or more of loss [5].

In this paper, we wish to recreate the design procedure used by [4], to create a prototype design for a metamaterial lens. We have elected to design the lens for use at 20GHz using layers of RO3003 PCB material and copper, with the intention that the lens will be fabricated using existing well-known PCB fabrication techniques, rather than the more complicated photolithography and layering techniques used by [4]. It is hoped that this will provide a cost-effective method for prototyping and testing our design process before we move on to create lenses to operate at higher frequencies. Our lens is also designed with a focal length to diameter ratio of f/0.37, as compared to the longer-focal length f/3 design achieved previously, which we believe would be easier to design around for the purposes of CubeSat deployment.

II. Materials and Methods

A. Theory

The design for the lens is based primarily on quasi-optics and, in particular, the physics of Gaussian beams. The purpose of our metamaterial lens, like any remote-sensing lens, is to focus an incoming plane wave to a beam which converges at some focal distance *f*. In order to accomplish this, it is necessary for the phase of the incoming plane wave to be transformed to that of a converging Gaussian beam. This phase transformation is given by [4, 7] as

$$\phi(r) = -\frac{\pi r^2}{\lambda R}$$

where r is the distance on the lens plane from the lens center, and R is the radius of curvature of the phase front, given by

$$R(f) = f + \frac{\left(\frac{\pi w_0{}^2}{\lambda}\right)^2}{f}$$

and w_0 is the waist of the focused beam at the focal plane.

The way in which the lens then produces this desired phase shift may be described theoretically by transmission line (TL) theory. The lens may be subdivided in the plane of the lens into many TLs, each of which ideally produces a single phase shift determined by their distance from the center of the lens. This is realized by designing the embedded metal elements in the lens to lie on a grid pattern, aligning with these transmission lines, and choosing the size of the metal elements accordingly [4].

B. Design Procedure

Our design procedure closely follows that of [4] In order to perform these optimizations to find optimal TLs, we first require, as noted earlier, S-parameters of individual elements to be used in the TLs. There are two types of these: substrate layers, of g by g by l; and embedded metal element layers, which were modeled as metal squares within an area of vacuum. The dimensions of the vacuum area were g by g by m, where g is the grid distance between the centers of adjacent metal elements, m is the thickness of the metal element, and l is the thickness of a substrate layer. The dimensions of the metal element are b by b by m. The parameter b may be changed for every individual element in a TL according to the requirements of the optimizer.

Given these models, S-parameters were calculated using finite element methods in the commercially available software HFSS. To perform these calculations, Floquet ports were used, implying that our S-parameters represent the output transmission magnitude and phase, relative to the input, of a plane wave which goes through an infinite periodic array of identical models. The periodicity was given length scale g. S-parameters were calculated in this manner from both the substrate model and the embedded element model, and de-embedded to the appropriate dimensions of each. For the embedded elements, S-parameters were calculated across a sweep of 48 different b values, with b going from $0.02g$ to $0.98g$.

This set of S-parameters was exported and loaded into Matlab, and converted into ABCD matrices. Matlab's nonlinear optimizer was used to optimize for sets of b values which produce maximal transmission and the desired output phase. Note that we choose to calculate the S-parameters of individual elements, rather than entire transmission lines, because in general, optimizations may take thousands of iterations or more. To do thousands of iterations in HFSS could take days to optimize for a single transmission line, whereas pre-calculating S-parameters for individual elements and combining them with transmission line theory means optimizations take seconds. The goal function to accomplish this optimization was defined as follows: given a set of b values, choose the ABCD matrix for the metal elements with those b values. Cascade them together

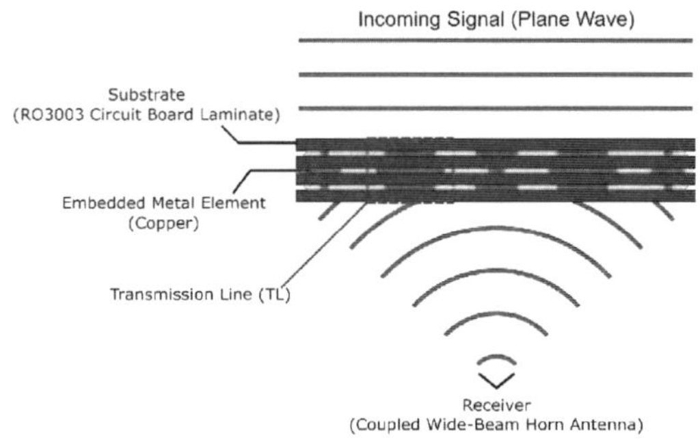

Fig. 1. Schematic diagram of metamaterial lens, not to scale

in the appropriate order, as in [8], with substrate ABCD parameters in between. Then find the S_{21} of the resulting transmission line. The function to be maximized is

$$\text{mag}(S_{21}) \cdot \cos(\phi - \text{ang}(S_{21}))$$

where ϕ is the desired output phase and $\phi - \text{ang}(S_{21})$ will be referred to here as the phase error. In our case, this optimization was only performed for a single frequency, though it can in principle be extended to optimization over a particular bandwidth.

C. Design Specifics

With this procedure, we designed a lens according to the following specifications: the substrate dielectric is RO3003. The embedded metal elements are copper. The intended operating frequency is 20GHz. The grid spacing g is a tenth of a wavelength, equal to 1.5mm. The metal layer thickness is 35 microns, corresponding to 1oz copper cladding. The thickness of the substrate layers was constrained to 0.50mm (20 mil), which is a standardly available thickness of RO3003. The lens is composed of 10 layers of metal elements embedded within 11 substrate layers, as using fewer layers did not seem to provide sufficient transmission. The lens is a 30cm x 30cm square, with an 11cm focal distance. It is designed to couple to an antenna with a HPBW of 60 degrees: given our choice of focal length, this ensures it should capture about 99% of the beam power of the antenna. It was decided that 120 TLs would be optimized to be incorporated into the lens, which gives 1 TL

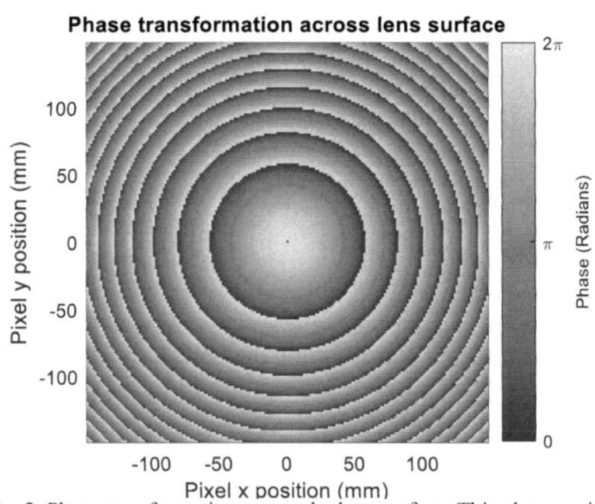

Fig. 2. Phase transformation across the lens surface. This phase varies rapidly at the edges of the lens due to its short focal distance.

for every 3 degrees of phase shift between 0 and 360. In Fig. 2, the resulting phase design of the lens is depicted.

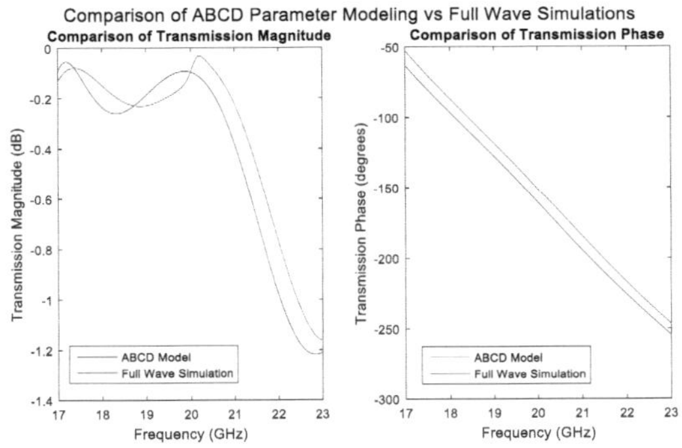

Fig. 3. Comparison of TL theory model vs. full wave simulation

III. Results

Given that our design relies upon the use of TL modelling in place of full wave simulations to determine the properties of our TLs, it was important to test is how well modelling the layering of substrate and metal elements with TL theory works. In **Error! Reference source not found.**, we show results comparing the S-parameters of our TL model vs. the S-parameters for the same structure obtained in HFSS via a full wave simulation. A number of optimized sets of b values were tested in this way, and this particular one exhibited the worst phase error. In spite of this, the phase error between the two is 8 degrees. We note also that the transmittances of the two models do not match perfectly, but are very close, generally off by no more than 0.2dB.

We also wanted to determine how high a transmission coefficient could be obtained across the range of desired output phases of our TLs. In Fig. 4, we demonstrate the transmittances of the 120 TLs which were optimized across an output phase range of 360 degrees. Across much of the range, the transmittance was above 98.5%; it was above 95% for the entire range. We note also that there is a distinct range of phases between ~100-150 degrees where the transmittance and phase-weighted transmittance drop sharply. This indicates that there is effectively a trade-off within this range between maximizing transmittance and minimizing the phase error. We found also, that, when using TLs with fewer than 10 layers, the reachable transmittance would in general drop lower. For example, with 8 layers of embedded elements, the transmittance was reduced to around 80%.

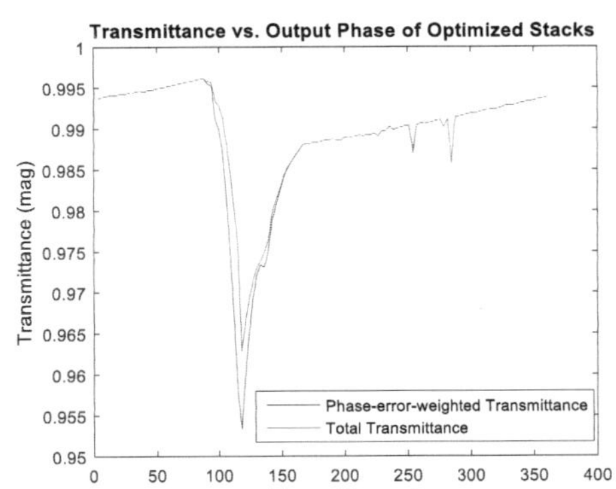

IV. Conclusion

We have here successfully designed TLs, using 10 layers of embedded metal elements, to be incorporated into a 20GHz lens design which have transmittances of over 95%. This falls somewhat short of the findings of [4], who were able to optimize for an average transmission of 97.5% across an entire 30% bandwidth. It may be that the optimization procedures previously used were more rigorous, and the procedures used here did not find global maxima within our parameter ranges. We also have no explanation for the significant drop-off in optimized transmittances between 100 and 150 degrees, and it is suggested that this could be studied in the future to determine if it is caused by non-rigorous optimization or some more inherent difficulty.

We do note that our design successfully constrained the thickness of the substrate layers to a single, standardly-available value, while the method in [4] required the optimization of the thicknesses of these layers. Our design has further been engineered to be simple to fabricate with standard PCB printing methods.

We also note that our simulation analysis of the transmission lines assumed periodicity in the lens plane of the structures which were tested. This means that our model may fail to take into account potential couplings between adjacent TLs which are different from each other.

Finally, we were unable to perform a full-wave simulation of the entire modeled lens, owing to unrealistic memory requirements. It is possible such a simulation could be performed by a supercomputer. It is suggested that future work will entail building a physical prototype based on the design procedure presented here, and measuring it to determine whether or not the 95% expected transmittance is realizable in practice.

Acknowledgments

This work was supported in part by JPL award SP.17.0001.003.

References

[1] E. Peral et al., "Radar Technologies for Earth Remote Sensing From CubeSat Platforms," in Proceedings of the IEEE, vol. 106, no. 3, pp. 404-418, March 2018. doi: 10.1109/JPROC.2018.2793179

[2] Pina, L., et al. "X-ray monitoring for astrophysical applications." Advances in X-Ray/EUV Optics and Components IX. Vol. 9207. International Society for Optics and Photonics, 2014.

[3] Babuscia, Alessandra, et al. "Inflatable antenna for cubesats: Motivation for development and antenna design." Acta Astronautica 91 (2013): 322-332.

[4] Pisano, Giampaolo, et al. "Dielectrically embedded flat mesh lens for millimeter waves applications." Applied optics 52.11 (2013): 2218-2225.

[5] Rodriguez, J. M., Hristo D. Hristov, and Walter Grote. "Fresnel zone plate and ordinary lens antennas: Comparative study at microwave and terahertz frequencies." Microwave Conference (EuMC), 2011 41st European. IEEE, 2011.

[6] Pisano, G., et al. "Metal mesh based metamaterials for millimetre wave and THz astronomy applications." Millimeter Waves and THz Technology Workshop (UCMMT), 2015 8th UK, Europe, China. IEEE, 2015.

Fig. 4. Over 95% transmittance is demonstrated over the lens surface

[7] Goldsmith, Paul F., Institute of Electrical and Electronics Engineers, and Microwave Theory and Techniques Society. Quasioptical systems: Gaussian beam quasioptical propagation and applications. New York: IEEE press, 1998.
[8] Pozar, David M. "Microwave Engineering 3e." Tramsmission Lines and Waveguides (2005): 143-149.

Tolerancing of the Submillimeter Array Optics using Physical Optics Simulations

Paul K. Grimes[1*], Scott N. Paine[1], Ramprasad Rao[2], Tirupati K. Sridharan[1] and Lingzhen Zeng[1]
[1]*Harvard-Smithsonian Center for Astrophysics, Cambridge, MA 02138, USA*
[2]*Institute of Astronomy and Astrophysics, Academia Sinica, Submillimeter Array Observatory, Hilo, HI 96720, USA*
[*]Contact: pgrimes@cfa.harvard.edu

Abstract—The Submillimeter Array (SMA) is an 8 element interferometer on Maunakea, Hawaii, operated by the Smithsonian Astrophysical Observatory and the Academia Sinica Institute of Astronomy and Astrophysics. Each of the 6m antennas houses a single cryostat containing four heterodyne receivers, operating between 180 GHz and 418 GHz. Following recent upgrades to the SMA's correlator system, routine observations with the SMA now use two receivers at the same time with each receiver receiving one of the two linear polarization signals from the sky. Observations may be made either with the two receivers tuned to different frequencies, or with both receivers tuned to the same frequency and either combining the outputs for greater sensitivity or forming the full Stokes parameters for polarimetric observations.

In order to achieve maximum efficiency in observations in any of the observing setups, the receivers must all be co-aligned, so that the peaks of the primary beam patterns from each antenna for each receiver are all pointed at the source being observed. At the same time, each receiver must illuminate the dish correctly, to maximize the aperture efficiency. Each receiver has its own feedhorn and focusing lens in the cryogenic receiver insert, and an ambient temperature optics insert used to couple the local oscillator signal into the sky signal and to transfer the sky signal from the moveable wire grid and mirror that separate the sky signal into the two linear polarizations and direct it to the selected receivers. We would like to quickly measure and adjust the pointing and aperture illumination of each SMA antenna for each receiver to ensure that each receiver and its associated optics is correctly aligned.

We have carried out a set of TICRA GRASP simulations of the SMA beam waveguide optics, in which the positions of the optical elements are varied in order to predict the expected changes in the pointing, beam pattern, and aperture illumination, due to misalignments in the receiver optics. These simulations are used in conjunction with 1- and 2-dimensional holography using a beacon overlooking the SMA site to prescribe adjustments to elements within the receiver optics to maximize the antenna efficiency and beam overlap of the SMA receivers. In this paper we present the physical optics simulation set up within GRASP, the key results of the effects of alignment tolerances of various elements of the SMA beam waveguide on the antenna efficiency and pointing alignment, as well as measurement results from holography that validate the physical optics simulations.

Wafer-bonded Antireflection Layers for Silicon optics

F. Defrance[1*], G. Chattopadhyay[2], J. Connors[3], S. Golwala[1], M. I. Hollister[4], C. Jung-Kubiak[2], E. Padilla[5], S. Radford[3], J. Sayers[1], E. C. Tong[3], and H. Yoshida[1]

[1]*California Institute of Technology, Pasadena, 91125, US*
[2]*NASA Jet Propulsion Laboratory, Pasadena, 91109, US*
[3]*Harvard–Smithsonian Center for Astrophysics, Cambridge, 02138, US*
[4]*Fermi National Accelerator Laboratory, Batavia, 60510, US*
[5]*Cal State San Bernardino, San Bernardino, 92407, US*
[*]Contact: fdefranc@caltech.edu

Abstract— Many applications in astronomy from tens of GHz to THz frequencies, on the ground and in space, would benefit from silicon optics because silicon's high refractive index and low loss make it an ideal optical material at these frequencies. Silicon can also be used for ambient temperature vacuum windows, however, it's large refractive index necessitates an antireflection coating. Moreover, multilayer antireflection treatments are necessary for wide spectral bandwidths, with wider bandwidths requiring more layers. To this end, we are developing multilayer coatings for silicon by bonding together wafers individually patterned with deep reactive ion etching (DRIE).

While a standard approach to antireflection coating is to deposit or laminate dielectric layers of appropriate refractive index, it is difficult (but not impossible) to find low loss dielectrics with the correct refractive index and other properties to match silicon well, especially if more than one layer is required, operation up to THz frequencies is desired, and/or the optic will be used cryogenically. Textured surfaces are an attractive alternative to dielectric antireflection coatings. For millimeter wavelengths, multi-layer antireflection textures with up to 4:1 bandwidths have been cut successfully into silicon lens surfaces with a dicing saw, but this technique becomes unusable at frequencies of 300 GHz and higher given the saw dimensions. Laser machining is being explored but demonstrations are not yet available. DRIE works well on flat surfaces (and has been demonstrated for narrowband windows to THz frequencies), but there are limits to the depth and aspect ratio of the features it can create. Furthermore, etching has not been adapted to large, curved optics.

We are pursuing a hybrid approach to this problem: construct a silicon optic by stacking flat patterned wafers. The starting point is a multilayer optical design incorporating both an axial gradient in the refractive index for antireflection and a radial index gradient for focusing. For each optical layer, a hole or post pattern is used to achieve the required effective index of refraction. Using a novel multilayer etching procedure, several layers of the optical structure are fabricated on a flat wafer. Several individually patterned wafers are stacked and bonded together to produce the completed optic. This approach can thus address the aspect ratio limitations of DRIE, and it obviates etching on curved surfaces.

We present our results to date, which include simulations, fabrication and measurements of 2- and 4-layer coatings with wafer-bonding, on high resistivity silicon wafers, at 75-330 GHz. The good agreement between the simulations and the test results validates the fabrication and test setup, and allows us to continue the development of larger bandwidth and more efficient coatings. Our near-term goal is to produce a 10-cm lens with a 7-layer coating providing 5.5:1 bandwidth from 75 to 420 GHz, with less than 1% reflection, eventually scaling up to 15-cm, 30-cm, and larger elements.

Single-Chip 135-160 GHz Doubler with more than 150 mW Output Power based on Discrete Schottky Diodes

D. Moro-Melgar[1*], O. Cojocari[1], I. Oprea[1], M. Rickes[1] and H. Hoefle[1]
[1]ACST GmbH, Hanau, 63457, Germany. (http://www.acst.de/)
*Contact: diego.moro-melgar@acst.de

***Abstract*— Single chip doubler based on novel diamond-based discrete Schottky diodes is presented here. CVD Diamond technology has been integrated by ACST GmbH in the discrete diode structure to implement in high power multipliers development. A single chip 135-160 GHz doubler able to handle up to 400 mW input power and provide more than 150 mW output power is reported. Conversion efficiency higher than 30 % is performed in band and up to 40% at some frequencies point.**

I. Introduction

Schottky diode technology has demonstrated during last decades to be a good option to achieve THz frequencies. This technology is particularly interesting because of the performance provided at room temperature. Schottky modules do not require any cryogenic systems to work and it makes them very compact and robust. These properties have encouraged the use of Schottky diodes in space-born applications like MIRO [1], HERSEL [2] and more recently in JUICE-SWI [3]. The THz modules proposed in the submillimeter wave instrument for JUICE is completely based on Schottky multipliers and mixers to cover the frequency range from 540-640 GHz and 1080-1280 GHz [4]. However, the power handling capabilities of Schottky technology at THz frequencies is the main limiting factor to provide higher power. The local oscillator power requirements of the 1.2 THz mixer in JUICE-SWI was address using power combining techniques at 135-160 GHz and 270-320 GHz multiplication stages [4]. There parallelization of multipliers [5] or/and chips [6] are the typical approaches proposed for power-combining applications. However, power-combining techniques increase both the complexity and the amount of resources implemented in every Schottky multiplier. Additionally, power combining usually requires the implementation of power splitters that increase the complexity of the design and fabrication costs. The interest of the JUICE-SWI frequency range and the demand of more powerful Schottky multipliers has motivated the work reported in this paper. ACST GmbH [7] has developed a new kind of Schottky diodes to address power handling capabilities of Schottky multipliers. The new Schottky diodes developed at ACST integrate CVD diamond in the structure. A 135-160 GHz doubler based on these novel diodes has been developed to handle at least twice the input power reported by other authors at similar frequencies [8], [9] while avoiding power combining techniques. We report in this paper a Schottky doubler able to provide even more power than the reported in [5] and [10] using power combining, and also a state-of-the-art efficiency comparable to MMIC-based doublers [10].

II. Doubler Design

An illustrative image of the diamond-based Schottky diode developed at ACST is presented in Fig. 1 (left side). The Schottky diode structure integrates a CVD diamond substrate. The diamond substrate is physically connected to the anodes to more efficiently distribute the heat to the contact pads used to place the diode on the chip. Two discrete diodes with three anodes each have been used in the 135-160 GHz doubler design. The design of the chip is based on the doublers used in SWI [4] and [11]. However, the MMIC design has been modified in this case to place two discrete diodes in antiparallel balance configuration. The chip architecture has been defined in AlN substrate to further improve thermal distribution of the heat generated in the diodes.

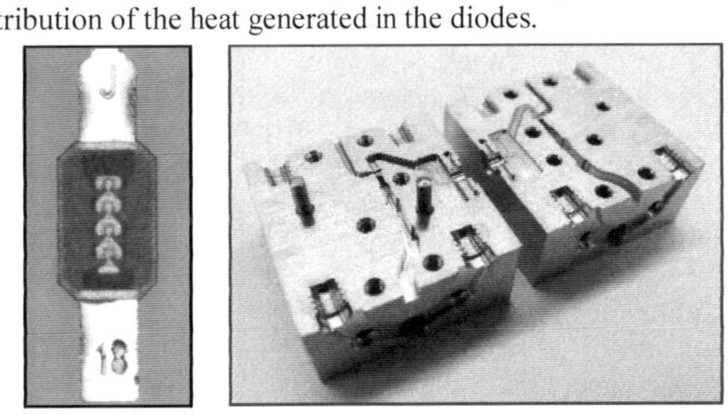

Fig. 1. Representative structure of diamond-based Schottky diodes developed by ACST (left). Mechanical part of the single chip 135-160 GHz doubler (right).

The mechanical design is shown in Fig. 1 right, and it is possible to notice a single waveguide module with WR12 input waveguide flange and WR6.5 output flange. The ACSTs varactor Schottky diamond-diodes used in this doubler have a breakdown voltage between -13.5 V and 14 V per anode. The doubler has been designed to optimally handle up to 400 mW (26 dBm) input power. The bias is provided via an SMA connector and the chip does not require more than -15 V bias.

III. Results

A. Experimental Setup

The experimental setup is shown in Fig. 2 and it consists of a signal generator used to provide an 8.125-10 GHz input signal to an 8x active multiplication chain (AMC). Then the 65-80 GHz signal provided by the AMC is used to pump the 135-160 GHz doubler. The output power of the doubler is measured using a VDI PM4 through the corresponding WR6.5 to WR10 waveguide transition taper. An E-band isolator is placed after the AMC to reduce the presence of standing waves and a standard E-band directional coupler is placed between the 135-160 GHz doubler and the isolator. A E8486A Keysight power sensor head is placed in the third port of the directional coupler to monitor the input power in the 135-160 GHz doubler.

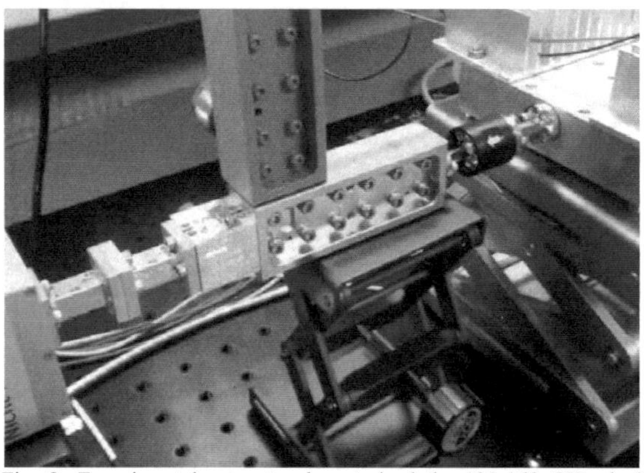

Fig. 2. Experimental setup to characterized the 135-160 GHz doubler developed at ACST.

B. Experimental Results

The experimental output power and efficiency in the 130-160 GHz frequency range are plotted in Fig. 3. The doubler has been characterized from 65-79.5 GHz with 24 dBm and 26 dBm fixed input power. The bias of the doubler has been optimized for every frequency in accordance with the input power. An output power between 19 dBm and 20 dBm is typically performed by this doubler with 24 dBm input power. More than 21 dBm is provided in band with 26 dBm input power and a peak value of 154 mW is achieved by this particular unit at 157 GHz. The efficiency at every frequency point shows a very flat performance of the doubler along the frequency band and also at different input power levels. More than 30 % efficiency is provided by the doubler in the specified frequency range and more than 35 % is achieved at certain frequencies. A peak efficiency of 38 % is performed by this unit at 140 GHz and 157 GHz with 24 dBm and 26 dBm input power, respectively.

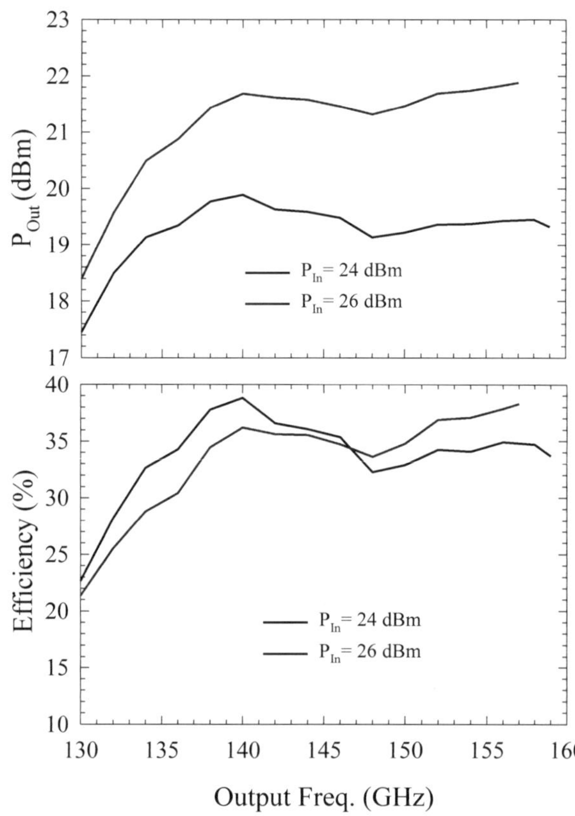

Fig. 3. Experimental output power (a) and efficiency (b) of the doubler from 130 to 160 GHz. The results are obtained with 24 dBm (black line) and 26 dBm (blue line) fixed input power.

The results obtained by the reported discrete-diode-based doubler are in agreement with the state-of-the-art efficiency achieved even by MMIC based doublers [8]. However, it is able to efficiently handle twice more power than other reported doublers at similar frequencies [8], [9]. Similar power handling capabilities with 25 % conversion efficiency have been previously reported by Dr. J. V. Siles at 170-200 GHz using Schottky MMIC chips [12]. These results indicate a state-of-the-art discrete-diode-based doubler in both power handling capabilities per chip and conversion efficiency.

Conclusions

A 135-160 GHz doubler based on a novel discrete Schottky diamond diodes is reported in this paper. The new type of diode developed at ACST GmbH has been implemented in the development of a single chip doubler able to efficiently handle up to 400 mW input power and provide more than 150 mW output power. The conversion efficiency is higher than 30 % in band and achieve up to 38 % in some frequency points. These are state-of-the-art results stablished by MMIC devices but using discrete diode technology.

References

[1] S. Gulkis, M. Frerking, J. Crovisier, G. Beaudin, P. Hartogh, P. Encrenaz, ... & R. Irigoyen. "MIRO: microwave instrument for Rosetta Orbiter". *Space Science Reviews*, vol. 128, no 1-4, p. 561-597, Nov. 2007.

[2] L. A. Samoska, T. C. Gaier, A. Peralta, S. Weinreb, J. Bruston, I. Mehdi, ... & H. Wang. "MMIC power amplifiers as local oscillator drivers for FIRST". In *Astronomical Telescopes and Instrumentation*. International Society for Optics and Photonics, p. 275-284, 2000.

[3] JUICE Jupiter Icy Moons Explorer SWI Submillimetre Wave Instrument (SWI), Science Concept Document, JUI-MPSSWI-PL-002, 13/05/2013.

[4] A. Maestrini, L. Gatilova, J. Treuttel, F. Yang, Y. Jin, A. Cavanna, D. Moro-Melgar,... & F. Dauplay. (2016, April). 1200GHz and 600GHz Schottky Receivers for JUICE-SWI. In *27th International Symposium on Space Terahertz Technology*, (April, 2016).

[5] B. Thomas, H. Gibson, A. Walber, I. Oprea, O. Cojocari & T. Nahri. "Power-combining of 186-210 GHz frequency doublers and W-band power amplifiers with more than 100 mW peak output power". *26th International Symposium on Space Terahertz Technology, Cambridge, MA,* 16-18 March, 2015.

[6] J.V. Siles, K. B. Cooper, L. Choonsup, L. Robert, G. Chattopadhyay, & I. Mehdi. "Next Generation of Room Temperature Broadband Frequency Multiplied LO Sources with 10 times Higher Output Power in the 100 GHz-1.9 Thz Range. In *29th International Symposium on Space Terahertz Technology*, (March, 2018).

[7] [DB/OL], http://www.acst.de/

[8] T. Waliwander. M. Fehilly and E. O'Brien. "An ultra-high efficiency high power Schottky varactor frequency doubler to 180–200 GHz". In *Millimeter Waves (GSMM) & ESA Workshop on Millimetre-Wave Technology and Applications, 2016 Global Symposium on* (pp. 1-4). IEEE. June 2016.

[9] C. Viegas, B. Alderman, C. G. Pérez-Moreno, J. Powell, C.I. Duff and R. Sloan. "Design and thermal analysis of a high-power frequency doubler at 160 GHz". In *Microwave and RF Conference (IMaRC), 2016 IEEE MTT-S International* (pp. 1-3). IEEE, December 2016.

[10] C. Viegas, H. Liu, J. Powell, H. Sanghera, A. Whimster, L. Donoghue, P.G. Huggard and B. Alderman. "A 180-GHz Schottky Diode Frequency Doubler With Counter-Rotated E-Fields to Provide In-Phase Power-Combining". *IEEE Microwave and Wireless Components Letters*, 2018.

[11] J. Treuttel, L. Gatilova, A. Maestrini, D. Moro-Melgar, F. Yang, F. Tamazouzt, T. Vacelet, Y. Jin, J. Mateos, A. Feret, C. Chaumont, C. Goldstein. "A 520–620-GHz Schottky Receiver Front-End for Planetary Science and Remote Sensing with 1070 K–1500 K DSB Noise Temperature at Room Temperature". *IEEE Trans.Terahertz Science and Tech.*, vol. 6, no 1, p.148-151, Jan, 2016.

[12] J. V. Siles, E. Schlecht, L. Robert, L. Choonsup and I. Mehdi. "High-efficiency planar Schottky diode based submillimeter-wave frequency multipliers optimized for high-power operation." In *Infrared, Millimeter, and Terahertz waves (IRMMW-THz), 2015 40th International Conference on*, pp. 1-1. IEEE, 2015.

Broadband Frequency Triplers for Radar Applications in the THz Range

Carlos G. Pérez-Moreno* and Jesús Grajal
Information Processing and Telecommunications Center, Universidad Politécnica de Madrid, E.T.S.I. Telecomunicación, Av. Complutense 30, 28040 Madrid, Spain
*Contact: carlos@gmr.ssr.upm.es

***Abstract*—We report the design and characterization of Schottky diode-based frequency triplers at 75-105 GHz and 225-315 GHz for the development of frequency multiplied signal sources to be used in a space-borne radar prototype for orbital debris detection at 94 GHz, a ground-based cloud profiling radar at 94 GHz, and a 3-D imaging radar at 300 GHz. The frequency triplers are designed with the aim of optimizing the bandwidth and conversion efficiency for a wide range of input powers, allowing the configuration of versatile sources able to fulfill the bandwidth and power requirements of diverse applications. For the 75-105 GHz frequency tripler, the measured room-temperature conversion efficiency is over 5% between 75 GHz and 102 GHz and over 8% across the 87-99 GHz band for an input power of 100 mW. For the 225-315 GHz frequency tripler, measurements across the 252-303 GHz band show a room-temperature conversion efficiency above 3% for an input power of 100 mW.**

I. Introduction

Millimeter-wave radars are being developed at the Universidad Politécnica de Madrid for diverse applications: A space-borne orbital debris radar at 94 GHz [1], a ground-based cloud profiling radar at 94 GHz [2-3], and a 3-D high-resolution imaging radar at 300 GHz [4]. In all these systems, the transmitted signal is generated by direct digital synthesis and several stages of frequency multiplication and power amplification. This work focuses on the design and characterization of two frequency triplers at 75-105 GHz and 225-315 GHz based on Schottky diodes for the transmitter chains of these radars.

Since self-heating is a significant limiting factor for the performance of high-power millimeter-wave frequency multipliers, the design procedure focuses on thermal management. Section II describes the design methodology, where an analytical electro-thermal model implemented into commercial circuit simulation software has been added to the design methodology presented in [5]. Section III displays the design and characterization of the 75-105 GHz frequency tripler, whereas Section IV presents the design and characterization of the 225-315 GHz.

II. Design Methodology

A physics-based numerical self-consistent electro-thermal model is developed to optimize the Schottky diode electrical and geometrical parameters together with the multiplier circuit performance from a joint electrical and thermal point of view [6]. For the purpose of improving the power handling capability, the multiplier circuit layouts are optimized via 3-D thermal modeling to maintain low operating temperatures. To complement the numerical approach, an analytical electro-thermal model is implemented into Keysight ADS to optimize the overall multiplier circuit performance accounting for the thermal effects. Fig. 1 shows the iterative design process, which can be divided into three steps [5]. The incorporation of thermal aspects into the design procedure provides an integrated thermal management approach that allows optimizing the device and circuit performance and designing for reliability.

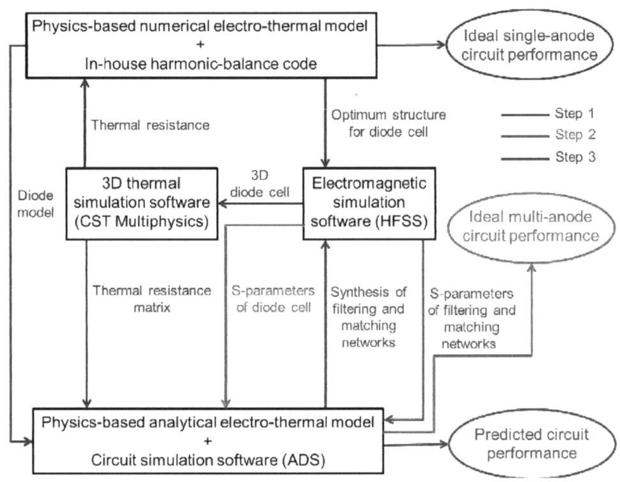

Fig. 1. Iterative design process utilized to optimize circuit performance.

III. Design and Characterization of a 75-105 GHz Frequency Tripler

For the 75-105 GHz frequency tripler, the predicted room-temperature conversion efficiency is over 9% across the 87-102 GHz band for an input power of 100 mW, with a peak efficiency above 12% at 99 GHz.

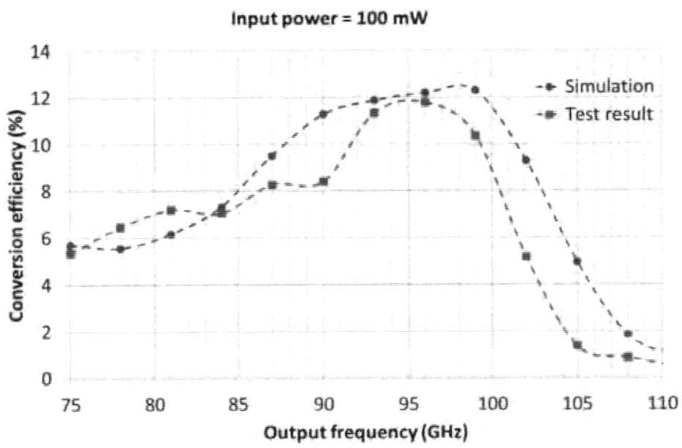

Fig. 2. 75-105 GHz frequency tripler: Conversion efficiency as a function of output frequency for an input power of 100 mW.

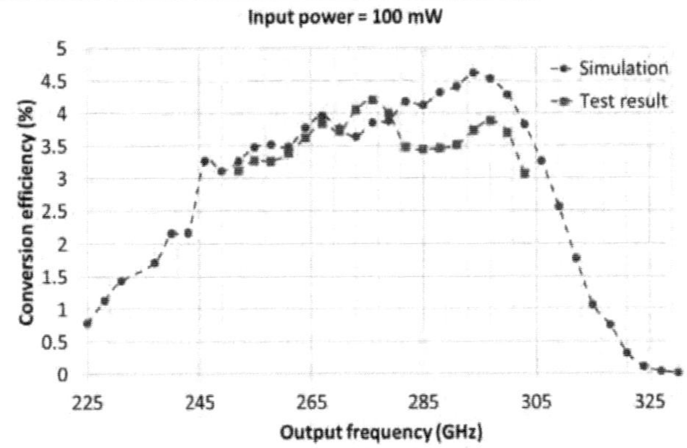

Fig. 3. 225-315 GHz frequency tripler: Conversion efficiency as a function of output frequency for an input power of 100 mW.

The designed frequency tripler is fabricated and assembled by Teratech Components Ltd. (UK) and uses a discrete GaAs diode chip with six planar Schottky varactors in a series configuration. The measured room-temperature conversion efficiency is over 5% between 75 GHz and 102 GHz and over 8% across the 87-99 GHz band for an input power of 100 mW (Fig. 2).

IV. Design and Characterization of a 225-315 GHz Frequency Tripler

For the 225-315 GHz frequency tripler, the simulated room-temperature conversion efficiency is over 3% across the 246-306 GHz band for an input power of 100 mW, with a peak efficiency of 4.5% at 294 GHz.

The designed frequency tripler is fabricated and assembled by Teratech Components Ltd. (UK) and features four anodes integrated into a GaAs membrane. Measurements across the 252-303 GHz band show a room-temperature conversion efficiency above 3% for an input power of 100 mW (Fig. 3).

Conclusions

Two broadband frequency triplers at 75-105 GHz and 225-315 GHz for radar applications have been designed and characterized. The incorporation of thermal aspects into the design procedure provides an integrated thermal management approach that allows optimizing the device and circuit performance and designing for reliability.

Acknowledgment

The authors would like to thank Teratech Components Ltd. for the fabrication, assembly and preliminary test of the designed frequency triplers. In particular, the authors would like to thank Dr. Jeff Powell and Dr. Byron Alderman for very helpful discussions along the design stage. This work was supported by the Spanish National Research and Development Program under projects TEC2014-53815-R and TEC2017-87061-C3-1-R, and by the Madrid Regional Government under project S2013/ICE-3000 (SPADERADAR-CM).

References

[1] M. Ramírez-Torres, M. Ferreras, C. Hernández, C. García-de-la-Cueva, M. Barba, G. Pérez-Palomino, J. A. Encinar, M. Sierra-Castañer, J. Grajal, J.-L. Vázquez-Roy, and E. Rajo-Iglesias, "Technological developments for a space-borne orbital debris radar at 94 GHz," in *IEEE Radar Conf.*, Oklahoma City, OK, USA, April 2018.

[2] G. Rubio-Cidre, C. Rodríguez, M. Ramírez-Torres, J. Grajal, and O. Rubiños, "Understanding cloud dynamics using a ground-based radar at 94 GHz," in *IEEE Radar Conf.*, Seattle, WA, USA, May 2017.

[3] G. Rubio-Cidre, J. Grajal, A. García-Pino, and O. Rubiños, "Design and preliminary results of a ground-based cloud profiling radar at 94 GHz," in *IEEE Radar Conf.*, Philadelphia, PA, USA, May 2016.

[4] J. Grajal, A. Badolato, G. Rubio-Cidre, L. Úbeda-Medina, B. Mencia-Oliva, A. García-Pino, B. González-Valdés, and O. Rubiños, "3-D high-resolution imaging radar at 300 GHz with enhanced FoV," *IEEE Trans. Microw. Theory Techn.*, vol. 63, no. 3, pp. 1097–1107, Mar. 2015.

[5] C. G. Pérez-Moreno and J. Grajal, "Frequency triplers at 94 GHz and 300 GHz for millimeter-wave radars," in *Proc. 28th International Symposium on Space Terahertz Technology*. Cologne, Germany, March 2017.

[6] C. G. Pérez-Moreno and J. Grajal, "Physical electro-thermal model for the design of Schottky diode-based circuits," *IEEE Trans. THz Sci. Technol.*, vol. 4, no. 5, pp. 597–604, Sep. 2014.

Pure-Rotational Molecular Spectroscopy with a Low-Power CMOS-Based W-Band Transmitter

Deacon J. Nemchick[*1], Brian J. Drouin[1], Adrian J. Tang[2], Yanghyo Kim[2], Gabriel Virbila[3] and Mau-Chung Frank Chang[3]

[1]*Laboratory Studies and Atmospheric Observations, Jet Propulsion Laboratory, Pasadena, CA 91109, USA*
[2]*Submillimeter Wave Advanced Technology, Jet Propulsion Laboratory, Pasadena, CA 91109, USA*
[3]*Department of Electrical Engineering, University of California at Los Angeles, Los Angeles, CA 90025 USA*
[*]Contact: Deacon.J.Nemchick@jpl.nasa.gov

Abstract— The distinct rotational signatures of gas-phase molecular species in the millimeter (mm) and sub-millimeter (sub-mm) spectral regions have long assisted remote sensing communities in the interrogation of atmospheric and astrophysical media. *In situ* studies employing highly-mobile instrumentation have not been able to reproduce the success of their remote-based counterparts largely due to the unaccommodating size and power requirements of traditional mm and sub-mm wave hardware. The Laboratory Studies and Atmospheric Observations group at JPL has embraced the marriage of novel custom-designed CMOS source and heterodyne detection electronics, which often leverage advances in the mobile phone industry, and traditional cavity enhanced laboratory techniques to combat the issues that have plagued the deployment of *in situ* mm wave sensors.

One device emerging from these efforts is a freestanding CMOS-based transmitter tunable to sub-500 Hz resolution over the operational bandwidth of 90 – 105 GHz. For prototyping purposes this transmitter, the output of which can be both frequency and amplitude modulated, has been deployed as the radiation source in a high-resolution sub-Doppler (Lamb-dip) absorption spectrometer. The presented experimental findings have shown that this device, which effectively functions as a USB powered/controlled W-band source, has sufficient output power (~2 mW peak) to perform spectral-hole burning saturation experiments and a phase-noise floor low enough to determine spectral line positions with a precision of 1 part in 10^9 and accuracy within the error of measurements made with traditional millimeter-wave sources. These findings highlight the promise of exploiting CMOS architectures for use in gas specific, low-power, and potentially low-cost *in situ* sensors.

Low power consumption quantum cascade lasers at 2.7 THz for compact and sensitive heterodyne detectors

François Joint[1,2], Gregory Gay[1], Thibaut Vacelet[1], L. Li[3], E. Linfield[3], Yan Delorme[1],

Raffaele Colombelli[2]

[1]*LERMA, Laboratoire d'tudes du Rayonnement et de la Matire en Astrophysique et Atmosphères*
[2]*Centre de Nanosciences et de Nanotechnologies, CNRS, Univ. Paris-Sud, Université Paris-Saclay, C2N – Orsay, 91405 Orsay cedex, France*

[3]*School of Electronic and Electrical Engineering, University of Leeds, Leeds LS2 9JT, UK*

Email: francois.joint@u-psud.fr

Terahertz (THz) technology is demonstrating more and more attractive applications in different fields, from astronomy, medicine and biology to material science and security. Superconducting heterodyne THz receivers combine very high spectral resolutions with sensitivity approaching the quantum noise limit. HEB (Hot Electron Bolometers) mixers are currently the most sensitive for operating frequencies beyond the THz and QCL (quantum cascade lasers) sources offer a great potential as local oscillators for heterodyne detection in the THz range.

In order to realize compact and sensitive heterodyne THz receivers combing superconducting HEB mixers and QCLs, currently the most promising and optimal receiver configuration for frequencies beyond 2 THz, we have undertaken the development of the QCL operating around 2.7 THz with the following specifications : single mode emission at a specified target frequency, low power consumption and near Gaussian beam emission.

We have chosen the distributed-feedback (DFB) architecture, in particular the 3rd-order DFB approach that can provide single mode emission as well as small beam divergence [1]. The DFB is implemented by introducing a deeply-etched lateral corrugation along the laser ridge that provides the necessary distributed feedback.

To obtain single mode operation at the desired frequency we have fabricated several devices with different grating periods and/or grating duty cycle. Based upon electro-optical characterizations of the lasers, the devices that best suit the application can be selected. During the electromagnetic modelling and simulation for the design of the lasers, special care has been taken to reduce the overall size as much as possible in order to minimize the power dissipation. We obtained devices with electrical consumption lower than 100 mW, which is very interesting for integrating QCLs to HEB mixers for embedded applications.

We have studied different quasi-optical configurations to reshape the QCL's output beam into a Gaussian beam for an efficient coupling with the HEB mixer. In particular, systems using Pyrex dielectric hollow waveguides [2] [3] and parabolic mirrors with short focal length have been studied with the aim of building a fully integrated heterodyne receiver. We'll present the simulation and experimental results.

Discrete GaAs Schottky beamlead mixer diodes for space-borne receiver applications

Peter J. Sobis*, Vladimir Drakinskiy[2], Tony Pellikka[1], Slavko Dejanovic[1], Anders Emrich[1], and Jan Stake[2]

[1]*Omnisys Instruments AB, Västra Frölunda, SE-421 32, Sweden*
[2]*Chalmers University of Technology, Göteborg, SE-412 96, Sweden*
*Contact: peter.sobis@omnisys.se

Abstract— This paper presents the design and implementation of a anti-parallell GaAs Schottky beamlead mixer diode suitable for use in discrete diode type subharmonic mixer applications up to about 500 GHz. The particular chip has been developed for the MWI instrument 118 GHz receiver channel which is part of the METOP Second Generation weather and climate research satellite programme. The main objective of the activity has been to develop a fabrication route for "pick and place" type antiparallel diode chip structures employing beamlead interconnects as a part of an ESA supervised lot acceptance testing (LAT) program. This has included qualification of the Chalmers E-beam based GaAs Schottky diode membrane process and statistical on-wafer yield analysis, assembly methods for beamlead type diodes, and pre-qualification activities. The results so far include 96 hours of biased and 500 hours of unbiased high humidity high temperature tests conducted at 85 degrees C and 85 % relative humidity, high temperature tests, total dose tests of up to 2 MRad and bondpull strength tests on representative carriers. The MetOp 118 GHz receiver development is currently in a pre-LAT phase with final LAT program scheduled to the end of 2018.

CubeSat Interferometry for THz Astrophysics, Planetary Science and Earth Observing

Christopher E. Groppi[1*], Paul F. Goldsmith[2], Jose V. Siles[2], Philip Mauskopf[1], Christopher K. Walker[3], Daniel C. Jacobs[1], Adrian Tang[2] and Paul Scowen[1]

[1]*Arizona State University, Tempe, 85287, USA*
[2]*NASA JPL, Pasadena, 91109, USA*
[3]*University of Arizona, Tucson, 85721, USA*
[*]Contact: cgroppi@asu.edu

Abstract—While great strides have been made in far-infrared astrophysics with the NASA Spitzer and ESA Herschel missions, sub-arcsecond spatial resolution from space is still beyond the reach of current technologies. The Atacama Large Millimeter Array has produced stunning images from the ground of planetary systems in the process of formation, but cannot observe two key molecules, water (H_2O) and oxygen (O_2), due to the presence of Earth's atmosphere. The concept proposed here will enable interferometric imaging with sub-arcsecond resolution of water and other key far-infrared molecular species from space at a cost far lower than the flagship class interferometric missions previously proposed (e.g. ESA's ESPRIT). We propose to study the concept of a far-infrared interferometer based on a constellation of CubeSat antenna elements with a central ESPA-class correlator satellite optimized for the imaging of water in protoplanetary systems. Such a mission would produce groundbreaking images of newly forming planetary systems in a key astrophysical and astrobiological tracer, the 557 GHz ground state line of water. By leveraging recent developments in CubeSat technology, inflatable reflectors, miniaturized receiver systems and low power CMOS digital electronics, such a mission could be implemented at an Explorer level budget. In addition to the proposed astrophysics application, the developments proposed here could also find application in planetary science (FIR spectroscopy of comets and small bodies) and Earth observing (high resolution imaging of Earth from geostationary orbit).

Integrated Schottky Receiver for Small Satellite Deployment

Jonathan Hoh[1], Christopher Groppi[1], Choonsup Lee[2], Robert Lin[2], Philip Mauskopf[1], Phil Putman[3], and Jose Siles[2]

[1]Arizona State University School of Earth and Space Exploration, 85287, Tempe AZ, USA
[2]NASA Jet Propulsion Laboratory, 91109, Pasadena CA, USA
[3]Sierra Lobo Inc., 43429, Fremont OH, USA
[*]Contact: jrhoh@asu.edu

Abstract—The terahertz portion of the electromagnetic spectrum is of intense interest to astrophysics. In this wavelength regime, we find the spectral signatures of both water vapor and molecular oxygen. Specifically, water lines at 557 GHz and the 1100-1200 GHz band are excellent diagnostics of water vapor in the interstellar medium, the Earth's atmosphere and the atmospheres of other planetary bodies. Recent advances in small satellite technology now allow us to consider their use for remote sensing these lines in ways that were before impossible.

Here we will be presenting the preliminary results of a low-mass, low-power, highly integrated Schottky diode based coherent receiver system suitable for deployment on cubesat or other small satellite platforms. Currently, coherent Schottky receivers are far too large to be considered for deployment on any smaller forms of space-based satellites. Using an already existing design for a modular 520-600 GHz receiver designed at JPL, we have used novel packaging methods to condense this receiver into an integrated system. This integrated receiver has shown to have a volume and power consumption significantly smaller than the current state of the art. We further present the designs of a similar integrated receiver for the first excited state of water vapor operating at the 1040-1200 GHz range. Further research will be spent exploring whether we can use passively cooling technologies to better enhance the performance of these Schottky receivers.

All Solid-State Receiver Designs at 2 THz for Atmospheric Sounding

J. Treuttel[1&2], D. Hayton[1], J-V. Siles[1], E. Schlecht[1], C. Lee[1], B. Thomas[3], R. Lin[1] and I. Mehdi[1]
[1]Jet Propulsion Laboratory California Institute of Technology, Pasadena, CA, 91109 USA
[2] LERMA, Observatoire de Paris, PSL Research University, UMR CNRS 8112, Paris France, Jeanne.treuttel@obspm.fr
[3] Radiometer Physics GmbH, Meckenheim, Germany

Abstract— **Measurement of the atomic oxygen emission at 2060 GHz has been identified as a key measurement to improve space weather predictions. In this paper we report on an all solid-state 2 THz receiver system. The mixer circuit is based on an anti-parallel pair of diodes in a sub-harmonic bias-able configuration. The design requires no cooling and is tolerant to nominal LO power variations. A compact multiplied source at 1.03 THz with more than 1 mW output power has been demonstrated and mixer chips have been designed.**

I. Introduction

Thermospheric density and atmospheric motion perturbations from space weather events influence higher atmosphere and upper mesosphere dynamic processes. These processes are involved in the climate evolution and play a major role in the trajectory of low-orbit space vehicles and debris during reentry. It is therefore desirable to build a novel instrument that provides observational data to characterize tropospheric winds at an altitude of 80 to 150 km [1]. One of the identified tracers for the neutral gas winds is the oxygen line (OI) which can be measured via doppler at 2060 GHz. The two main critical goals to develop a state-of-the-art 2 THz radiometer fulfilling the requirements of very small platforms include the development of a mixer and its 1 THz local oscillator source.

This paper describes an on-going effort to build the first ever all-solid-state receiver at 2 THz. While receivers in this frequency range have been demonstrated, see Table I, they have all utilized bulky lasers for the local oscillator. We report on mixer chip designs that are purposely optimized both for performance as well as reduced LO requirements. A mixer chips designed with anti-parallel pair topology provides sensitivity and performance that will enable measurement of the oxygen line. Finally, a compact LO chain that produces more than 1 mW at 1 THz is briefly described.

II. System Requirements

The implementation requirements for a 2 THz radiometer system compatible with small-sats are 4 kg mass, 10U volume, and power below 15W. This will be addressed by a Schottky based heterodyne spectrometer working at terahertz frequencies and operating at room temperature with large instantaneous bandwidth and high spectral resolution. The two main critical system parameters to fulfill all criteria while keeping state-of-the-art performances include: 1) the configuration of the mixer, its epitaxial layer definition, matching transmission lines and housing, 2) the local oscillator sub-system part and its calibration. Table I provides context for the current design work. Best recorded systems [2] and [3] use fundamental balanced mixers with respectively 5 mW and 10-12 mW of local oscillator power at 2 THz, comfortably provided by a CO2-pumped methanol gas laser. The choice of fundamental mixers is justified by the fact that they can theoretically reach better noise performance over subharmonic mixers [4]. However, the sub-harmonic topology relaxes the local oscillator (LO) source by cutting its operating frequency by a factor of two. This configuration also avoids the requirement to use a bulky CO2 laser that is far from fulfilling the mass/volume/power criteria and doesn't offer spectral tunability easily achievable by Schottky local oscillator sources [5] [6]. The proposed receiver utilized planar Schottky diodes for the mixer and multiplied LO.

III. Mixer Development

In this part we describe the development of a 2060 THz mixer including the mixer topology, the Schottky junction fabrication parameters, and the circuit matching constraints. We describe comparison between two chips with distinctive topologies that have been fabricated, namely, *Chip-A* originally designed for 1.8 THz and described previously in [7] and *Chip-B which has been designed to optimize performance at 2.06 THz .*

A. Mixer configuration

Recent designs of 1-2 THz MMIC sub-harmonic biasable mixers are based on successful sub-harmonic 1200 GHz

Frequency	Mixer type	Local Oscillator type	Epitaxy (cm^{-3})	Receiver noise temp. DSB (K)	Conversion gain (dB)	Ref.	Flight instrument
2.5 THz	Fundamental	CO2 Laser > 5 mW	1 *10e18	16500 K	- 17 dB	[2]	MLS AURA
2.5 THz	Fundamental	CO2 Laser > 10 mW	1 *10e18	6000 K	- 12.5 dB	[3]	-
1.96 THz	Subharmonic (chip A)	Schottky multipliers – 1 mW	1 *10e18	7000 K (simulated)	- 13 dB (simulated)	This work	-
2.06 THz	Subharmonic (chip B)	Schottky multipliers – 1 mW	1 *10e18	2000 K (simulated)	-10 dB (simulated)	This work	-

Table 1: State-of-the-art performance of Schottky based 2 THz heterodyne room temperature radiometers.

mixer topology [8] and have been design to give - 15 dB conversion gain with 1.5 mW of LO power [5]. The preferred mixer circuit topology at submillimeter and terahertz frequencies features a minimum of two-anode in order to provide balancing, implemented in a series-balanced [7] or antiparallel-balanced configurations, both featuring antiparallel topology at LO and IF frequencies and named after their anode positioning across the waveguide channel. In both cases a bias favor flexibility in the operative system, especially when the local oscillator source remains a critical sub-system. Indeed, if great LO power (P_{ol}) is usually preferred at THz to relax the mixer efficiency, an optimum in the coupled parameter (P_{ol}, mixer bias voltage) can be found during the optimization of the mixer junction and its matching network so that limited hot-electron noise is added to the mixer conversion noise. In the series-balanced topology the DC biasing is made possible by decoupling the DC path to the RF and LO transmission lines with very compact on-chip capacitors with very well-behaved mode confinement near the diode. The antiparallel configuration is generally used under unbiased operation [9] due to the difficulty to isolate the two anode DC paths. For example, the bias can be applied through the IF port and uses a multilayer transmission line [10]. This solution relies on having a good quality dielectric without pin-hole defects. Recent work has converged lately on a novel mixer configuration [11] that combines both on chip capacitor and a planar transmission line to differentiate the RF, LO and IF port directly at the diode cell level. This alleviates the problem of DC biasing of the anti-parallel configuration mixer in which the extremely secure balance allowing to reach very pure even LO excitation of the anodes for AM noise rejection. In Fig 1. we illustrate two designed chips of seire-balanced and antiparallel configuration that are fabricated at JPL-MDL: *Chip-A* originally designed for 1.8 THz [7]; *Chip-B* [this work].

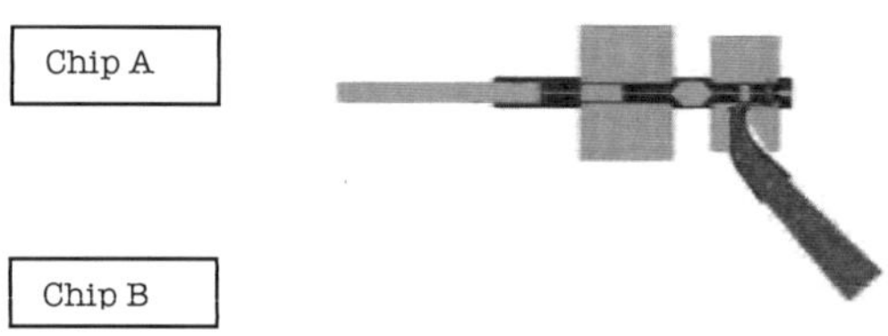

Fig. 1: Top – *Chip A is a series-balanced mixer configuration that has been described in* [7]. Chip B shown in the bottom of the figure is an anti-parallel balanced configuration and provides enhanced performance.

B. *Schottky junction parameters*

The doping and epitaxial layer thicknesses of the GaAs-Schottky junction are defined as a trade-off to optimize the junction mixer rectification efficiency for a same anode size. Firstly if we assume equivalent bandgap and electron mobility of the GaAs fabricated material, monte-carlo dedicated studies [12] typical doping initial value range for THz GaAs-based Schottky mixer-operation is located between $3e^{17}$ and $1e^{18}$ cm $^{-3}$, also confirmed by the state-of-the-art heritage listed in Table I. Secondly the thickness of the epitaxial level defines the reverse current excursion, which is maximized in a case of a multiplier operation and minimized in a forward mixer operation in order to reduce series resistance related losses. In this study *Chip-A* [7] and *Chip-B* both feature a thin epilayer thickness and high epitaxial doping suited to mixer operation. The junction parameters used in the harmonic balance electrical simulation are respectively for *Chip-A* [7] and *Chip-B*: a series resistance values of 100 ohms, a junction capacitance of 0.39 and 0.32 fF. Both are simulated with a saturation current I_{sat} = 1.6 pA, an ideality factor η = 1.3, and a built-in potential V_{bi} = 0.718.

C. *Transmission line*

The mixer sensitivity and conversion loss are related to the diode parasitics at RF and LO frequencies and to how well the RF and LO signals are coupled from the feed-horn and waveguide input ports to the anodes. Diode topology including mesa-to-mesa and finger-to-finger distances are optimized to limit diode parasitic coupling and harmonic trapping. *Chip-B* features a photolithographic backside process in the diode cell vicinity used in order to reduce the substrate parasitic loss by avoiding mode confinement in the substrate. For both chips, the transmission lines are designed to minimize loss while providing effective impedance matching. *Chip-A* [7] uses a very thin 3-μm thick GaAs membrane to support the diode devices while reducing the dielectric loading of the chip. The sections of the transmission lines are suspended by gold beam leads on both sides of the chip. The anti-parallel pair configuration *Chip-B* provides wider channel dimensions not affordable by *Chip-A* [7] configuration and thus introduces a larger range of low loss transmission lines for matching the RF and LO signals to the anodes. Its transmission lines are defined by a suspended central gold line so that the signal propagates on the edge of the gold only and the membrane underneath the gold does not significantly affect the mixer performance. *Chip B* features two different channel sections. The channel dimensions are defined to ensure that the RF and LO signals are propagated with the quasi-TEM mode of the central channel line, and that no unwanted transmission mode coupling occurs. The first matching section is designed so that the RF signal is confined as close as possible and coupled efficiently to the diode anodes. The second section supports the LO signal only, therefore its dimensions can be relaxed to minimize the losses coming from the LO source. The transmission sections feature two low (60 ohms) and medium (100 ohms) impedance lines with respectively 140, 160 dB/m loss at RF frequency, and around 110 and 100 dB/m losses at LO frequency.

D. *Mixer performance comparison*

Some key features of the two mixer chips (A and B) are compared in Table II. RF and LO signal coupling in percent is defined as the percentage of the input RF and LO signals from the waveguide input ports coupled to both Schottky anodes. These are distinct from the *RF and LO return losses* simulated at RF and LO waveguide input port as they also depend on the transmission line losses. The *RF and LO lost signal* indicates how efficient the coupling is and illustrates the possible differentiation in the two designs. *Chip-B* features a higher RF return loss compared to *Chip-A*. However, its RF coupling signal is higher by 25 % and lost RF signal lower by more than 20% thanks to its reduced RF transmission line losses.

The expected mixer performance as a function of design frequency for the two chips is shown in Fig. 2. The improved sensitivity expected from Chip B makes it the baseline design. *Chip-B* features a LO return loss that is higher compared to *Chip-A* and its LO signal coupling it higher by more than 30%. This has an impact on the operation mode of the junction requiring lower DC bias voltage level, and where *Chip-A* requires 1 V of DC voltage, *Chip-B* requires only 0.4 V. Both conversion loss and sensitivity of Chip B are drastically improved. These results are for room temperature and further improvement can be expected if the mixers can be cooled even to 120 K.

Table 2 : *Comparison performance of the two mixer chips (A and B) shows the expected enhancement of the new topology.*

	Chip-A [7]	*Chip-B*
RF Return loss	**< -15 dB**	**< -12 dB**
RF signal coupling	**50 %**	**75%**
Lost RF signal	**47 %**	**19 %**
LO Return loss	**< -12 dB**	**< -15 dB**
LO signal coupling	**40 %**	**75 %**
Lost LO signal	**50 %**	**22 %**
Pol **Bias Per 2 anodes**	**1 mW** **1 V**	**1 mW** **0,4 V**

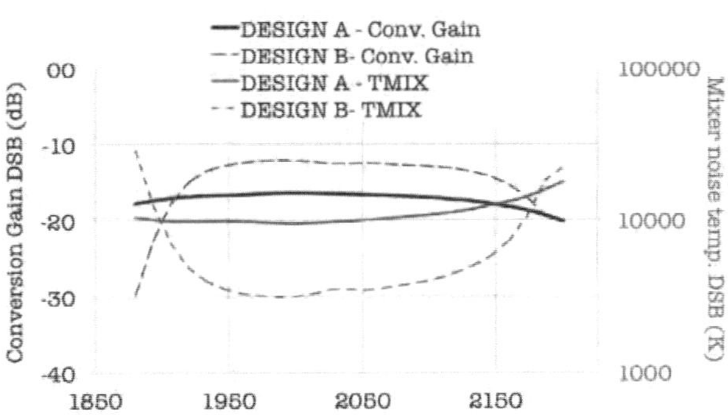

Fig. 2 Simulated performance of the two mixer chips as a function of frequency is shown. Identical anode parameters are selected for both designs.

IV. LOCAL OSCILLATOR DEVELOPMENT

Another key hurdle towards developing an all-solid-state 2 THz receiver system is the associated local oscillator source. For implementation on small satellites it is important that the LO source is compact and highly efficient. The local oscillator consists of a synthesized signal at 38 GHz which is amplified with power amplifiers to >1.5 Watt of power at Ka-band. The signal is then fed into two separated branches each featuring one dual chip tripler at 114 GHz based on [14] that are recombined [15] to produce > 300 mW at the input of the 350 GHz stage. This allows us to generate a 350 GHz spectrally pure signal of > 35 mW which is then used to pump the last stage tripler. The mixer is subharmonically pumped and thus requires the LO to be at 1.03 THz. The LO schematic is consolidated and secured with an E-H tuner to allow mismatch correction over frequency, and thermally secured to operate the 110 GHz at their maximum input power. The LO chain is shown in Fig. 3. The output power was recorded with an Erickson-VDI PM5 meter. The measurement plot demonstrates a record power of 1.75 mW at 1.02 THz (Fig. 3). Two LO chains were assembled and their performance is found to be fairly comparable, see Fig. 3. This LO chain has already demonstrated that it is capable of pumping the mixer as reported in [13].

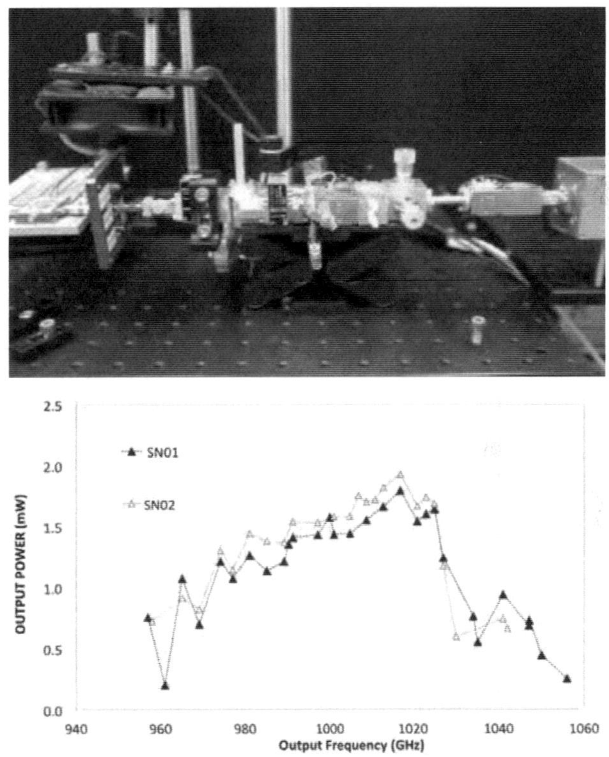

Fig. 3: A compact 1 THz LO chain has been built and demonstrated. The measured power should be sufficient to pump Schottky diode mixers in the 2 THz range.

V. CONCLUSION AND PERSPECTIVES

All-solid-state receivers capable of making spectroscopic measurements in the 2 THz range are being developed to observe the neutral oxygen line in the troposphere and measure wind velocities. Winds at higher altitudes are directly linked to understanding key heliophysics questions such as energy balance. A solid state LO chain that provides more than 1 mW of output power at 1.03 THz has been demonstrated. A novel mixer chip has been designed and simulated to show enhancement in performance from existing state of the art. The chip is currently being fabricated and will be packaged and characterized in the future.

VI. Acknowledgment

The authors acknowledge many helpful discussions and insightful comments from their colleagues at the Jet Propulsion Laboratory. This work was carried out at the Jet Propulsion Laboratory, California Institute of Technology, Pasadena, CA, under a contract with National Aeronautics and Space Administration (NASA). © 2018. California Institute of Technology. The first author was funded via the NASA Post-doc office during her visit to JPL.

References

[1] D. Wu, J-H(S). Yee, E. Schlecht, I. Mehdi, J. Siles, J. Treuttel and R. Lin, "*THz Limb Sounder (TLS) for Lower-Thermospheric Wind, Oxygen Density, and Temperature*", USNC-URSI National Radio Science Meeting, Jan. 6, 2017, University of Colorado, Boulder.

[2] P. Siegel, R. Smith, M. Gaidis, and S. Martin, "*2.5-THz GaAs monolithic membrane-diode mixer*", IEEE Trans. MTT, vol. 47, no. 5, pp. 596–604, May 1999.

[3] H.-W. Hübers , T. W. Crowe , G. Lundershausen , W. C. B. Peatman , H. P. Roserl , 4th ISSTT, 1993.

[4] A. Kerr, "*Noise and Loss in Balanced and Subharmonically Pumped Mixers: Part II—Application*", IEEE Transaction on Microwave Theory and Techniques, Vol 27, N0.12, December 1979

[5] A. Maestrini, J.S. Ward, J.J. Gill, C. Lee, B. Thomas, R.H. Lin, G. Chattopadhyay, and Imran Mehdi, "*A Frequency-Multiplied Source with more than 1 mW of Power across the 840-900 GHz Band*", IEEE Transactions on MTT , Vol. 58, no. 7, pp. 1925-1932, July 2010.

[6] I. Mehdi, B. Thomas, R. Lin, A. Maestrini, J. Ward, E. Schlecht, J. Gill, C. Lee, G. Chattopadhyay, and F. Maiwald, "*High-power local oscillator sources for 1- 2 THz*", in Proc. SPIE, 7741, 774112 (2010).

[7] B. Thomas, A. Maestrini, J. Ward, E. Schlecht, G. Chattopadhyay, J. Gill, C. Lee, R. Lin, and I. Mehdi, "*Terahertz cooled sub-harmonic Schottky mixers for planetary atmospheres,*" 5th ESA Workshop Millimeter Wave Tech. Apps, 3/30/2015.

[8] E. Schlecht, et al, "*Schottky diode based 1.2 THz receivers operating at room-temperature and below for planetary atmospheric sounding*" IEEE Trans. TST, Vol 4 , No. 6, Nov. 2014.

[9] J.Treuttel, L.Gatilova, A.Maestrini, D.MoroMelgar, F.Yang, F.Tamazouzt, T.Vacelet, Y.Jin, A.Cavanna, J. Matéos, A. Féret, C. Chaumont, and C. Goldstein, "*A 520–620-GHz Schottky Receiver Front-End for Planetary Science and Remote Sensing With 1070 K–1500 K DSB Noise Temperature at Room Temperature*", IEEE TST, Vol 6, No.1, Nov. 2015.

[10] J. Treuttel, B.Thomas, A.Maestrini, J.V-Siles, C.Lee and I.Medhi. "*A Novel 330 GHz Sub-Harmonic Mixer with Independently Biased Schottky Diodes*", International Symposium on Space Terahertz Technology, Tokyo, April 2012.

[11] A.Maestrini, L.Gatilova, J.Treuttel, F.Yang, Y.Jin, A.Cavanna, D. MoroMelgar, F.Tamazouzt, T.Vacelet, A. Féret, F. Dauplay, J-M. Krieg and C. Goldstein, "*1200 GHz and 600 GHz receivers for JUICE-SWI*", International Symposium on Space Terahertz Technology, Nanjing April 2016.

[12] D. Moro-Melgar, A. Maestrini, J.Treuttel, L. Gatilova; T. González; B. García Vasallo, J. Mateos, "*Monte Carlo Study of Two-Dimensional Capacitance Fringing Effects in GaAs Planar Schottky Diodes*", IEEE Transactions on Electron Devices, Vol. 63, issue 10, p. 3900-3907, 2016.

[13] J. Treuttel, D. Hayton, E. Schlecht, C. Lee, J-V. Siles, A. Maestrini, B. Thomas, R. Lin and I. Mehdi, "*A Cryogenic 2 THz Schottky Solid-State Heterodyne Receiver for Atmospheric Studies*", In Proceeding of IRMMW, Cancun, August 2017.

[14] J. V. Siles, C. Lee, R. Lin, G. Chattopadhyay, T. Reck, C. Jung-Kubiak, I. Mehdi, and K. B. Cooper, "*A High-Power 105–120 GHz Broadband On-Chip Power-Combined Frequency Tripler*", IEEE MCWL Vol. 5, No 3, March 2015.

[15] A. Maestrini, J. Ward, C. Tripon-Canseliet, J. Gill, C. Lee, H. Javadi, G. Chattopadhyay, and I. Mehdi, "*In-Phase Power-Combined Frequency Triplers at 300 GHz*", IEEE MCWL, Vol. 18, no. 3, pp. 218-220, March 2008.

Session W1: **Invited IV**

HEterodyne Receiver for OST (HERO)

Martina C. Wiedner*[1], Andrey Baryshev[2], Victor Belitsky[3], Yan Delorme1, Vincent Desmaris[3], Anna Di Giorgio[4], Brian Ellison[5], Juan-Daniel Gallego[6], Maryvonne Gerin[1], Paul Goldsmith[11], Christophe Goldstein[7], Frank Helmich[8], Fabrice Herpin[9], Jean-Michel Huet[10], Willem Jellema[8,2], Jean-Michel Krieg[1], Philippe Laporte[10], André Laurens[7], Imran Mehdi[11], Gary Melnick[12], Benjamin Quertier[9], René Plume[13], Christophe Risacher[14], Russel Shipman[8], OST STDT and NASA Goddard Engineering Team[15],

[1] *LERMA, Observatoire de Paris, ENS, PSL Research University, CNRS, Sorbonne Universités, UPMC Univ. Paris 06, avenue de l'Observatoire, 75014 Paris, France*
[2] *Kapteyn Astronomical Institute, University of Groningen, P.O. Box 800, 9700 AV, Groningen, NL*
[3] *Group for Advanced Receiver Development, Chalmers University of Technology, Gothenburg, SE 41296,*
[4] *Istituto Fisica Spazio Interplanetario INAF, via Fosso del Cavaliere 100, 00133 Roma, Italy*
[5] *Rutherford Appleton Laboratory, Space Department, Harwell Campus, Didcot, OX11 0QX, UK*
[6] *Observatorio de Yebes, CDT-IGN, Apdo. 148, Guadalajara 19080, Spain*
[7] *CNES, 18 Avenue Edouard Belin, 31400 Toulouse, France*
[8] *SRON Netherlands Institute for Space Research, Landleven 12, 9747 AD Groningen, The Netherlands & Kapteyn Astronomical Institute, University of Groningen, Groningen, The Netherlands*
[9] *Laboratoire d'Astrophysique de Bordeaux, Pessac, 33615, France*
[10] *GEPI, Observatoire de Paris, PSL Research University, CNRS, Paris,75014, France*
[11] *JPL, California Institute of Technology, 4800 Oak Grove Drive, Pasadena CA 91109, USA*
[12] *Harvard-Smithsonian Center for Astrophysics, 60 Garden Street, Cambridge, MA, 02138, USA*
[13] *University of Calgary, T2N1N4, Canada*
[14] *Max-Planck-Institut für Radioastronomie, Auf dem Hügel 69, 53121, Bonn, Germany*
[15] *NASA Goddard Space Flight Center, 8800 Greenbelt Rd., Greenbelt, MD 20771, USA*
*Contact: martina.wiedner@obspm.fr

Abstract— The Origins Space Telescope (OST) is one of the four science and technology definition (STDT) studies selected by NASA Headquarters for the 2020 Astronomy and Astrophysics Decadal survey. OST is designed to a) Chart the Rise of Metals, Dust, and the First Galaxies, b) Unveil the Growth of Black Holes and Galaxies Over Cosmic Time, c) Trace the Signatures of Life and the Ingredients of Habitable worlds, and d) Study the Solar System in Context. To enable these studies OST is equipped with five instruments including the HEterodyne Receiver for OST (HERO).

HERO is based upon proven heterodyne receiver technology and encompasses the frequency range of 0.468 to 4.752 THz with up to 128 pixels per band. The major design challenges were to accommodate the large number of receiver components and minimize power dissipation at 4K and the overall payload power consumption.

HERO's optical system is designed to be compact and light-weight. A two-axis Offner Relay configuration within the optics allows the selection of the individual frequency bands and the internal hot and cold calibration loads. The Offner relay also allows pointing agility and fast scanning of the beam on the sky without moving the whole heavy satellite.

The mixers are divided into six frequency bands each having two polarizations, five bands for the frequency range from 0.47 to 2.7 THz and the sixth from 4.54 to 4.75 THz. For easy packaging and compactness on the sky, the mixer arrays are in square geometries. The lowest two frequency bands have 2x16 SIS mixers, the upper four bands 2x64 HEB. All mixers posses 8 GHz IF bandwidth and have state-of-the-art noise temperature.

The local oscillators for HERO are critical as they need to be tunable over a wide RF band and must provide sufficient power to 'pump' up to 2x64 pixels. Amplifier multiplier chains provide the LO for all bands, including that at 4.7 THz. The LO is divided in waveguide to the appropriate number of pixels. Subsequent optical transport and alignment of the LO signal from the space bus to the mixers located approximately 8m away poses an additional engineering difficulty and requires a path length compensation system and tip-tilt mirrors. These compensate for any vibrations or relative changes between the position of the space bus and the mixers.

Low power dissipation, low noise cryogenic SiGe amplifiers amplify the signal at 4K and 20K, and are followed by compact CMOS amplifiers in the warm space bus. The signal then passes into 128 (+2) extremely low power (1W/ 8GHz) digital spectrometers built either using CMOS ASIC technology or FPGAs.

The instrument control is located in 3 subunits and is fully redundant, while most other components are internally redundant.

An international team of experts has designed HERO and the OST STDT team and NASA engineering team for OST have substantially supported the concept. HERO builds on the successful Herschel/HIFI mission payload heritage, but exploits latest technical developments and very considerably extends HIFI in terms of sensitivity, and spectral and spatial coverage.

Session W2: **Large Systems and Applications**

Millimetron Space Observatory – Large-Aperture and Cooled Space Telescope

A. Smirnov[1*], E. Golubev[1], M. Arhipov[1], V. Pishnov[1], T. Kosmovich[1], E. Filina[1], A. Baryshev[1,2], Thijs de Graauw [1], S. Pilipenko[1], S. Likhachev[1] and N. Kardashev[1] on behalf of the Millimetron team

[1]*Astro Space Center of P.N. Lebedev Physical Institute, Moscow, Russia*
[2]*University of Groningen, Kapteyn Astronomical Institute, Groningen, the Netherlands*
*Contact: asmirn@asc.rssi.ru

Abstract— Millimetron Space Observatory (MSO) is the next generation space instrument based on 10-meter deployable telescope designed for cryogenic (< 10K) operation at Lagrangian L2 orbit, which decrease all sources of thermal emission from the telescope and atmosphere. It is planned that MSO will cover an extremely wide wavelengths band of from about 50μm up to 20mm. Despite that the ground instruments, e.g., ALMA, has got the big collecting area and angular resolution, MSO will have a great potential for covering the far-infrared band which cannot be matched by any ground, balloon or airborne observatory. Uniqueness of studies in the far infrared range are driven by the fact that the earliest stages of star formation, during the time when gas and dust clouds are collapsing and planets forming can only be observed in it. Another MSO science goal are the compact objects in the Universe, such as the black holes and its surroundings, pulsars and gamma-ray bursts. It's requiring from instrument sub-microarcsecond angular resolution that can be realized only by using Space-Earth interferometry technic at millimeter or submillimeter wavelength ranges. The suite of the state of the art instruments on-board, like a bolometer cameras, imaging spectrometers and heterodyne instruments will provide MSO an unprecedented imaging, spectroscopy and high-resolution spectroscopy capability at the same time as an extremely high angular resolution. We will provide an overview and a current progress in the development of MSO.

The Gas And Ice Spectrometer/Radar (GAISR): A Millimeter/Submillimeter-Wave Multi-Functional Instrument for Cometary Jet Observations

Ken B. Cooper, Raquel R. Monje, Robert J. Dengler, Corey J. Cochrane, Stephen L. Durden, Adrian Tang, and Mathieu Choukroun

Jet Propulsion Laboratory, California Institute of Technology, Pasadena, 91109, USA
Contact: ken.b.cooper@jpl.nasa.gov

Abstract— An instrument prototype combining a small-particle Doppler radar and a dual-channel submillimeter-wave spectrometer is being developed to enable new observations of cometary jets and outer-moon plumes. Dubbed GAISR (Gas And Ice Spectrometer/Radar), its medium-range W-band (95 GHz) radar will operate in a frequency-modulated continuous-wave (FMCW) mode with 1 Watt of transmit power to achieve ultra-high sensitivity detection of the range and velocity distribution of 0.1-10 mm sized ice and dust particles released by jets and plumes. The radar's primary aperture also serves as an antenna for two passive heterodyne spectrometer channels at 270 and 560 GHz for detecting the abundance, temperature, and velocity of water vapor and its isotopes (including HDO), as well other major cometary volatiles. The GAISR instrument design focuses on being low mass and power instrument for compatibility with future planetary missions. This is accomplished by leveraging recent innovations in W-band signal generation using ultra-low power silicon integrated circuits, state-of-the art III-V semiconductor (GaN, InP, and GaAs) devices for signal amplification and detection, compact quasioptical duplexing, and modern low-power digital processors. A new signal processing algorithm for FMCW Doppler radar detection out to the maximum range ambiguity limit has also been developed. This talk will summarize GAISR's system design, focusing on the engineering tradeoffs in building a multi-functional flight-like instrument, and will describe very promising initial performance testing including gas cell isotopic detection and the range/Doppler mapping of rain.

Measuring Volcanic Eruption Dynamics with WAMS, a Millimeter-wave Radar and Imager

Sean Bryan[1*], Amanda Clarke[1], Loÿc Vanderkluysen[2], Christopher Groppi[1], Scott Paine[3], Daniel W. Bliss[1], James Aberle[1], and Philip Mauskopf[1]

[1]*Arizona State University, Tempe, AZ, 85281, USA*
[2]*Drexel University, Philadelphia, PA, 19104, USA*
[3]*Smithsonian Astrophysical Observatory, Cambridge, MA, 02138, USA*
[*]Contact: sean.a.bryan@asu.edu

Abstract—Millimeter-wave remote sensing technology can significantly improve measurements of volcanic eruptions. Forecasts of drifting volcanic ash for aviation safety would improve with direct measurements of high-altitude volcanic ash, and with better understanding of internal eruption processes. Here we present an update on the development of the WAMS (Water and Ash Millimeter-wave Spectrometer) instrument, a radar/radiometer system to measure water vapor and ash throughout an entire eruption cloud. The radiometer system will measure millimeter-wave water lines (24 GHz, 183 GHz, 220 GHz) to image the water vapor density and temperature inside eruption clouds, improving on existing measurements with infrared cameras that are limited to measuring the outer cloud surface. The radar system will measure the 3D mass distribution of volcanic ash inside eruption plumes and their nearby drifting ash clouds. The wavelength of the 220 GHz radar is optimized to be well matched to typical ash particle sizes, offering better sensitivity than longer wavelength existing weather radar measurements. This proposed mm-wave system would acquire data that no existing instruments can obtain: direct imaging of water vapor path integrated concentration, corresponding water vapor temperature, and 3D radar mapping of ash concentration. The system is built with off the shelf components, except for some millimeter-wave components our team commonly builds for radio astronomy applications.

High Frequency High Spectral Resolution Focal Plane Arrays for ATLAST

A.M. Baryshev[*1], R. Hesper[1], A. Khudchenko[1], K. Rudakov[1,2]

[1]*NOVA/Kapteyn Astronomical Institute, University of Groningen, Groningen, the Netherlands*
[2]*Institute of Radioengineering and Electronics, Moscow, Russia*
[*]Contact: a.m.baryshev@astro.rug.nl

Abstract—Large collecting area single dish telescope such as ATLAST will be especially effective for medium (R~1000) and high (R~50000) spectral resolution observations. Large focal plane array is a natural solution to increase mapping speed. For medium resolution direct detectors with filter banks (KIDs) and or heterodyne technology can be employed. We will analyze performance limits of comparable KID and SIS focal plane array taking into account quantum limit and high background condition of terrestrial observing site. For large heterodyne focal plane arrays, a high current density AlN junctions open possibility of large instantaneous bandwidth >40%. This and possible multi frequency band FPSs presents a practical challenge for spatial sampling and scanning strategies. We will discuss phase array feeds as a possible solution, including a modular back-end system, which can be shared between KID and SIS based FPA. Finally we will discuss achievable sensitivities and pixel counts for a high frequency (>500 GHz) FPAs and address main technical challenges: LO distribution, wire counts, bias line multiplexing, and monolithic vs. discrete mixer component integration.

Session W3: **Sources and Local Oscillator Systems II**

Next Generation of Room-Temperature Broadband Frequency Multiplied LO Sources with 10 times Higher Output Power in the 100 GHz – 1.9 THz Range

Jose V. Siles, Ken B. Cooper, Choonsup Lee, Robert Lin, Goutam Chattopadhyay and Imran Mehdi

Jet Propulsion Laboratory, California Institute of Technology, Pasadena CA 91109, USA
*Contact: Jose.V.Siles@jpl.nasa.gov

Abstract— Building upon the tremendous success of the Heterodyne Instrument for the Far Infrared (HIFI) on board the Herschel Space Observatory, high-resolution submillimeter-wave receivers able to map galaxies at unprecedented speeds have been proposed for a number of future NASA missions to address key unanswered questions about the stellar life cycle. Almost 10 years after Herschel was launched, Schottky diode frequency multiplied local oscillator sources are still the preferred local oscillator technology for these receivers. Unlike other competing technologies, Schottky based LO sources are compact, broadband, frequency stable, temperature stable, and can operate at room temperature.

To meet the requirements of the next terahertz heterodyne array instruments, it was necessary to push the limits of the current Schottky technology even further to achieve unprecedented output power levels while preserving or even improving the form factor of the LO sources and reducing the overall dc power consumption to make them suitable for suborbital or space missions.

Here we are presenting the newest generation of compact THz LO sources recently demonstrated at JPL, which exhibit output power levels more than 10 times higher than the previous state-of-the-art. These are based on a novel JPL-patented circuit topology called "on-chip power-combining". This concept, together with a precise optimization of the devices for high-power operation, yields an improvement in power-handling capabilities by one order of magnitude without using multiple-chips power-combined structures or diamond substrates. These results represent a major breakthrough in power generation at terahertz frequencies, and contribute enormously to closing the so-called "terahertz gap". Prototypes at 180 GHz, 240 GHz, 340 GHz, 530 GHz, 1 THz and 1.6 THz have been designed, fabricated, assembled and tested at JPL all with world record performances. For example, output power levels of 550 mW (25% efficiency) at 180 GHz, 130 mW (15% efficiency) at 230 GHz, 30 mW (7% efficiency), 2 mW (5% efficiency) and 0.8 mW (3% efficiency) have been measured at 300K. This also translates to world record input power handling capabilities. The conversion efficiency for all these stages is also state-of-the-art and pretty close to the theoretical limits predicted using physics-based numerical simulators. Further improvements could still be achieved by cooling the sources down to 120K. With this new generation of LO sources available, large arrays of high-resolution terahertz receivers can now be driven with a single broad-band compact source.

Broadband Metasurface External-Cavity THz QC-Lasers

Christopher A. Curwen[*,1], John L. Reno[2], Tatsuo Itoh[1], Benjamin S. Williams[1]

[1]*Department of Electrical Engineering, University of California, Los Angeles, CA 90095, USA*
[2]*Sandia National Laboratories, Center of Integrated Nanotechnologies, MS 1303, Albuquerque, NM 97185, USA*
[*]Contact: ccurwen@ucla.edu

Abstract— Terahertz quantum-cascade lasers are high-power, compact sources of THz radiation that have potential to operate as local oscillators for THz heterodyne receivers. However, practical implementation has proved challenging due to difficulties in THz-QCL power extraction, poor beam patterns, and limited tunability – all resulting from the subwavelength, metallic waveguides used as laser resonant cavities. A new architecture has been demonstrated that solves many of these problems; the quantum-cascade vertical-external-cavity surface-emitting-laser (THz QC-VECSEL) [1, 2]. The THz QC-VECSEL uses an amplifying, reflective metasurface composed of a subwavelength array of antenna-coupled microcavities loaded with quantum-cascade gain material. When feedback to the metasurface is provided by an external mirror, the resulting Fabry-Perot cavity forms an external cavity laser that has demonstrated a near diffraction limited beam with very high power and slope efficiency. Furthermore, the QC-VECSEL is naturally suited for tunability by adjusting the length of the external cavity. By using piezoelectric actuation, we have demonstrated single-mode tuning of a QC-VECSEL over 250 GHz at ~3.4 THz [3]. However, this tuning range is limited by the bandwidth of the metasurface which is made up of a single resonant microcavity. Here, we further demonstrate metasurfaces with enhanced bandwidth made up of coupled microcavity resonators. We demonstrate continuous tuning over more than a terahertz, between ~2.85-3.9 THz, however, multimoding is consistently observed as a result of the external-cavity free-spectral-range being smaller than the gain bandwidth of the metasurface. A narrow beam is emitted (~5x5°), and a peak power of ~15 mW is observed at 77 K.

[1] L. Xu, C. A. Curwen, P. W. C. Hon, Q.-S. Chen, T. Itoh, and B. S. Williams, "Metasurface external cavity laser," Appl. Phys. Lett. 107, 221105 (2015).
[2] L. Xu, C. A. Curwen, D. Chen, J. L. Reno, T. Itoh, and B. S. Williams, "Terahertz metasurface quantum-cascade VECSELs: Theory and Performance," IEEE J. Sel. Top. Quantum Electron. 23, 1200512 (2017).
[3] C. A. Curwen, L. Xu, J. L. Reno, T. Itoh, and B. S. Williams, "Broadband continuous tuning of a THz quantum-cascade VECSEL," 2017 Conference on Laser and Electro-Optics (CLEO).

Diamond-Substrate Schottky Diodes for high-power MM-wave Multipliers

Oleg Cojocari*, Diego Moro-Melgar, and Ion Oprea.

ACST GmbH, Hanau, 63457, Germany
*Contact: oleg.cojocari€@acst.de

Abstract— Relatively high power, in the range maybe of 1W, is desired for many applications at MM/SubMM waves, including active imaging, high bit-rate data transmission, and Inferior chains for LO-sources in THz heterodyne receivers. Moreover, most of "real-live" applications impose severe requirements to THz sources like compactness, reliability, and price. Many R&D groups worldwide address these requirements but the current state-of-the-art devices still do not fulfil all requirements.

The most practicable approach nowadays to generate power above 100GHz is frequency multiplication approach based on Schottky diodes. However, power-handling capability of multiplier modules is still a severe limiting factor for obtaining high power at MM/SubMM-waves by frequency multiplication. For instance, state-of-the-art MM-wave doubles perform about 25%-30% efficiency, but their maximum input power-handling capability is usually limited in the range of 200mW. This limitation basically defines maximal achievable output power of about 50-60mW.

The limiting factor for maximal power-handling capability is twofold. On one hand, this is limited electrically, basically due to breakdown voltage of Schottky varactors. On another hand, power-handling capability is thermally limited by inferior heat dissipation approaches.

ACST has recently modified their Film-Diode process for fabrication of high-breakdown varactor diodes on transferred Diamond substrate. The breakdown voltage of ACST varactor diodes is highest-possible for any particular doping concentration, and is close to theoretical limit. On another hand, CVD-diamond is known to be one of the best thermal conductive dielectrics in the nature. Therefore, combination of these two aspects in ACST varactor diodes suggest considerably-increase of maximal power-handling capability, which is key-factor for achieving high-efficiency high-power MM-wave frequency multipliers.

This contribution will address the above challenges and solutions for MM-wave frequency multipliers. Finally, preliminary results shall be presented on a 135-160GHz doubler, which survives input power of up to 750mW and provides output power more than 150mW. On Authors knowledge this is the highest power achieved from a single device without use of power combining approach.

High Power Discrete Schottky Diodes Based 275-305 GHz Transceiver for FMCW-Radar

D. Moro-Melgar[1*], O. Cojocari[1], I. Oprea[1], H. Hoefle[1] and M. Rickes[1]
[1]ACST GmbH, Hanau, 63457, Germany. (http://www.acst.de/)
*Contact: diego.moro-melgar@acst.de

***Abstract*— A heterodyne 275-305 GHz transceiver intended for frequency modulated continuous wave radar applications is presented here. A commercial E-band active multiplication chain is used to provide the required power and frequency band of the transceiver. All modules beyond E-band are solid-state discrete Schottky diodes based developed by ACST GmbH. The transceiver consists of a 135-160 GHz and a 270-320 GHz doublers to increase the frequency up to Y-band for the transmitter part. The receiver part consists of a 270-320 GHz sub-harmonic mixer pumped by the 135-160 GHz doubler. A D-band directional coupler distributes the power at 137.5 GHz to 152.5 GHz between the transmitter and receiver part. The transmitted power is higher than 20 mW in band and 8.5 dB average doubler side band noise figure is performed by the receiver.**

I. Introduction

Nowadays the market volumes for submm-wave sources and receivers are very limited. Moreover, specification requirements may vary from one particular application to another. Under these conditions, the use of discrete diode structures for frequency multipliers and mixers is of significant advantages in comparison to monolithically-integrated diode circuits, concerning flexibility and price. However, the implementation of discrete diodes approach becomes difficult beyond W band due to the size reduction as the frequency increases. Dedicated high frequency diodes structures and substrates are required to go up in frequency and provide the maximum performance of the chip. Discrete diodes approach has been successfully demonstrated up to 200 GHz using doublers in [1], [2] with equivalent performance provided by MMIC like in [3]. A 270-320 GHz discrete diodes based doubler and mixer are demonstrated by ACST in this work as part of the presented 275-305 GHz transceiver. These modules are combined together to define the transmitter and receiver part. The aim of this work is to provide as much power as possible from the transmitter part using discrete diodes approach and single chip modules. Additionally, it is of high interest to obtain as low noise as possible from the receiver. The maximization of the transmitted power and the minimization of the receiver noise improves the radar range equation of the transceiver [4]. A new type of diodes has been developed by ACST for high power applications. These diodes integrate a CVD diamond substrate to improve heat dissipation and have been implemented in the doublers of the transmitter part. Standard ACST mixer diodes of type 2MAF1.5 have been used in the mixer development. These discrete diodes are fabricated using ACSTs Film-Diode process in a similar way described in [5]. The discrete diodes approach allows both flexibility in the design and the assembly of diodes. On one hand, the same diodes properties can be used in different designs with different frequency range. On the other hand, different diodes properties can be used in the same design to enhance the performance at different frequencies. The use of proper discrete diodes structures and substrate chips have allowed to demonstrate a transceiver with more than 10 mW transmitted power and less than 9 dB average double side band (DBS) noise figure (NF) of the receiver between 275-305 GHz.

II. Transceiver Architecture

An image of the transceiver is shown in Fig. 1. The transceiver consists of an initial active multiplication chain (AMC) and a high power amplifier (HPA) commercially available. This AMC is able to multiply six times the initial signal and provide more than 250 mW output power from 68.7 GHz to 76.3 GHz. The transmitter part consists of a 135-160 GHz (D150) and a 270-320 GHz (D300) doublers completely developed by ACST [6]. The receiver part consists of a 270-320 GHz subharmonic mixer (SHM300) developed at ACST and a commercial 46 dB gain low noise amplifier (LNA).

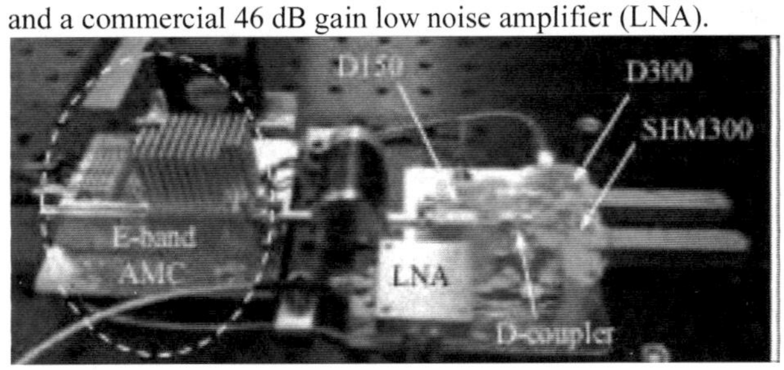

Fig. 1. Discrete Schottky diodes based 275-305 GHz transceiver developed by ACST.

The transmitter and receiver parts are integrated together using a dedicated D-band coupler (D-coupler) developed by ACST. The D-band coupler distributes the power provided by the 135-160 GHz doubler between the 270-320 GHz doubler and mixer. An E-band tuneable attenuator is placed between the 135-160 GHz doubler and the E-band AMC to calibrate the ensemble of the transceiver in the required frequency band. The transmitter and the receiver use a diagonal antenna each.

A dedicated power supply unit (PSU) is included in the system to bias the different modules.

III. Diodes Technology & Modules

A. Diodes Technology

An illustrative image of ACSTs diamond-based varactor and film varistor Schottky diodes is shown in Fig. 1. The Schottky diode structure integrates a CVD diamond substrate (Fig. 1 left). The diamond substrate is physically connected to the anodes to more efficiently distribute the heat to the contact pads, used later to place the diode on the chip. The film diodes (Fig. 1. right) feature a film substrate as part of the fabrication process. Both diode structures are based on a quasi-vertical fabrication process [7].

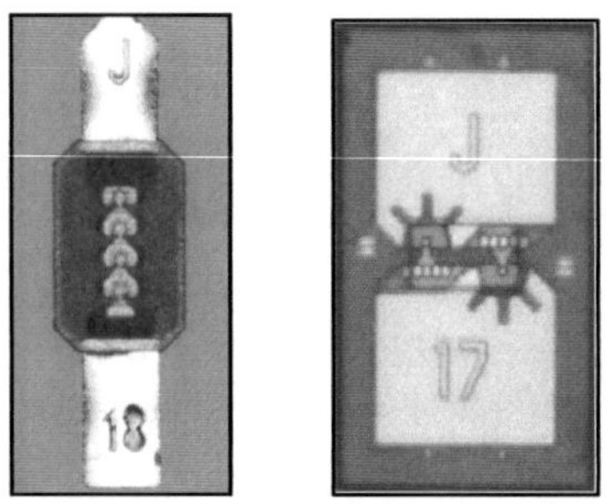

Fig. 2. Representative structures of diamond-based varactor (left) and film varistor (right) Schottky diodes developed by ACST.

B. 135-160 GHz Doubler

Two discrete diamond diodes are placed in the chip in anti-series balanced configuration. The discrete diode features three Schottky anodes with -13.5 to -14 V breakdown voltage each. The bias is applied using a SMA connector with no more than -16 V required. A single chip is used in this multiplier based on the same architecture proposed in [8]. The mechanical design of the module features a WR12 and WR6.5 input and output waveguide flange, respectively.

C. 270-320 GHz Doubler

Two discrete diamond diodes are placed in the chip in anti-series balanced configuration. The discrete diode features two Schottky anodes with -10 to -10.5 V breakdown voltage each. The bias is applied using a SMA connector with no more than -9 V required. A single chip is used in this multiplier based on the same architecture used in the D150. The mechanical design of the module features a WR6.5 and WR3.4 input and output waveguide flange, respectively.

D. 270-320 GHz Mixer

A single discrete film diode is placed in the chip featuring two Schottky anodes in anti-parallel configuration, like in [9]. No bias is required for this mixer. A WR6.5 and WR3.4 input waveguides lead the local oscillator (LO) and the radiofrequency (RF) signals to the chip. The chip is also defined on film-substrate technology [5] and the architecture is based on [10]. The output intermediate frequency (IF) signal is extracted through a 50 Ohm SMA connector featuring an 18 GHz frequency band. The DSB noise figure of this mixer is lower than 6 dB and it requires only 2.1 mW LO input power along the frequency band to correctly work.

E. D-band Coupler

The D-band coupler is specially design to work in the frequency range of the developed transceiver. The mechanical design features a WR6.5 waveguide input and two WR6.5 outputs in parallel configuration. The internal waveguide lengths are defined to have similar electrical path of the D-band signal to the D300 and SHM300 modules. A -15 dB to -16 dB coupling factor between the input of the coupler and the output for the SHM300 is defined. Less than 0.5 dB transmission loss factor between the input of the coupler and the output for the D300 is defined.

IV. Results

A. Transmitted Power

The experimental setup is described in Fig. 1. A signal generator is used to provide the initial signal at 11.45-12.71 GHz. The signal is multiplied and amplified up to E-band and the high power multipliers developed in this work increase the signal up to Y-band. The antenna of the transmitter is replaced by a power meter PM4 in this case. The bias required by the AMC, multipliers and LNA is provided by a dedicated PSU. It is important to note that the bias of multipliers is fixed along the frequency band. The experimental transmitted power between 275-305 GHz is plotted in Fig. 3. The receiver part was working during the measurement of transmitted power.

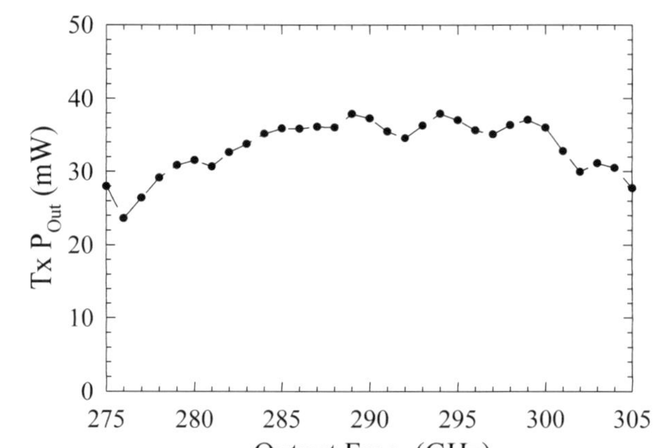

Fig. 3. Experimental transmitted power by the 275-305 GHz transceiver developed at ACST.

The transmitted power varies between 25 mW and 39 mW in the 275-305 GHz frequency range. It is important to note that this power is provided by single chip discrete diode based doublers. This power level is higher than the power provided by some authors using MMIC approach [5], [8] and the highest value reported using discrete Schottky diodes technology for doublers [6], [11]. Similar output power is obtained by JPL using MMIC Schottky technology [12]

B. Receiver Noise Figure

The experimental setup is shown in Fig. 1. The power delivered at D-band by the D150 doubler is split by the D-band coupler to properly provide the LO power to the SHM300 mixer. The calibration of the receiver part is achieved using a tunable attenuator between the E-band AMC and the D150 doubler. The D150 doubler is powerful enough to modify its output between 20-22 dBm and find the suitable output to correctly pump both the transmitter and the receiver parts. The LNA amplifies 46 dB the IF signal between 10 KHz and 200 MHz and perform ~1.9 dB noise figure. A diagonal antenna is used to receive the RF input signal. The noise figure of the receiver, while the transmitter is also transmitting, is plotted in Fig. 4. The noise figure has been obtained with the Y-factor method [13]. Liquid nitrogen has been used as cold load and a RF absorber at room temperature has been used as hot load. Each value of noise figure plotted in Fig. 4 is calculated from the Y-factor of the receiver obtained from 10 KHz to 200 MHz of the IF frequency range.

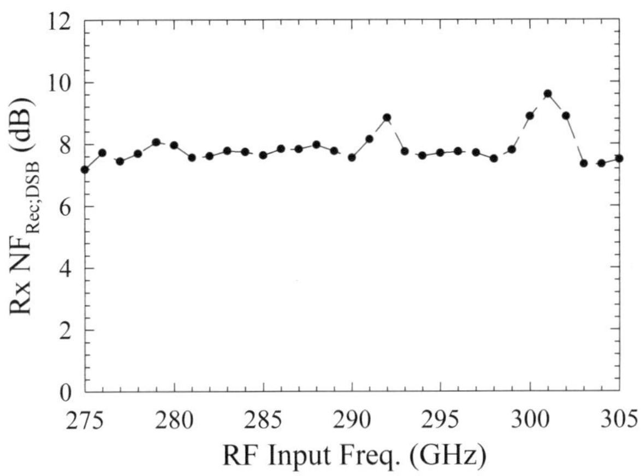

Fig. 4. Experimental double side band noise figure of the receiver part at 275-305 GHz.

The average DSB noise figure of the receiver part is ~8.5 dB in the 275-305 GHz band. The higher values at 294 GHz and 302 GHz are associated to a lack of LO power due to the standing waves between the mixer and the D-band coupler.

Conclusions

A discrete Schottky diode based 275-305 GHz transceiver for FMCW-Rada applications has been demonstrated with more than 20 mW transmitted power and 8.5 dB average DSB noise figure. A 39 mW peak value of transmitted power is provided by the transceiver system. The implementation of diamond diodes in the transmitter modules demonstrates high power handling capabilities to avoid power combining techniques. To authors knowledge, the obtained transmitted power is the highest value reported by a single chip 300 GHz source based on discrete Schottky diodes.

References

[1] T. Waliwander. M. Fehilly and E. O'Brien. "An ultra-high efficiency high power Schottky varactor frequency doubler to 180–200 GHz". In *Millimeter Waves (GSMM) & ESA Workshop on Millimetre-Wave Technology and Applications, 2016 Global Symposium on* (pp. 1-4). IEEE. June 2016.

[2] B. Thomas, H. Gibson, A. Walber, I. Oprea, O. Cojocari & T. Nahri. "Power-combining of 186-210 GHz frequency doublers and W-band power amplifiers with more than 100 mW peak output power". *26th International Symposium on Space Terahertz Technology, Cambridge, MA,* 16-18 March, 2015.

[3] C. Viegas, H. Liu, J. Powell, H. Sanghera, A. Whimster, L. Donoghue, P.G. Huggard and B. Alderman. "A 180-GHz Schottky Diode Frequency Doubler With Counter-Rotated E-Fields to Provide In-Phase Power-Combining". *IEEE Microwave and Wireless Components Letters*, 2018.

[4] S. W. Henriksen. „Radar-range equation". *Proceedings of the IEEE, 63*, no. 5, pp.813-814. May, 1975.

[5] O. Cojocari, I. Oprea, H. Gibson and A. Walber. "Submm-wave multipliers by film-diode technology." In *Microwave Conference (EuMC), 2016 46th European*, pp. 337-340. IEEE, 2016.

[6] D. Moro-Melgar, O. Cojocari, I. Oprea. "High Power High Efficiency 270-320 GHz Source Based on Discrete Schottky Diodes", In *Microwave Conference (EuMC), 2018 48th European*, (Submitted/Accepted). IEEE, 2018.

[7] O. Cojocari, C. Sydlo, H. L. Hartnagel, S. Biber, J. Schür, and L. P. Schmidt. "Schottky-structures for THz-applications based on quasi-vertical design-concept." In *International Symposium on Space Terahertz Technology, Göteborg, Sweden*. 2005.

[8] D. Jiangqiao, A. Maestrini, L. Gatilova, A. Cavanna, S. Shengcai and W. Wu. "High Efficiency and Wideband 300 GHz Frequency Doubler Based on Six Schottky Diodes." *Journal of Infrared, Millimeter, and Terahertz Waves* 38, no. 11, 1331-1341, 2017.

[9] I. Oprea, A. Walber, O. Cojocari, H. Gibson, R. Zimmermann and H. L. Hartnagel. "183 GHz mixer on InGaAs Schottky diodes." *Proc. ISSTT* (2010): 159-160.

[10] A. Maestrini, L. Gatilova, J. Treuttel, F. Yang, Y. Jin, A. Cavanna, D. Moro-Melgar, ... & F. Dauplay. (2016, April). 1200GHz and 600GHz Schottky Receivers for JUICE-SWI. In *27th International Symposium on Space Terahertz Technology,* (April, 2016).

[11] C. Yao, M. Zhou and Y. Luo. "A High Power 320–356GHz Frequency Multipliers with Schottky Diodes". *Chinese Journal of Electronics*, *25*(5), pp.986-990, 2016.

[12] J.V. Siles, K. B. Cooper, L. Choonsup, L. Robert, G. Chattopadhyay, & I. Mehdi. "Next Generation of Room Temperature Broadband Frequency Multiplied LO Sources with 10 times Higher Output Power in the 100 GHz-1.9 Thz Range. In *29th International Symposium on Space Terahertz Technology,* (March, 2018).

[13] A. R. Kerr. "Suggestions for revised definitions of noise quantities, including quantum effects". *IEEE transactions on microwave theory and techniques*, vol. 47, no 3, p. 325-329, March 1999.

Session W4: **Invited V**

Terahertz Astronomy from Near Space and Beyond

Christopher K. Walker

University of Arizona, Tucson AZ 85721, USA
*Contact: cwalker@as.arizona.edu

Abstract—With the advent of ultra-long duration ballooning and new innovative approaches to CubeSats and large space-based telescopes, the prospects are bright for a new generation of powerful terahertz observatories. To reach their full potential these observatories will push beyond the boundaries of current terahertz technology, necessitating advancements in detectors, sources, amplifiers, optics, and cryogenics. During the presentation plans and concepts for this new generation of observatories and their instrumentation will be discussed.

Session W5: **Superconducting Heterodyne Detectors II**

Planar-integration of Arrayed SIS Receiver Frontends

Wenlei Shan*, Shohei Ezaki, Shinichiro Asayama, Takashi Noguchi, and Satoru Iguchi

National Astronomical Observatory of Japan, Tokyo, 181-8588, Japan
*Contact: wenlei.shan@nao.ac.jp

Abstract—We have been conducting studies on planar integrated arrayed SIS receiver for wide field-of-view observation. Along with this study, our design ideas have gradually established, and were presented consecutively in previous ISSTTs ("Concept Design of a Dual-Polarization Sideband-Separating Multi-Pixel SIS Receiver," ISSTT 2016; "Design and Fabrication of a Dual-Polarization," ISSTT 2017). This approach has the following features: (1) independent coupling of signal and LO into a planar integrated mixer chip through membrane-based waveguide probes, (2) on-chip RF devices that include OMTs, hybrid bridges, and LO couplers, and (3) a much simplified LO metal waveguide distribution network that is embedded in the mixer mount.

For proof-of-concept, we have prototyped a single-pixel integrated SIS receiver, aiming to be readily extend to an array. The prototype receiver is designed at 2 mm wavelength to be capable of dual-polarization and balanced mixing. The mixer ICs are fabricated from silicon-on-insulator (SOI) wafers, which are locally thinned by a deep ion-etching process where LO and signal probes locate, with remaining silicon membranes of 6-micrometer thick. Orthogonally placed polarization-sensitive probes are adopted for signal coupling with polarization separation.

The mixer ICs were assessed in a 4 K cryostat. A corrugated feed horn is attached to the mixer mount for low cross-polarization signal coupling. The cross-polarization level of the horn/OMT combination was measured with a standard near-field beam scanner. Results of initial measurement show a cross-polarization level as low as -30 dB and a reasonable receiver noise. These results provide clear evidence of the feasibility of this planar integration approach.

A 1x4 Focal Plane Array Using 230 GHz SIS Mixers

John Garrett[1,*], Jamie Leech[1], Faouzi Boussaha[2], Christine Chaumont[2], Brian Ellison[3], and Ghassan Yassin[1]

[1]*University of Oxford, Oxford, U.K.*

[2]*GEPI, Observatoire de Paris, Paris, France*

[3]*Rutherford Appleton Laboratory, U.K.*

*Contact: john.garrett@physics.ox.ac.uk

***Abstract*—A new 1x4 focal plane array centered around 230 GHz is presented in this paper. The size of the array was limited to 4 pixels due to the space available in the test cryostat; however, we can expand the design in the future. On the front of the array block there are 4 waveguides flanges for the RF feed horns, while the local-oscillator signal enters through a separate waveguide on the side. The local-oscillator power is multiplexed using cascaded E-plane power dividers and then combined with the RF signals using directional couplers. Preliminary tests of the array block have now been completed. They show reasonable local-oscillator distribution and excellent RF signal isolation. Future work will involve testing the noise properties of the array block and improving the local-oscillator distribution.**

I. Introduction

A common application of millimeter-wave receivers is observing molecular transitions in star forming regions. We would like to map these regions to understand their physical and chemical properties; however, from our perspective, these regions can extend over very large angular diameters (~0<θ<~100 arcmin). Large surveys are slow using single dish antennas and even more difficult with interferometers due to their inherently narrow beam widths. Unfortunately, increasing the mapping speed of heterodyne receivers around 230 GHz is very challenging because the current state-of-the-art Superconductor/Insulator/Superconductor (SIS) receivers are already very close the quantum limit sensitivity. This means that any further improvements to sensitivity (and therefore mapping speed) will be modest.

An alternative technique to improve the mapping speed is to increase the number of receivers in the focal plane. The mapping speed of a focal plane array is proportional to

$$\frac{N}{T_n^{\ 2}} \qquad (1)$$

where N is the number of pixels in the array, and T_n is the system noise temperature. The challenge then is to add more pixels without increasing the noise temperature. Two general approaches are to (a) create a large number of pixels with relatively poor noise properties, or (b) create a modest number of pixels with very good noise properties. The latter is more desirable due to the lower requirements on the backend of the receiver.

Perhaps the most obvious challenge in creating a focal plane array is the complexity since more pixels require more feed horns, low noise amplifiers (LNAs), magnetic coils, bias wires, etc. All of these components are required to fit into a very compact space, and for older telescopes, this often means fitting into a space that was only ever intended for a single pixel receiver. (A side effect of the complexity is also the added thermal load due to the additional wiring and LNAs.)

Another major challenge comes from efficiently pumping all of the SIS devices. The local-oscillator (LO) signal has to be evenly divided amongst the pixels (i.e., multiplexed) and then injected into each SIS device (i.e., diplexed) without interfering with the RF signal path. SIS mixers are relatively insensitive to LO power (compared to HEBs), but it is important to keep Eqn. (1) in mind to balance the cost/benefit of adding more pixels. Most of the other focal plane arrays around 230 GHz use waveguide power splitters to divide the LO signal (e.g., HERA [1] and SuperCam [2]). Alternatively, the NRAO 8-beam array used cascaded wire grids [3], DesertStar used a reflective phase grating [4] and HARP-B uses a freespace meander line [5], but these techniques either become very complex above ~16 pixels or they limit the RF bandwidth of the receiver. In order to combine the LO and RF signals, SuperCam, HARP-B and DesertStar all use Mylar beam splitters; however, this complicates the telescope's optics since they require additional alignment. At frequencies below ~400 GHz, waveguide directional couplers, such as those used by HERA, are likely a better option.

In this paper, we present a new 1x4 focal plane array centered around 230 GHz. The array uses waveguide power splitters to divide the LO signal and directional couplers to inject the LO signals into the RF waveguides. These techniques facilitate the optical system since no freespace components are required for the LO. This should allow the design to be expanded into a much larger format in the future.

II. Design

A 3D render of the 1x4 array is shown in Fig. 1. The waveguide flanges for the four pixels are seen on the front of the array block and one flange is visible on the side for the LO signal (Fig. 1a). Spindles for the magnetic coils are also visible on top of the block with soft iron pole pieces to direct the magnetic field across the SIS devices. The size of the new focal plane array was limited to 1x4 due to the space available in our test cryostat. This array will act as a demonstrator to test our new design and then it can be expanded in the future. This may include expanding the array into a larger linear strip, such as 1x8, and then stacking the strips to form an 8x8 focal plane array, similar to SuperCam.

(a) The array block fully assembled.

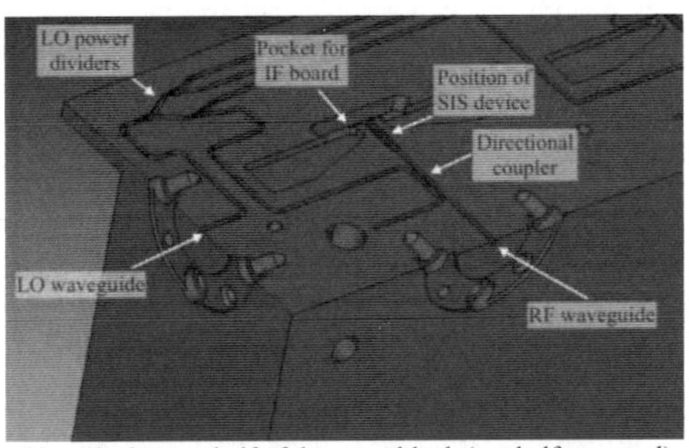

(b) The bottom half of the array block (top half removed).

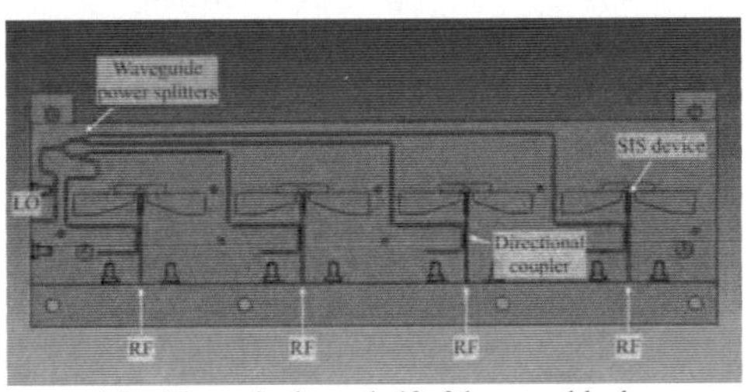

(c) The entire lower half of the array block.

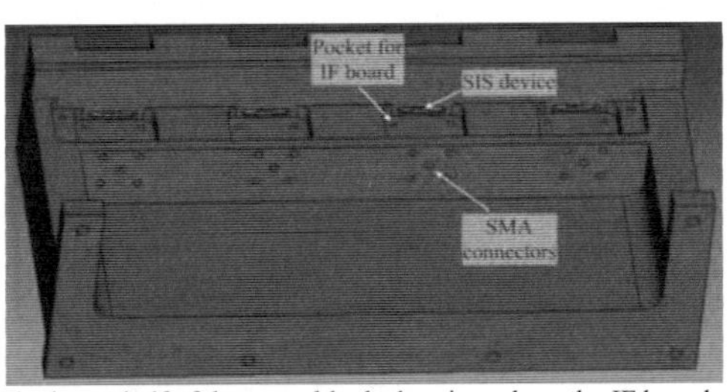

(d) The lower half of the array block showing where the IF boards are placed to route the IF signal downwards.

Fig. 1. Design of the 1x4 focal plane array block.

The waveguides were machined using split-block fabrication. In Fig. 1b, the upper half is removed to display the inner waveguides. As the LO signal enters the array block from the left-hand side, it is first divided in four using cascaded E-plane power splitters, and then the LO signal is injected into each RF waveguide using a direction coupler. The entire lower half of the array block is shown in Fig. 1c. Since the RF and LO waveguides are in the same plane, the IF signal is routed downwards to avoid any crossover, as shown in Fig. 1d. An IF tuning board sits in the pocket on the back of the array block, which connects the SIS device to an SMA connector (not shown in Fig. 1d).

A. Bow-tie couplers

The directional couplers were designed using bow tie antennas (Fig. 2a). These couplers are similar to the couplers that were previously used by the GUBBINS system [6]. The coupling was optimized using HFSS simulations to be approximately -17 dB (Fig. 2b); although, past experience from the GUBBINS system suggests that the actual coupling could be closer to -13 dB.

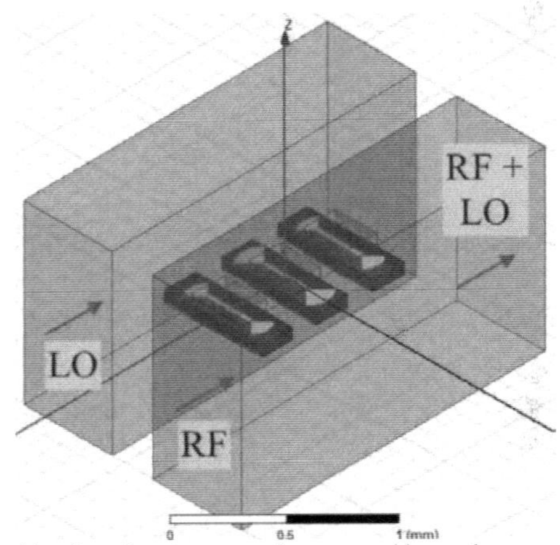

(a) Bow-tie antennas used as cross-guide couplers.

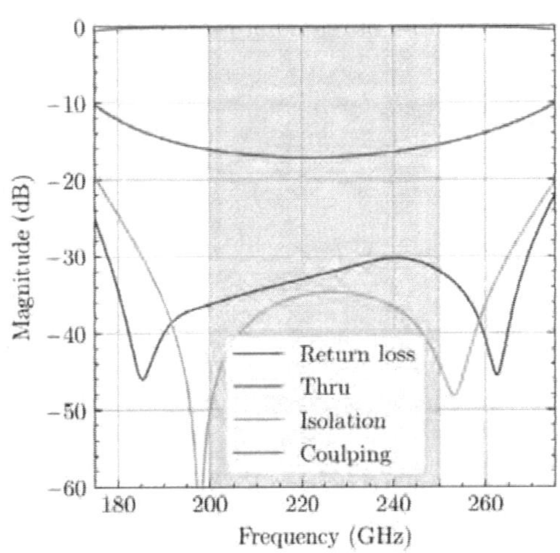

(b) Simulated performance from HFSS.

Fig. 2. Waveguide directional coupler using bow-tie antennas.

B. *E-plane waveguide power splitters*

As seen previously in Fig. 1b, the LO power is divided using cascaded E-plane power splitters. The 3D model of one of these power splitters is shown in Fig. 3a. The design was optimized to provide less than -15 dB of return loss from 200 GHz to 280 GHz (Fig. 3b).

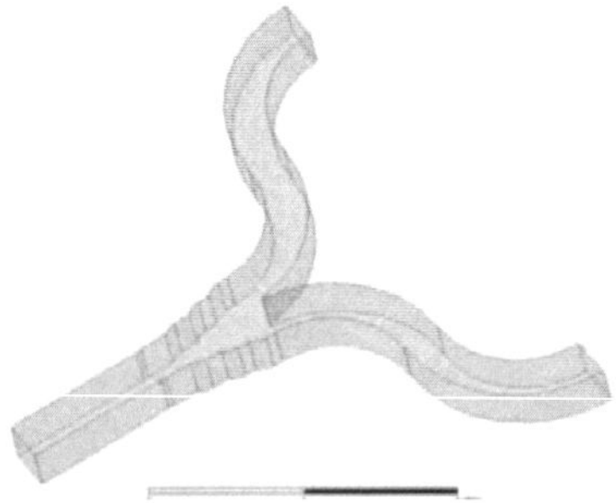

(a) 3D rendering of one of the power splitters.

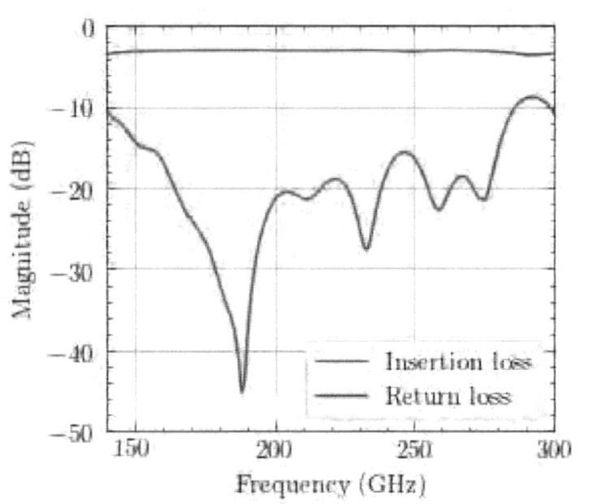

(b) Simulated performance of the power splitters.

Fig. 3. -3dB E-plane power splitters.

III. 230 GHz Mixer Device

The SIS device that was installed inside the array block has already been reported in [7,8]. It is a single-ended device that uses a finline transition to couple the RF and LO signals from the waveguide to the planar circuit. The device was fabricated on a 100 μm quartz substrate with a 1.5 μm^2 Nb/Al-AlO$_x$/Nb SIS junction. It was designed to operate from 140 to 260 GHz with an IF range from DC to +12 GHz.

This SIS device has been tested extensively in a single pixel mixer block. The SIS junction exhibits excellent DC properties with a typical quality factor of $Q = R_{sg}/R_n > 30$ and a current density of $J_c \sim 13.7$ kA/cm^2. Using an LO with a 213-257 GHz tuning range, excellent RF performance was found from 215 to 245 GHz, with a noise temperature close to 40 K (Fig. 4). The IF response spans from DC to 12 GHz, although the noise temperature quickly degrades past ~8 GHz.

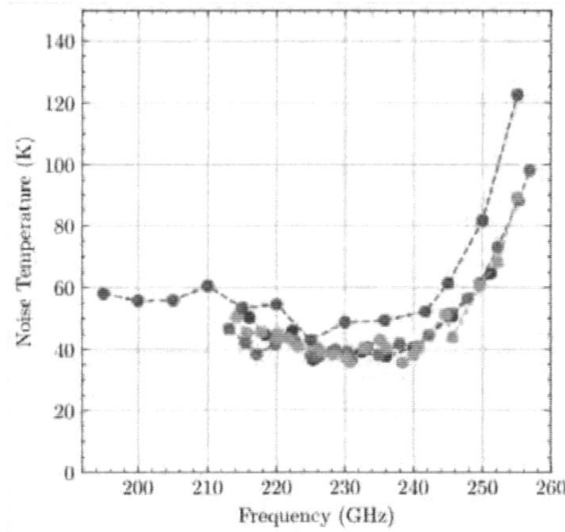

Fig. 4. Measured RF performance from the 230 GHz SIS device in a single mixer block. The different colors represent different devices. The device represented by the red line was tested with a different LO that has a wider tuning range but worse noise properties, which is why the noise temperature of this device is ~10 K higher than the others.

IV. Fabrication and Assembly

The array block was manufactured at the Rutherford Appleton Laboratory (RAL) Space Precision Design Facility. All of the components were machined from high quality copper except for the magnetic pole pieces which were machined from soft iron. The finished array can be seen in Fig. 5. Note that highly precise machining was required for the power dividers (Fig. 5b).

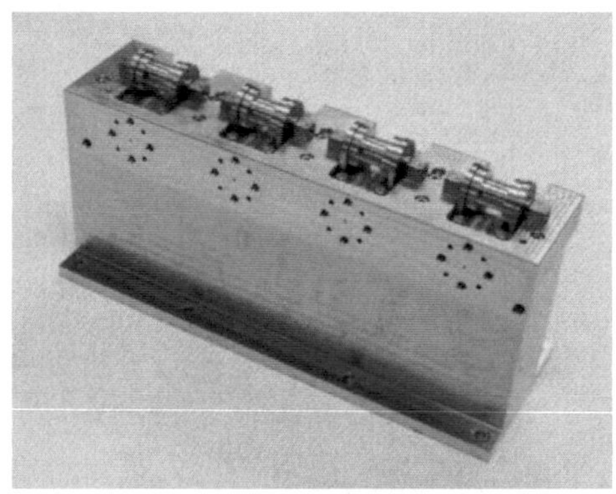

(a) The assembled array block.

(b) The waveguide power dividers.

Fig. 5. Photos of the finished array block.

A. IF boards

In the original design, the IF boards were meant to be installed vertically at a 90° angle to the SIS device (see Fig. 1d). In practice, however, it was too difficult to attach the bond wires over the 90° corner. Instead, the IF board was printed onto a flexible substrate (DuPont™ Pyralux® TK1810018R), so that it could bend over the corner. This allowed one end of the IF board to be in the same plane as the SIS device, making the wire bonding much easier. Although the flexible substrate has higher attenuation than Duroid®, the insertion loss was measured to be less than 1 dB at 20 GHz (for a 3 cm long 50 Ω microstrip).

B. Directional couplers

When the array block was first fabricated, the directional couplers were machined in the wrong direction (see Fig. 1b). To fix this error, the old waveguides were drilled out and then filled back in with copper slugs. This allowed the directional couplers to be re-machined. Unfortunately, at the time, no CNC milling machines were available. The new waveguide bend then needed a new machining technique that only required straight cuts. The solution, shown in Fig. 6a, was to leave a small amount of material in the outside corner. HFSS simulations showed excellent performance from 150 to 270 GHz (Fig. 6b).

(a) A cross-section of the new waveguide bend design.

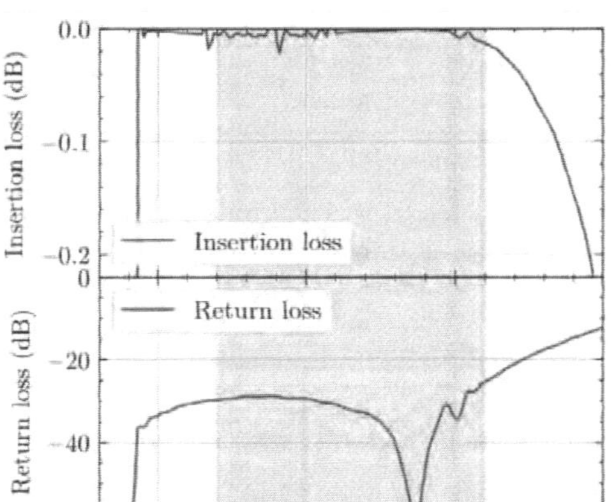

(b) Simulated performance from HFSS. The gray region is the standard frequency range for aWR4.3 waveguide (170-260 GHz).

Fig. 6. An E-plane waveguide bend that can be machined using a drill bit that is the diameter of the waveguide's minor axis and using a milling machine that can only make straight cuts (i.e., not a CNC milling machine).

V. Preliminary Test Results

For the preliminary tests, the array block was installed into the open-cycle cryostat shown in Fig. 7. Since this cryostat has had difficulties cooling in the past, only two windows were installed on the front of the cryostat to limit the radiative heat load. This allowed for freespace coupling for the LO signal and one RF pixel. Since each pixel is identical, each pixel should also have the same noise properties, provided that the LO pumping level is similar. Even though only one pixel could see outside the cryostat, all of the other pixels were still connected to DC bias supplies and current sources for the magnetic coils. This allowed us to measure the I-V curves from all of the pixels, from which we could recover the pumping levels. This setup allowed us to test the LO distribution and RF signal isolation.

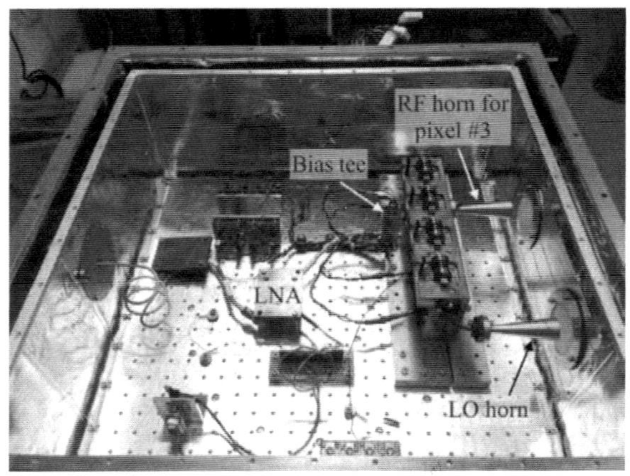

Fig. 7. The 1x4 focal plane array installed in the test cryostat. Only two windows were used for the initial tests. This allowed for freespace coupling for the LO signal and pixel #3.

A. LO distribution

In any focal plane array, it is important that the LO power is evenly distributed between the pixels in order to pump each SIS device to the optimum level. Otherwise, if the pumping level of some of the pixels is lower than others, the conversion gains of those pixels will be lower, resulting in higher IF noise contributions and therefore lower sensitivities.

To test the LO distribution, the LO signal was injected into the LO port of the array block, and then the pumped and unpumped I–V curves were measured from each pixel. The junction drive level and the power delivered to each SIS junction was then recovered by comparing the pumped and unpumped I-V curves. With the LO set to 230 GHz, the recovered junction drive levels were $\alpha_1 = 0.80$, $\alpha_2 = 0.72$, $\alpha_3 = 1.07$, and $\alpha_4 = 0.55$ (where $\alpha = V_J/V_{ph}$, V_J is the voltage across the junction, and V_{ph} is the equivalent photon voltage). Based on how the noise temperature changes with the drive level in the single pixel mixer block (Fig. 8), these drive levels roughly correspond to noise temperatures of $T_{n1} = 43.1$ K, $T_{n2} = 46.2$ K, $T_{n3} = 40.2$ K, and $T_{n4} = 57.4$ K. Although, this distribution is not ideal, it does result in adequate noise properties. The noise temperatures could be improved by over-saturating some of the pixels (i.e., driving some of the pixels past $\alpha \sim 1$) since the noise temperature does not deteriorate very much past the saturation point.

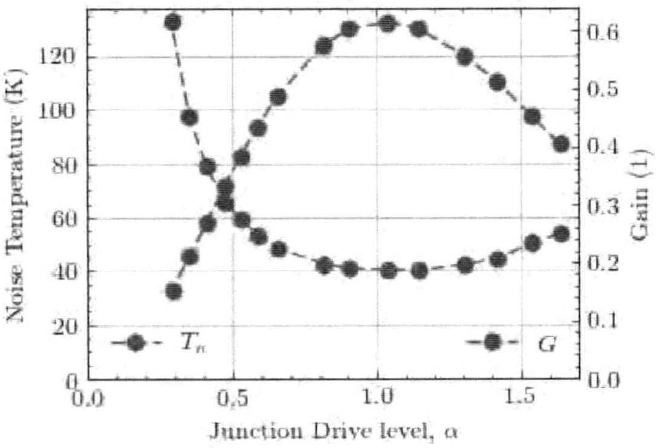

Fig. 8. Noise and gain performance compared to the junction drive level. This data was measured using a single pixel mixer block.

B. RF signal isolation

The RF signal isolation is a measure of the signal leakage between adjacent pixels. Typically isolation is measured by injecting a strong signal into one of the pixels, and then comparing the IF outputs from the surrounding pixels. However, in the current test setup, none of the pixels are connected to an IF backend. Instead, the signal isolation was measured by injecting a strong LO tone into pixel #3, and then measuring the pumped and unpumped I-V curves from the other pixels. Similar to the LO distribution test, this allowed us to recover the junction drive levels.

When this test was performed at 230 GHz, pixel #3 was pumped to $\alpha_3 = 1.624$. The pump levels of the surrounding pixels were then measured to be $\alpha_1 = 0.045$, $\alpha_2 = 0.048$ and $\alpha_4 = 0.046$. Since the junction drive level is proportional to the voltage across the junction, the signal isolation between the 3rd pixel and the ith pixel was then calculated by

$$I_{i,3} = -20 \cdot \log_{10}\left({}^{\alpha_i}/_{\alpha_3}\right).$$

For the drive levels listed above, this corresponds to isolation levels of $I_{1,3} = 31.1$ dB, $I_{2,3} = 30.6$ dB, and $I_{4,3} = 31.0$ dB. This is a very good level of isolation since less than 1/1000th of the signal power of pixel #3 leaked into the adjacent pixels.

VI. Future Work

The next step for this project will involve connecting the focal plane array to an IF measurement chain and then measuring the response to hot and cold black body loads to characterize the noise temperature and gain. The LO distribution could potentially be improved by repositioning some of the bow-tie antennas.

Conclusions

A new 1x4 focal plane array has been presented. It uses waveguide power splitters to divide the LO signal and a waveguide coupler based on bow-tie antennas to combine the LO and RF signals. Preliminary results show reasonable LO distribution and very good LO isolation. Future work will involve testing the noise properties of the array, and potentially expanding the array design to a larger format, such 8x8.

Acknowledgment

We would like to thank Mat Beardsley (RAL Space) for fabricating the array block, Byron Alderman (STFC/Teratech Components Ltd.) for fabricating the bow tie antennas, and Nick Brewster (STFC) for his assistance installing the bow-ties.

References

[1] K.-F. Schuster, *et al.*, "A 230 GHz heterodyne receiver array for the IRAM 30 m telescope," *Astronomy & Astrophysics*, vol. 423, no. 3, pp. 1171–1177, Sep. 2004.

[2] C. Groppi, *et al.*, "SuperCam: A 64 pixel heterodyne array receiver for the 350 GHz Atmospheric Window," in *Proc. 20th Int. Symp. Space THz Tech.*, Charlottesville, 2009, pp. 90–96.

[3] J. Payne and P. Jewell, "The Upgrade of the NRAO 8-beam receiver," in *ASP Conf. Series*, vol. 75, 1995, pp. 114–148.

[4] C. E. Groppi, *et al.*, "DesertSTAR: a 7 pixel 345 GHz heterodyne array receiver for the Heinrich Hertz Telescope," in *Proc. of SPIE (vol. 4855)*, Marseille, France, 2003, p. 330-337.

[5] H. Smith, *et al.*, "HARP: a submillimeter heterodyne array receiver operating on the James Clerk Maxwell Telescope," in *Proc. of SPIE (vol. 4855)*, Marseille, France, 2003, pp. 338–348.

[6] P. Grimes, *et al.*, "GUBBINS: A novel millimeter-wave heterodyne interferometer," in *Proc. 20th Int. Symp. Space THz Tech.*, Charlottesville, VA, Apr. 2009.

[7] J. Garrett, B.K. Tan, F. Boussaha, C. Chaumont, and G. Yassin, "A 220 GHz Finline Mixer with Ultra-Wide Instantaneous Bandwidth," in *Proc. 26th Int. Symp. Space THz Tech.*, Cambridge, MA, Mar. 2015.

[8] J. Garrett, F. Boussaha, C. Chaumont, B.K. Tan, and G. Yassin, "A 230 GHz finline SIS receiver with wide IF bandwidth," in *Proc. 27th Int. Symp. Space THz Tech.*, Nanjing, China, Mar. 2016.

Towards the Improvement of the Heterodyne Receiver Sensitivity beyond the Quantum Noise Limit

Denis Sych[1,2], Michael Shcherbatenko[2,3], Michael Elezov[2], Gregory N. Goltsman[2,4*]
[1] P. N. Lebedev Physical Institute of the Russian Academy of Sciences, 119991 Moscow, Russia
[2] Department of Physics, Moscow State University of Education, 29 Malaya Pirogovskaya St, Moscow, 119435, Russia
[3] Moscow Institute of Physics and Technology, Moscow Oblast, Russia
[4] National Research University Higher School of Economics, Moscow 101000, Russia
*Contact: goltsman@mspu-phys.ru

***Abstract*—Noise reduction in heterodyne receivers of the terahertz range is an important issue for astronomical applications. Quantum fluctuations, also known as shot noise, prohibit errorless measurements of the amplitude of electro-magnetic waves, and introduce the so-called standard quantum limit (SQL) on the minimum error of the heterodyne measurements. Nowadays, the sensitivity of modern heterodyne receivers approaches the SQL, and the growing demand for the improvement of measurement precision stimulates a number of both theoretical and experimental efforts to design novel measurement techniques aimed at overcoming the SQL. Here we demonstrate the first steps towards the practical implementation of a sub-SQL quantum receiver. As the principal resources, it requires a highly efficient single-photon counting detector and an interferometer-based scheme for mixing the signal with a low-power local oscillator. We describe the idea of such receiver and its main components.**

The ultimate fundamental limits of the measurement precision are studied in the fields of quantum physics, quantum radiophysics and quantum optics for many years. It is well understood that the quantum fluctuations of the signal, such as shot noise, lead to the finite measurement precision even in the absence of the technical noise and introduce the unavoidable signal discrimination error. Back in the 60s of the last century, the fundamental limit on the maximum accuracy of quantum measurements was calculated [1, 2], nowadays called the Helstrom bound, which turned out to be much smaller than the corresponding value provided by the ideal heterodyne receiver (called the standard quantum limit, SQL). Since then, a number of efforts were made to design a realistic setup to achieve the Helstrom bound, or, at least, to overcome the SQL. The Helstrom bound depends on the number of circumstances, for instance, the quantum state of the signal and *a priori* knowledge about the signal parameters, such as polarization, spectrum, spatial profile, timing, photon statistics, *etc*. The basic idea of overcoming the SQL is to take into account this *a priori* information about the signal, and design a measurement scheme specifically suited for the given type of signal.

Despite the Helstrom bound is known for several decades, the problem of designing a realistic measurement apparatus able to achieve the Helstrom bound remains open. Over the past 50 years, several approaches to reduce the measurement error below the SQL were proposed [3, 4, 5]. The key components of the receivers that, in principle, can beat the SQL, include highly efficient single-photon counting detectors [6, 7] and various interferometer-based schemes for mixing the signal with a weak local oscillator, controlled by feed-back or feed-forward loops.

To date, implementation of heterodyne detection using a weak local oscillator (whose power is equal or slightly higher than that of the signal) has been demonstrated, with a low noise photon counters used as a detector. Photon-counting nature of the detector ensures quantum-limited sensitivity, provided 100% photon detection efficiency. In the IR-range, superconducting nanowire single photon detectors (SNSPDs) are widely used due to their remarkable characteristics. Such detectors have quantum efficiency close to 100%, and also high counting rate (limited by the dead time of $\sim$ 10 ns). In the direct detection mode the detector count probability is time independent (i.e., registration of a photon is equiprobable in any given time interval).

In contrast to that, in the heterodyne mode, the probability of detecting a photon periodically changes with time, which is the result of the interference beats of the signal and the local oscillator. Processing of statistics of photocounts allows reconstruction of the frequency response of the signal. A heterodyne receiver based on SNSPD was demonstrated in [8]. The receiver was assembled using standard optical fibers (Fig. 1). As the signal and the local oscillator (LO) we used 1550 nm DFB-lasers. The IF analog signal read out from the detector was processed with help of RF spectrum analyser. However, this method of processing was not optimal.

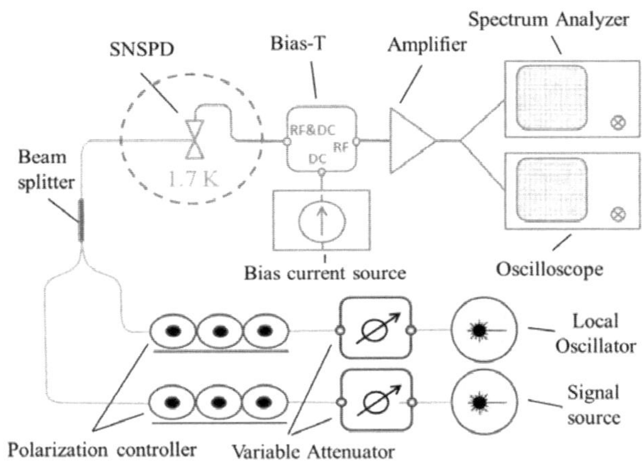

Fig. 1. A schematic view of the heterodyne receiver based on SNSPD

In [9], a similar measurement but implementing a digital post-processing of the signal was carried out. The voltage pulses from the detector were replaced by the delta pulses (since we are only interested in the registration of the photon arrival time moments), after which the Fourier transform was used to reconstruct the spectrum of the signal (Fig. 2).

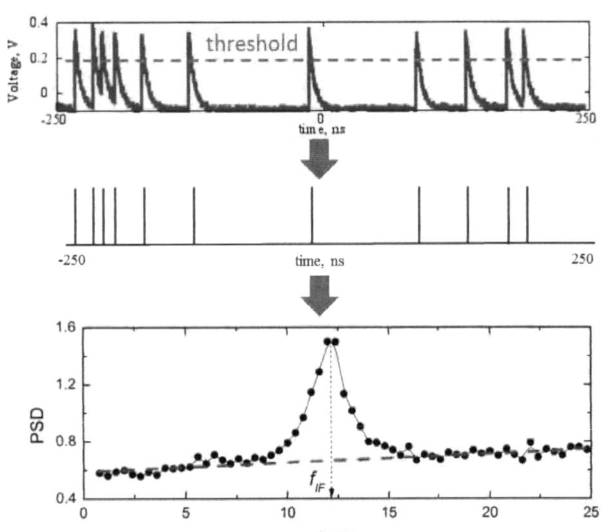

Fig. 2. Digital post-processing: first step – detection of photons, second step - the voltage pulses are replaced by delta-pulses, third step – Fourier-spectrum reconstruction.

The application of this approach makes it possible to get rid of the electrical component of noise, as well as the pulse spectrum contribution. Noise in such a system will be only dependent on the false response (dark counts), which provides a relatively insignificant contribution (dark count rate is 10^{-2} s^{-1}). Such a heterodyne receiver operates at the quantum noise limit, and the local oscillator power necessary for optimal operation is comparable to the signal power.

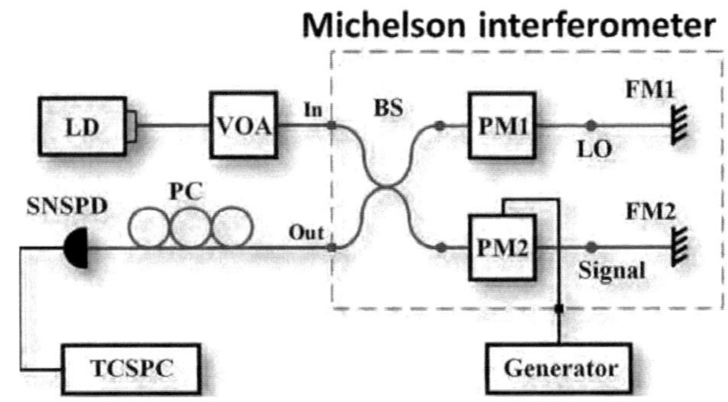

Fig. 3. Experimental setup: LD - Laser Diode at 1550 nm wavelength; PM1 and PM2 - phase modulators; TCSPC - Time-Correlated Single Photon Counting electronics, BS - beam splitter 50/50; FM1 and FM2 - Faraday's mirrors; VOA - variable optical attenuator; PC - polarization controller; SNSPD - Superconducting Nanowire Single-Photon Detector,

Recently, we proposed a novel approach to reduce the measurement error below the SQL [5]. With a new quantum receiver scheme based on the adaptive heterodyne technique, we can unconditionally suppress the measurement error below the SQL down to the Helstrom bound. In this work, we elaborate the idea further on and demonstrate the first steps of a proof-of-principle experimental realization of the sub-SQL receiver (see Fig. 3). Here we consider the simplest case – discrimination of a binary coherent signal, i.e. the task is to discriminate between two phase-conjugated states $|\alpha\rangle$ and $|-\alpha\rangle$. As a basis we use the Kennedy receiver scheme proposed in [3].

The main idea is the following. We interfere the signal and the LO on an almost transparent beam splitter (99:1 splitting ratio), such that almost all the power passes through the beam splitter. The LO power is chosen such that the reflected power of LO is comparable to the transmitted power of the signal (Fig. 4).

The result of interference between the LO and the signal can be tuned to be almost destructive, ideally leading to the completely nulling result. Such operation is called the nulling signal displacement. Ideal "nulling" occurs due to superposition of the signal with the LO of the same frequency, polarization, spatial profile, *etc*, but π phase difference. The ideally nulled signal provides no photon counts, thus the presence of photon counts after the interferometer is treated as 0 phase difference between the signal and the LO.

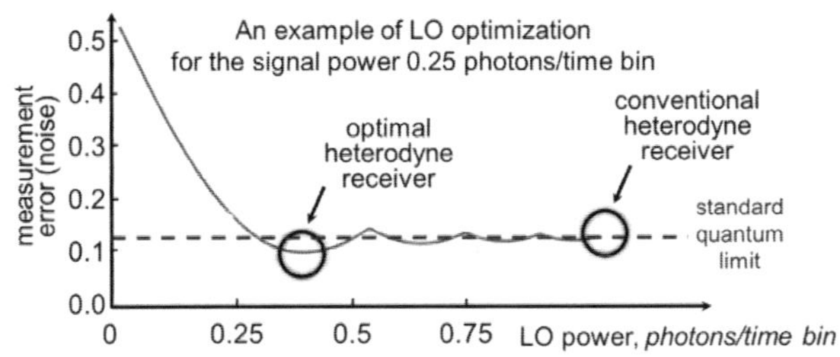

Fig. 4. Error rate as a function of the LO power for various heterodyne receivers. Optimized heterodyne receiver (global minimum) outperforms the conventional heterodyne receiver.

The optical part of the receiver is realized on the basis of the Michelson interferometer made of standard optical fiber components. We got an interference contrast of 25 dB with a temporal stability of a few seconds. As a source, a 1550 nm fiber-coupled DFB laser is used, serving as both the LO and the test signal, which is a typical technique used in similar receivers. The laser power is split by an optical fiber beam splitter. Then the signal is highly attenuated by a variable attenuator down to the single-photon level.

Using a fiber phase modulator installed in one arm of the interferometer, a binary test signal (periodically repeated "01" sequence at 100 kHz repetition rate) is produced. The result of the interference between the signal and the LO is either destructive or constructive, corresponding to zero photon counts (regarded as logical "0") and non-zero photon counts (regarded as logical "1", respectively). Photons are detected by the SNSPD, and the electronics registers the arrival time of the photon, counted from the beginning of the measurement.

To calculate the amount of errors, we send a signal with a known sequence of "0" and "1" to the receiver, and compare it with the measured sequence. After processing the results, the number of errors is normalized to the number of transmitted signal bins, and the error probability is compared to the SQL, which was calculated by the formula [1]:

$$e_{SQL} = \frac{1}{2}\left(1 - erf\left(\sqrt{2n}\right)\right), \qquad erf(x) = 2\int_0^x e^{-t^2} dt/\sqrt{\pi}$$

For the approximate average power of 1 photon per time bin, the observed state discrimination error (1.9%) was 27% below the SQL (2.6%). Thus, we were convinced that a receiver based on standard optical fibers and SNSPD can operate below the SQL, though we do uncorrected for the quantum efficiency of the detector (around 65%) and the losses in the interferometer. Based on the obtained results, in the future work aimed to attain the Helstrom bound, we plan to create a receiver with a tunable signal displacement and an adaptive feed-forward control of the LO. Development of the single-photon detector technology in the near- and mid-infrared range allows for realistic implementation of a new heterodyne receiver technique in this spectral range. Attempts are made to develop the single photon counting technology in a longer wavelength range [10], which opens good perspectives for the development novel receivers in far-infrared and THz ranges.

ACKNOWLEDGMENT

The research has been carried out with the support of the Russian Science Foundation (project No. 17-72-30036)

REFERENCES

[1] C. W. Helstrom, "Detection theory and quantum mechanics", *Inf. Control* 10, 254, 1967.

[2] P. A. Bakut and S. S. Shchurov, "Optimal detection of a quantum signal", *Probl. Peredachi Inf.* 4, 77, 1968.

[3] R. Kennedy, MIT Res. Lab. Electron. Quart. Prog. Rep. 108,219, 1973.

[4] S. Dolinar, MIT Res. Lab. Electron. Quart. Prog. Rep. 111,115, 1973.

[5] D. Sych and G. Leuchs, "Practical receiver for optimal discrimination of binary coherent signals", *Phys. Rev. Lett* 117, 200501, 2016.

[6] C. Wittmann, et al, "Discrimination of binary coherent states using a homodyne detector and a photon number resolving detector", *Phys. Rev. A* 81, 062338, 2010.

[7] C. Wittmann, et al, "Demonstration of coherent-state discrimination using a displacement-controlled photon-number-resolving detector", *Phys. Rev. Lett.* 104, 100505, 2010.

[8] M. Shcherbatenko, et. al., "Potential of a superconducting photon counter for heterodyne detection at the telecommunication wavelength," *Opt. Express* 24, 30474-30484, 2016.

[9] Y. Lobanov, et. al., "Superconducting Nanowire Single Photon Detector for Coherent Detection of Weak Signals" *IEEE Transactions on Applied Superconductivity* 27, 4, 2016.

[10] Korneev, et. al., "NbN nanowire superconducting single-photon detector for mid-infrared", *Physics Procedia*, 36, pp. 72-76, 2012.

Development of SIS Receivers with Ultra-wide Instantaneous Bandwidth for wSMA

C. Edward Tong[1,*], Lingzhen Zeng[1], Paul Grimes[1], Wei-Chun Lu[2], Tse-Jun Chen[2], Yen-Pin Chang[2] and Ming-Jye Wang[2]
[1]*Harvard-Smithsonian Center for Astrophysics, Cambridge, MA 02138, USA*
[2]*Institute of Astronomy & Astrophysics, Academia Sinica, Taipei, Taiwan*
*Contact: etong@cfa.harvard.edu

***Abstract*—We report on the development of SIS receivers with ultra-wide instantaneous bandwidth, intended for the new wSMA (wideband Submillimeter Array) instrumentation. The different factors which would put a limit on the IF bandwidth of an SIS mixer are first examined. In order to deliver up to 20 GHz of IF bandwidth, we have based our SIS mixer design on a 3-junction series array connected to a wideband cryogenic isolator. Special considerations are given to the grounding of the mixer chip, which can introduce significant grounding inductance. We present the test results for the prototype wSMA 240 GHz receiver. Low noise operation has been confirmed over the IF range of 3.5 – 19 GHz, for Local Oscillator frequencies between 210 and 270 GHz.**

I. Introduction

As the Superconductor-Insulator-Superconductor (SIS) mixer becomes established as the ultimate low noise heterodyne front-end at submillimeter wavelengths, it is recognized that its utility in astronomical instrumentation can see further enhancement by increasing its instantaneous operating bandwidth [1, 2]. At present, the majority of SIS mixer based receivers operate with an IF of 4 – 8 GHz, while an increasing number offer a wider IF band of 4 – 12 GHz [3, 4]. A number of SIS mixers have been tested with even higher IF bandwidths [5-7].

In this paper we will discuss the challenges of extending the IF bandwidth of SIS mixers. We will also present the design and performance of the wSMA 240 GHz receiver we are currently developing for the Submillimeter Array

The Submillimeter Array (SMA) is a radio interferometer located on Mauna Kea, Hawaii. It is jointly operated by the Smithsonian Astrophysical Observatory (SAO) and Academia Sinica Institute of Astronomy and Astrophysics (ASIAA), Taiwan. Since 2017, the SMA has been conducting routine nightly astronomical observations with dual SIS receivers operating between 200 and 420 GHz, delivering an on-sky bandwidth of 32 GHz using its 4 – 12 GHz double-side-band IF coverage. The total observation bandwidth will be further increased in the next phase of instrumentation upgrade. The new wideband SMA (wSMA) instrumentation [8] will operate with an IF of 4 – 20 GHz and will double the on-sky bandwidth of observation. This will lead to further increase in wideband continuum sensitivity of the array as well as further increase in spectral line observation capability.

II. IF Bandwidth Limits of SIS Mixers

At very high IF, Tucker's Quantum Theory of Mixing [9] stipulates that the upper side band signal behaves differently to the lower sideband signal because of quantization of the IF. Pan and Kerr [10] have conducted an investigation of the effect of very high IF in SIS. They have concluded that the zero IF approximation holds well if the IF is less than one tenth of the Local Oscillator (LO) frequency. Thus, for an SIS mixer operating above 200 GHz, there should be no fundamental physical reason that prevents it from delivering IF up to 20 GHz. In practice, however, there are a number of factors which would limit the IF bandwidth of SIS mixers. Here we discuss a few of the dominating factors.

A. Junction Capacitance

It has generally been recognized that a large junction capacitance in an SIS mixer may short the IF signal. In addition, the tuning circuit also presents additional capacitance to the IF in parallel with the junction capacitance, further limiting the IF bandwidth of the mixer. According to [11], the 3-dB IF bandwidth of an SIS mixer is given by:

$$F_{3\mathrm{dB}} = \frac{1}{2\pi(R_{\mathrm{out}}//R_L) * (C_j + C_{\mathrm{tune}})} \qquad (1)$$

where C_j and C_{tune} are the capacitances of the junction and the tuning circuit respectively; and R_L is the load resistance of the mixer, which is typically 50 Ω for wideband IF system; and R_{out} is the output resistance of the mixer, which is in parallel with R_L.

In order to reduce the junction capacitance, one can either use very small junction or use a series junction array [12]. Junctions with small dimensions are generally more difficult to fabricate, requiring electron beam lithography, and they have higher leakage current when compared to devices with similar critical current density. In addition, they also require higher magnetic field to operate. For these reasons, we have chosen to use series junction arrays with relatively large dimensions, which can be reproduced reliably using optical lithography.

B. Impedance Level

Equation (1) shows that the equivalent resistance of R_{out} in parallel with R_L is inversely proportional to the IF bandwidth. R_{out} is simply the dynamic resistance of the photon step, which is typically many times R_N, and at times infinite or even negative. As a result, a lower value of R_N can generally help to increase the IF bandwidth. This is the case for SIS mixers with very high critical current density [7]. Likewise, Parallel-Connected-Twin-Junction (PCTJ) type SIS mixers generally have very low values of R_N, although they often only offer moderate IF bandwidth as the junction capacitances appear in parallel, and so are additive.

C. Output Saturation

The dynamic range of an SIS mixer is limited by the phenomenon of output saturation in which the IF voltage swing becomes a significant fraction of the width of the photon step [13]. Kerr also pointed out that this situation is accentuated for noise-like signal which is characterized by a Gaussian amplitude distribution [14]. As the IF bandwidth increases, this IF voltage swing also rises.

Consider an SIS mixer operating with an IF bandwidth of 20 GHz, and a double-side-band (DSB) conversion gain of 0-dB. When the mixer input is terminated by an ambient (300 K) load, the IF output power is around -70 dBm, or 100 pW. Assuming a constant IF load impedance of 50-ohm, the RMS IF voltage swing works out to be 0.07 mV. For noise-like signal, the peak-to-peak voltage swing, taken to be 4σ, is 0.28 mV, which is a significant fraction of the 1 mV wide photon step of a single SIS junction, pumped by a local oscillator (LO) of 242 GHz. Clearly, SIS mixers, which operate with a high IF bandwidth, are more prone to saturation.

The solution to the saturation problem is to employ a series junction array, so that the output voltage swing is divided between the junctions. By employing a 3-junction array, one increases the power handling capacity of the SIS mixer by a factor of 9. Conversely, one can also consider that the same junction array increases the IF bandwidth of the mixer by a factor of 9.

D. Impedance Matching

For all SIS mixers, it is important that the down-converted power from the mixer be efficiently passed onto the LNA. This requires good impedance matching between the IF output of the mixer and the input of the LNA. Early attempts to integrate an SIS mixer with wideband low noise amplifier generally employed the direct interface method [5, 15]. It is now recognized that a careful design of the impedance matching circuit is essential [16, 17]. Tan [6] designed a six-section transformer circuit to match an SIS mixer to the LNA and measured reasonable response from a 650 GHz SIS mixer to IF of 15 GHz. Another approach is to use SIS device with very high critical current density [7] such that $R_{out} \sim 50\ \Omega$, to provide wideband matching to the LNA.

We have selected the more established method of using a wideband isolator between the SIS mixer and the LNA. This ensures that the LNA sees a constant 50 Ω match across the wide IF band, and on the other hand, the mixer IF port is terminated with 50 Ω. More details on this wideband isolator will be given in the next section.

E. Grounding of Mixer Chip

For a waveguide mixer chip, it is common to have the actual IF grounding point of the SIS junction to be located at some distance away from the junction itself. While beam-lead mixer chips provide the best grounding configuration, many chip designs have the grounding point located behind an RF choke filter that prevents RF power from leaking along the chip channel. This arrangement introduces a grounding inductance to the IF embedding impedance [18].

Fig. 1 shows the how the wSMA-240 mixer chip is grounded with a cylindrical metal contact over a low impedance section of the IF choke filter. The center of the contact point is 0.82 mm from the waveguide feed point. Also shown in Fig. 1 is the IF embedding impedance seen at the waveguide feed point. At low IF, the reactive part of this embedding impedance increases linearly with frequencies, corresponding to an added grounding inductance of L_{gnd} ~10 nH. The real part of the impedance starts around 50 Ω at low IF and first sees moderate increases followed by larger increases at IF > 10 GHz.

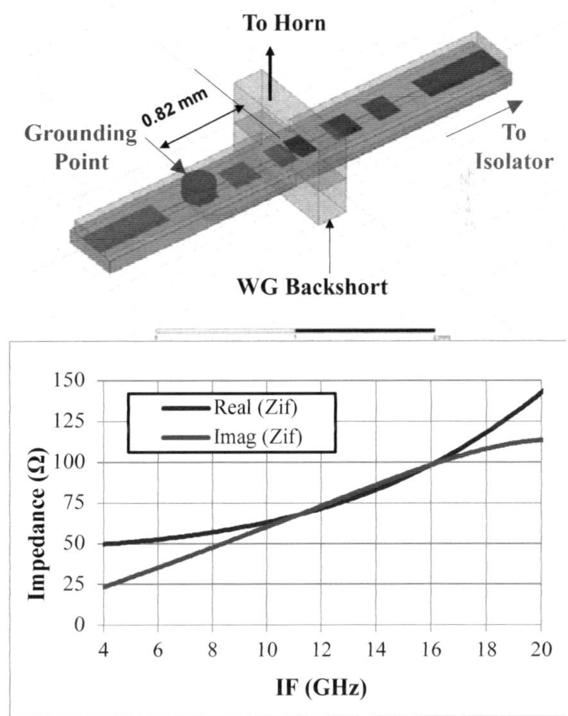

Fig. 1. (Upper) Grounding arrangement for the wSMA-240 mixer chip. (Lower) Simulated IF embedding impedance of the above mixer seen at the waveguide feed point.

This grounding inductance resonates with the junction and tuning capacitances, C_j and C_{tune} to form an impedance peak at an IF of f_R:

$$f_R = \frac{1}{2\pi\sqrt{L_{\text{gnd}}\left(C_j + C_{\text{tune}}\right)}} \qquad (2)$$

Beyond this resonant frequency, the conversion efficiency of the SIS mixer is expected to drop, marking the upper end of the usable IF bandwidth.

III. Wideband Cryogenic Isolator

In most SIS receivers, the cryogenic LNA is connected to the SIS mixer through a coaxial cable. A cryogenic isolator is usually inserted in front of the LNA to isolate reflections from the LNA from returning to the SIS mixer, which generally has high output impedance. Y-junction isolators can provide good isolation over an octave bandwidth. Edge-mode isolators are used when even wider bandwidths are required. At present, 4 – 12 GHz edge-mode isolators are commercially available.

We have developed a cryogenic edge-mode isolator which is usable between 4 and 22 GHz [19]. Its insertion loss is better than 1 dB from 4 to 17 GHz, rising to 1 – 1.5 dB from 17 to 22 GHz. The input return loss is about −15 dB from 4 to 22 GHz, and the isolation is better than 15 dB from 4 to 13 GHz, degrading to about 10 dB above 13 GHz.

The finite input return losses of the isolator, $|S_{11}|$, and the output reflection coefficient of the SIS mixer, Γ_{out}, drive a standing wave in the cable linking the isolator and the mixer. This standing wave ratio, SWR, is given by the following equation:

$$\text{SWR} = \frac{1 + |\Gamma_{\text{out}}||S_{11}|}{1 - |\Gamma_{\text{out}}||S_{11}|} \qquad (2)$$

Since the output impedance of an SIS mixer is generally high and at times infinite or even negative, $|\Gamma_{out}|$ can be considered to be around unity. For $|S_{11}| \sim$ -15 dB, equation (2) gives a SWR of 1.43 or 3.1 dB. Thus, the isolator shields the LNA from large source impedance variation, which can adversely affect the performance of the LNA. The isolator may help to prevent oscillation of the LNA when the SIS mixer exhibits negative output impedance.

IV. Tuning of SIS Junction

The discussion above highlights the importance of the output impedance of the SIS mixer for ultra-wideband operation. The other important parameter is its conversion gain. The insertion loss of the isolator means that the effective conversion gain of the mixer is reduced, especially towards the top end of the IF band. Furthermore, the noise temperature of an ultra-wideband LNA is unlikely to be as good as one with narrower band-width, and one also expects the noise of the LNA to rise with IF. Therefore, it is helpful if the SIS mixer can provide a net conversion gain so that the receiver noise temperature would remain flat across the wide IF band.

We present the result of a calculation based on Tucker's theory of quantum mixing [9] in Fig. 2. In this simulation, the source impedance presented to a tuned SIS junction is varied over the Smith chart, which is normalized to R_n. The capacitance of the junction is assumed to be cancelled out by a shunt inductor at the simulation RF frequency of 240 GHz. Both the output impedance of the SIS junction and its conversion gain are calculated.

In Fig. 2, the solid black contour gives the locus of the source impedance associated with infinite output impedance. The region of the Smith chart above this solid contour yields negative output impedance, while the region below yields positive output impedance. The output impedance is independent of the IF load impedance. Unity conversion gain contours are plotted in dotted lines for the cases of an IF load impedances of R_n and $2R_n$.

Several conclusions can be drawn from this simulation. To begin with, inductive source impedance is generally linked to negative output impedance, and the same can be said for conversion gain. A higher IF load impedance also expands the domain of conversion gain.

By employing an isolator at the IF port of the mixer, one can tolerate an infinite output impedance or even have the mixer operating slightly into the negative impedance region, without sacrificing stability. This allows us to use a small amount of conversion gain to counter-balance the IF roll-off, as mentioned above. The same holds true, if the load impedance rises with frequency. Referring to Fig. 1, this is actually the case for our mixer, as a result of the grounding arrangement.

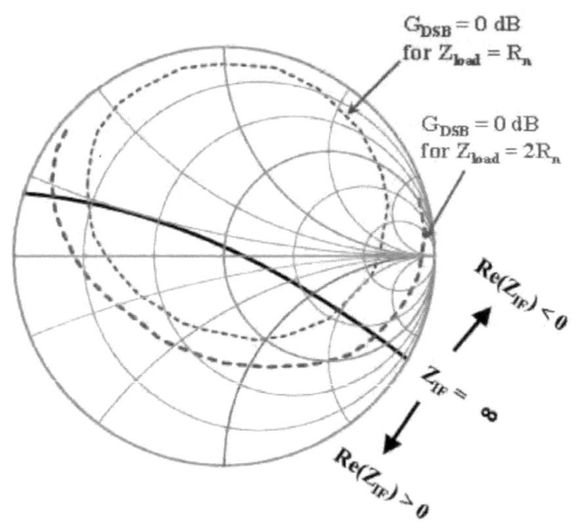

Fig. 2 Simulated conversion gain (G_{DSB}) and output impedance (Z_{IF}) of a Nb-based SIS mixer operating at an LO frequency of 240 GHz, and with α = 1. The source impedance of the SIS junction with its capacitance tuned out is varied over the Smith Chart normalized to R_n.

V. Design of the wSMA-240 Mixer

The wSMA-240 receiver will operate with an LO between 210 and 270 GHz, with a target IF coverage of 4 – 20 GHz. Based on the discussion above, we have chosen to use an SIS mixer based on a 3-junction array. The design parameters of the Nb/Al/AlO$_x$/Nb junction array are: nominal diameter of

1.6 μm, and R_nA product of 25 Ω-μm². The normal state resistance R_n of the array is ~40 Ω. Assuming a specific capacitance of 85 fF/ μm², the total capacitance of the junction array is ~57 fF, corresponding to an ωCR product of 3 – 3.5. This latter value implies that a tuning circuit can readily be designed to achieve a percentage band-width of ~30%. Considering that the IF coverage extends up to 20 GHz, the wSMA-240 receiver band is slated to provide sky coverage between 190 and 290 GHz, which corresponds to a percentage bandwidth in excess of 42%. Thus, our design goal is to provide good sensitivities to input frequencies between 205 and 275 GHz for a percentage bandwidth of 29%, with reduced performance beyond the band edges.

Equation (1) tells us that in order to maximize the IF bandwidth, not only does the junction capacitance have to be kept small, the capacitance of the tuning circuit, C_{tune}, has to be small as well. For this reason, we choose a simple transformer circuit to match the junction array to the waveguide feed point impedance. Fig. 3 shows the plot of the feed point impedance as a function of frequency as well as photos of the junction and the tuning transformer.

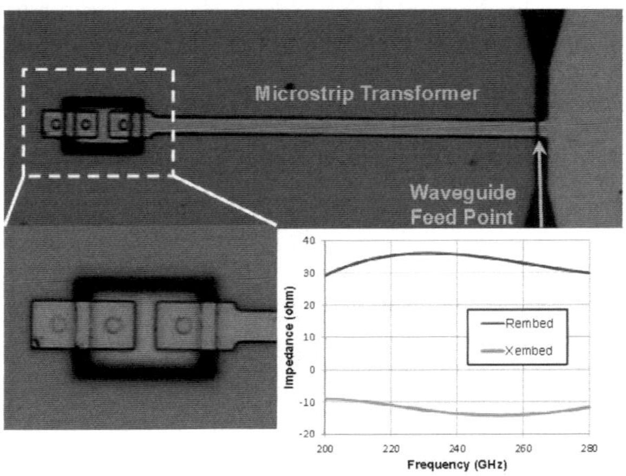

Fig. 3 (Upper) Photo of the SIS mixer chip showing the single section microstrip transformer linking the waveguide feed point and the SIS junction array. (Lower Left) Close up view of the junction array. Two of the junctions sit on top of a coplanar island. (Lower Right) Real (Rembed) and imaginary (Xembed) parts of the embedding impedance as a function of frequency seen at the waveguide feed point.

It should be noted that the embedding impedance at the waveguide feed point is capacitive. Consequently, the required length of the microstrip transformer is less than a quarter of a guided wavelength. This also helps to minimize C_{tune}, which amounts to ~58 fF, approximately equals to the junction capacitance. The sum of these 2 capacitances, C_{mixer}, is thus 115 fF. Using (1), one obtains a 3-dB IF bandwidth of 27 GHz if R_{out} is taken to be infinite.

However, when the grounding inductance, L_{gnd}, is taken into account, the resultant bandwidth is much reduced because L_{gnd} resonates with C_{mixer}. Equation (2) predicts a resonant frequency of 4.7 GHz. In practice, this frequency is affected by the real part of the impedance as well. By adding the susceptance due to C_{mixer} to the simulated complex IF load impedance plotted in Fig 1, we arrive at the total load impedance seen by the SIS mixer. This is plotted in Fig. 4.

As seen from the plot, the real part of the IF load impedance peaks at around 12 GHz, whereas the imaginary part is capacitive above 10 GHz. We, therefore, expect that the conversion gain of the mixer will also peak at 12 GHz and decline beyond it. Using this complex IF impedance data, we have performed simulation of the conversion gain of the wSMA-240 mixer at a number LO frequencies between 210 and 270 GHz for IF spanning 4 – 20 GHz. The results are presented in Fig. 5.

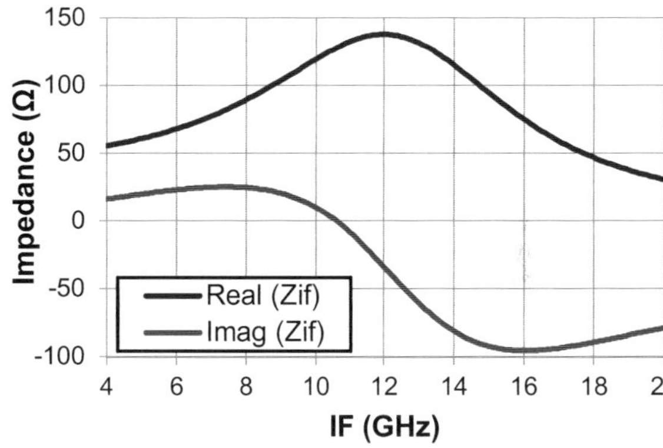

Fig. 4 Total IF load impedance, Z_{IF}, seen by the SIS mixer when both the grounding inductance, L_{gnd}, and mixer capacitance, C_{mixer}, are taken into account.

Fig. 5 confirms that the peak IF conversion gain is at approximately 12 GHz and drops above that frequency. Nevertheless, the single-side-band (SSB) conversion gain generally stays above -4 dB, except for the upper-side-band at the highest LO frequency. Adding in the 1 dB insertion loss from the isolator, the DSB conversion gain of the mixer is, therefore, generally above -2 dB. This means that the noise contribution of the isolator and the LNA is generally less than 1.6 times that of the inherent noise temperature of the LNA. We also note that the side-band ratio of the mixer is within 1 dB of unity ratio.

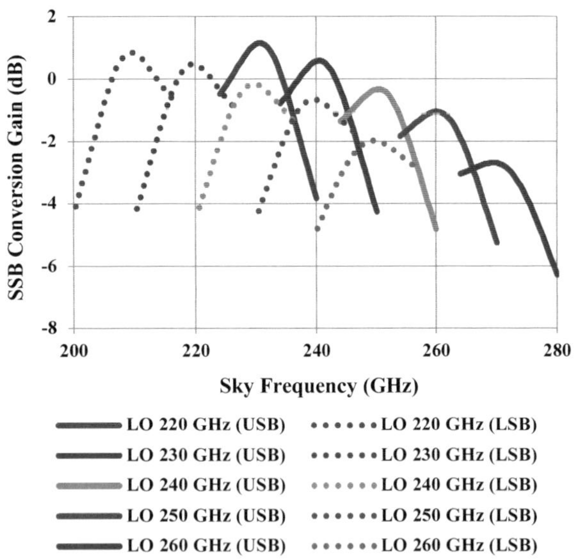

Fig. 5 Simulated SSB conversion gain of wSMA-240 mixer as a function of signal (sky) frequency for different LO frequencies when the IF is varied between 4 and 20 GHz. The upper-side-band (USB) responses are plotted in solid lines, while the lower-side-band (LSB) responses are in dotted lines.

VI. Performance of wSMA-240 Receiver

We have performed laboratory measurements of these wSMA-240 receivers. The SIS mixer was connected to the wideband isolator by a 15 cm long coaxial cable, with the junction array biased through a bias port on the side of the isolator. An LNF-LNC-6_20B amplifier was connected to the output of the isolator. From the data supplied by the manufacturer, the noise temperature of this amplifier is below 5 K between 6 and 14 GHz, and is below 10 K over 3.5 - 20 GHz.

Y-factor measurements using ambient load (295 K) and liquid nitrogen-cooled load (78.5 K) were conducted. Local oscillator was injected in front of the cryostat using a wire grid polarizer. DSB receiver noise temperature was calculated from the Y-factor directly without applying any corrections. The observed I-V and P-V curves of a test at an LO of 250 GHz were plotted in Fig. 6. The IF was 9 GHz and a maximum Y-factor of 2.93 was recorded, corresponding to a DSB noise temperature of 33.7 K, or around 2.8 times the photon noise at this LO frequency. We note from the pumped I-V curve that the photon step was bending downwards, indicating that the mixer had negative output impedance.

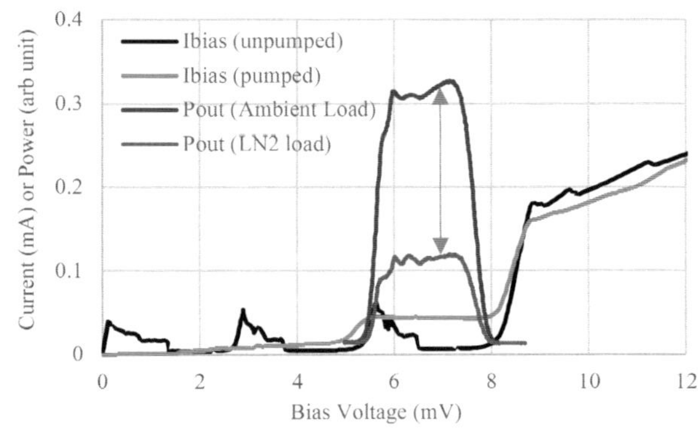

Fig. 6 Current Vs Voltage (I-V) and output Power Vs Voltage (P-V) curve of a wSMA-240 mixer when the mixer was pumped at an LO frequency of 250 GHz. The output power was measured at an IF of 9 GHz through a 30 MHz bandpass filter. A maximum Y-factor of 2.93 was obtained at 7.5 mV bias.

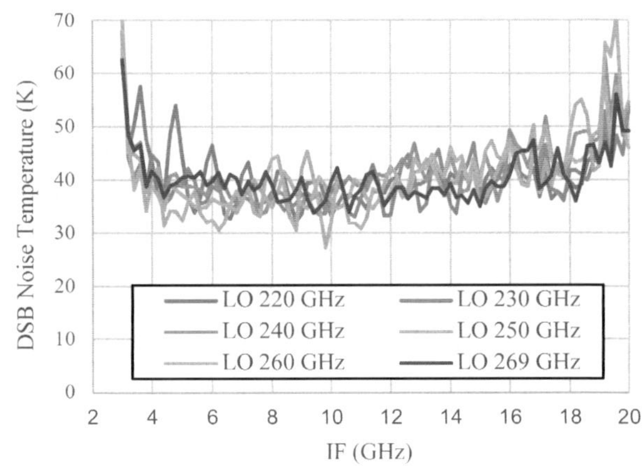

Fig. 7 Measured DSB receiver noise temperature of wSMA-240 mixer.

Using a YIG-tuned filter with a bandpass of 30 MHz, Y-factor was recorded between IF of 3 and 20 GHz, for various LO frequencies. The resulting DSB noise temperature is plotted in Fig. 7. On average, the noise temperature is around 40 K between IF of 3.5 and 19 GHz. A ripple with a period of 1 GHz is observed in the noise temperature plot Vs IF. This is caused by the standing wave given by (2) from the 15 cm long cable between the mixer and the isolator. One factor (2) does not account for is the finite isolation of the isolator, which couples with the finite input return loss of the LNA, to increase the magnitude of the standing wave.

VII. Conclusion

A 3-junction SIS mixer has been designed for the new wSMA-240 mixer, with very low mixer capacitance. By placing a wideband edge mode isolator between the mixer and the LNA, the receiver delivers low noise temperature for IF between 3.5 and 19 GHz, with LO spanning 210 and 270 GHz. The grounding inductance of the mixer is a limiting factor on its instantaneous bandwidth. Changes to the IF circuitry will be introduced to reduce this inductance.

References

[1] C.-Y.E. Tong, P.K. Grimes, J. Weintroub, and R. Blundell, "Ultra-wide IF bandwidth --- the next frontier for SIS receivers," in *Microwave Symp. (IMS), 2015 IEEE MTT-S Int.*, Phoenix, AZ, May 2015, doi 10.1109/MWSYM.2015.7166804.

[2] C.-Y.E. Tong, R. Blundell, K.G. Megerian, J.A. Stern, S.-K. Pan, and M. Popieszalski, "A distributed lumped-element SIS mixer with very wide instantaneous bandwidth," *IEEE Trans. Appl. Supercond.*, vol. 15, no. 2, pp. 490-494, June 2005.

[3] S.-K. Pan et al, "A fixed-tuned SIS mixer with ultra-wide-band IF and quantum-limited sensitivity for ALMA Band 3 (84-116 GHz) receivers," in *Proc. 15th Int. Symp. Space THz Techn.*, pp. 62-69, Amherst, MA, Apr. 2004.

[4] Y. Uzawa *et al*, "Development and testing of band 10 receivers for the ALMA project," *Physica C,* vol. **494**, pp. 189-194, Apr. 2013

[5] F. Rice, M. Sumner, J. Zmuidzinas, R. Hu, Henry G. LeDuc, A.I. Harris, and D. Miller, "SIS mixer design for a broadband millimeter spectrometer suitable for rapid line surveys and redshift determinations," in *Proc. SPIE 4855, Millimeter and Submillimeter Detectors for Astronomy*, Feb. 2003. doi 10.1117/12.459710.

[6] B.K. Tan, G. Yassin, and P. Grimes, "Ultra-wide intermediate bandwidth for high frequency SIS mixers," *IEEE Trans. THz Sci. Techn.*, vol. 4, no. 2, pp. 165–170, Mar. 2014.

[7] T. Kojima, M. Kroug, K. Uemizu, Y. Niizeki, H. Takahashi, and Y. Uzawa, "Performance and Characterization of a Wide IF SIS-Mixer-Preamplifier Module Employing High-Jc SIS Junctions," *IEEE Trans. on THz Sci. Techn.*, vol. 7, no. 6, pp. 694-703, Nov. 2017.

[8] P. Grimes, R. Blundell, S. Paine, C.-Y.E. Tong, and L. Zeng, "Next generation receivers for the Submillimeter array," in *Proc. SPIE 9914, Millimeter, Submillimeter, and Far-Infrared Detectors and Instrumentation for Astronomy VII*, 991424, July 2016. doi:10.1117/12.2232790.

[9] J.R. Tucker, and M.J. Feldman, "Quantum detection at millimeter wavelengths," *Rev. Modern Phys.*, vol. 57, no. 4, pp. 1055-1113, Oct. 1985.

[10] S.-K. Pan and A.R. Kerr, "SIS mixer analysis with non-zero intermediate frequencies," in *Proc. 7th Int. Symp. Space THz Techn.*, pp. 195-206, Charlottesville, VA, Mar. 1996.

[11] C.E. Tong, R. Blundell, K.G. Megerian, J.A. Stern, and H.G. LeDuc, "Quantum-limited mixing in a transformer-coupled SIS resonator for the 600 GHz frequency band," in *Proc. 13th Int. Symp. Space THz Technology,* pp. 23-32, Harvard University, March 2002.

[12] C.-Y.E. Tong, P.K. Grimes, A. Barbier, K. Schuster, and M.-J. Wang, "Design and performance of a 3-junction series distributed SIS mixer for wide IF applications," *IEEE Trans. Appl. Supercond.,* vol. 23, pp. 1400404, June 2013.

[13] L. R. D'Addario, "Saturation of the SIS mixer by out-of-band signals," *IEEE Trans. Microw. Th. Tech.*, vol. 26, no. 6, pp.1103-1105, Jun. 1988.

[14] A. R. Kerr, "Saturation by Noise and CW Signals in SIS Mixers," in *Proc. of the 13th Int. Symp. on Space THz Techn.*, Harvard University, pp. 11-22, 26-28 March 2002.

[15] S. Padin, D.P. Woody, J.A. Stern, H.G. LeDuc, R. Blundell, C.-Y.E. Tong, and M.W. Pospieszalski, "An integrated SIS mixer and HEMT IF amplifier," *IEEE Trans. Microw. Th. & Tech.*, vol. 44, no. 6, pp. 987-990, June 1996.

[16] A.R. Kerr *et al,* "Development of the ALMA Band-3 and Band-6 sideband-separating SIS mixers," *IEEE Trans. THz Sci. Technol.,* vol. 4, no. 2, pp. 201-212, March 2014.

[17] S. Montazeri, P.K. Grimes, C.-Y.E. Tong, and J.C. Bardin, "A 220-GHz SIS mixer tightly integrated with a sub-hundred-microwatt SiGe IF amplifier," *IEEE Trans. THz Sci. & Tech.,* vol. 6, pp. 133-140, Jan. 2016. DOI 10.1109/TTHZ.2015.2498041

[18] C.-Y.E. Tong, R. Blundell, K.G. Megerian, J.A. Stern, M. Popieszalski, and S.-K. Pan, "Gain enhancement in inductively-loaded SIS junction arrays," *IEEE Trans. Appl. Supercond.*, vol. 17, no. 2, pp. 371-374, June 2007.

[19] L. Zeng, C. E. Tong, R. Blundell, P. K. Grimes and S. N. Paine, "A Low-Loss Edge-Mode Isolator With Improved Bandwidth for Cryogenic Operation," *IEEE Trans. on Microwave Theory and Techniques*, vol. 66, no. 5, pp. 2154-2160, May 2018.

Session W6: **Optical Design, Systems, and Components II**

Vector Beam Pattern measurements of a 850 GHz wide field Microwave Kinetic Inductance Detector camera

Stephen J. C. Yates[1], Kristina K. Davis[2], Willem Jellema[3,4], Jochem J. A. Baselmans[4,5], Lorenza Ferrari[1], Ronald Hesper[3], and Andrey M. Baryshev[3]

[1] *SRON Netherlands Institute for Space Research, Groningen, 9747AD, The Netherlands*
[2] *Arizona State University, 781 Terrace Rd., Tempe, AZ, U.S.A*
[3] *Kapteyn Institute, University of Groningen, Groningen, 9747AD The Netherlands*
[4] *SRON Netherlands Institute for Space Research, Utrecht, 3584CA, The Netherlands*
[5] *Terahertz Sensing Group, Delft University of Technology, Delft 2628CD, The Netherlands*
*Contact: s.yates@sron.nl

Abstract—We present vector beam pattern measurements of a large field of view Microwave Kinetic Inductance Detector camera at 850 GHz. The angular and position dependent optical response of a receiver, its beam pattern, is given by the E-field of the detected radiation and as such is a vector with both amplitude and phase. Amplitude only measurements, even if taken in multiple optical planes, will be an incomplete description of the optical response. As such they will miss subtle phase errors and so not describe how a full instrument will operate in its final scientific operation configuration on sky. With the phase information, the full E-field is described, allowing the beam to be numerically propagated in either direction to investigate the optics or optical coupling between components. This allows testing at subcomponent level or in the near field of the full instrument, which can then be used to determine and understand the final deployed far field on sky performance. Phase and amplitude measurements are now standard procedure for phase sensitive heterodyne instruments, such as used in ALMA. However, vector beam pattern measurements have only recently been shown to be possible with direct, power-only sensitive, detector arrays: the lack of an intrinsic phase response, high pixel count and low detector speed make it more difficult. Additionally, with large field of view cameras the optics become more complex so making this technique more relevant. Measurements are presented from a wide field camera with a test array of lens-antenna coupled Kinetic Inductance Detectors. The vector beam patterns are measured using a dual optical source modulation scheme, using multiplexing electronics allowing around 400 pixels to be simultaneously characterized. Properties across the field of view can be investigated, including defocus and Gaussian beam coupling, that would not otherwise be available from an amplitude only beam pattern. An added advantage of the technique is that standing waves off the source can be corrected for, important for single frequency measurements. Finally, the dual source modulation scheme means the dynamic range for a given modulation depth is the square of that given by a single source amplitude only measurement. This has opened up much weaker features on the beam pattern for investigation, placing a limit on the residual in-detector-chip stray light and allowing characterization of the radiation scattering in the optical chain.

Integrated Micro-Lens Antennas for THz Heterodyne Receivers

Maria Alonso-delPino [*], Cecile Jung-Kubiak[1], Theodore Reck[1], Choonsup Lee[1] and Goutam Chattopadhyay[1]

[1]*Jet Propulsion Laboratory, Pasadena, 91109, USA*
[*]Contact: maria.alonso@jpl.nasa.com

Abstract—Most of astrophysics and planetary missions at terahertz frequencies require a multi-pixel array of receivers in order to reduce the acquisition time of the observations. At the same time, the power, mass and volume of the instrument needs to be reduced. Silicon micromachining is a technology that enables the integration of the heterodyne receiver front end in just a few silicon wafers. The volume, mass and loss reduction of these approach compared with split block metal machining technology is considerable. An antenna with high efficiency and that has seamless integration with a silicon wafer stack is necessary to successfully develop these heterodyne instruments.

This work presents the development of integrated micro-lens for frequencies up to 1.9 THz. The antenna is composed of a waveguide based feed with an iris and a Fabry Perot cavity to enhance the directivity of the feed. On top lies a shallow silicon lens. The aperture efficiency of this antenna is higher than 80% which makes them suitable for tight inter-pixel array stacking. The bandwidth is around 15%, sufficient for most Schottky based heterodyne receivers, and a cross-pol level below -25 dB. Moreover, the feed and lens can be fabricated using silicon micro-machining processes and can be vertically integrated with the rest of the receiver. Two prototypes have been built and evaluated at 1.9 THz: one with an aperture of 2.6 mm and a measured directivity of 33.2 dB and the other one with an aperture of 6.35 mm and a measured directivity of 41.2 dB. The technology development required to implement this antenna at 1.9 THz will be presented here, in terms of design fabrication, system integration and performance evaluation.

Sensitivity limits and design of THz absorption spectrometers

Jeffrey Hesler

Virginia Diodes Inc., Charlottesville, 22902, USA
Contact: hesler@vadiodes.com

Abstract—Absorption spectrometers are used to analyze the vibrational and rotational transitions of chemical species. This article will discuss the limits of spectrometer sensitivity, and the design of spectrometers to enable the measurement of the minimum detectable amount of a species. The classic Townes sensitivity limit was derived by assuming a square law detection method. However, the minimum detectable signal can be shown to be independent of the method of detection used, leading to useful insights into the optimum spectrometer design. In particular, the spectrometer can utilize the well-established techniques of homodyne detection of amplitude modulated signals. Analysis and measurements of different spectrometer architectures will be considered. The effect of various noise sources on the sensitivity will be presented, including Johnson noise, diode noise, source amplitude noise, and flicker noise. The dependence of the sensitivity on the available transmitter power will be analyzed with respect to these noise sources. Measurements of spectrometer performance will be presented at WR-5.1 (140-220 GHz) and WR-5.1 (500-750 GHz).

HiPTC: a compact, efficient and low vibration cooler for Terahertz detection

T. Wiertz[1], S. Carpentier[1], Y. Pennec[1], J-M Niot[1], W. Zhang[2], Q.J. Yao[2], K. Zhang[2], J. Li[2], S.C. Shi[2]

[1]*Air Liquide Advanced Technologies, Sassenage, 38360, France*
[2]*Purple Mountain Observatory, Nanjing,* 210008, *China*
*Contact: Thierry.wiertz@airliquide.com

Abstract— Since 2010, Air Liquide Advanced Technologies has developed a dual-stage pulse tube cooler for application to future space science missions. The target temperature for this cooler was 15K. The cooler is dubbed HiPTC, standing for Heat intercepted Pulse Tube Cooler.

Tests performed on one engineering model showed that below 9K temperature could be reached using this cooler. This makes it suitable for terahertz detector transition.

Preparatory work has been undergone with Purple Mountain Observatory to adapt the cooler for use in an instrument on board the future Chinese space station. In the frame of this work, a test has been done with coupling of the cooler with a SIS mixer built by PMO. Cryogenic tests have been performed and show very promising results.

The result of this positive testing experience is the availability of a compact, efficient and low vibration cryocooler for terahertz applications. The power consumption of the cooler is less than 500W. Vibration levels are very low thanks to a well balanced compressor and to pulse tube technology. Total cooler mass is below 25 kg including drive electronics and the cooler fits in a reduced volume.

This cooler is particularly well suited for space applications or other mobile application where mass, energy, volume and vibration are key factors.

The presentation will display cooler characteristics as well as test results and will discuss potential applications for this cooler.

Registered Symposium participants

Full Name	Email	Institution
Maria Alonso-delPino	maria.alonso@jpl.nasa.gov	JPL, California Institute of Technology
Olivier Auriacombe	olivier.auriacombe@stfc.ac.uk	RAL Space
Jan Barkhof	J.Barkhof@rug.nl	University of Groningen / NOVA
Andrey Baryshev	a.m.baryshev@astro.rug.nl	University of Groningen / NOVA
Victor Belitsky	victor.belitsky@chalmers.se	Chalmers University of Technology
Raymond Blundell	rblundell@cfa.harvard.edu	Harvard-Smithsonian CfA
Faouzi Boussaha	faouzi.boussaha@obspm.fr	Paris Observatory
Matt Bradford	bradford@caltech.edu	JPL, California Institute of Technology
Sean Bryan	sean.a.bryan@asu.edu	Arizona State University
Eric Bryerton	bryerton@vadiodes.com	Virginia Diodes, Inc.
Grigoriy Bubnov	bubnov@ieee.org	Institute of Applied Physics RAS
Goutam Chattopadhyay	Goutam.Chattopadhyay@jpl.nasa.gov	JPL, California Institute of Technology
Christine Chaumont	christine.chaumont@obspm.fr	CNRS-GEPI-Paris Observatory
Christine Chen	christine.p.chen@jpl.nasa.gov	JPL, California Institute of Technology
Sergey Cherednichenko	serguei@chalmers.se	Chalmers University of Technology
Oleg Cojocari	cojocari@acst.de	ACST GmbH
Jake Connors	jconnors@cfa.harvard.edu	Harvard-Smithsonian CfA
Ken Cooper	ken.b.cooper@jpl.nasa.gov	JPL, California Institute of Technology
Tom Crowe	Crowe@vadiodes.com	Virginia Diodes, Inc.
Daniel Cunnane	daniel.p.cunnane@jpl.nasa.gov	JPL, California Institute of Technology
Christopher Curwen	ccurwen@ucla.edu	UCLA
Kristina Davis	kdavis32@asu.edu	Arizona State University
Peter Day	Peter.K.Day@jpl.nasa.gov	JPL, California Institute of Technology
Fabien Defrance	fdefranc@caltech.edu	California Institute of Technology
Duccio Delfini	duccio.delfini@obspm.fr	Paris Observatory
Vincent Desmaris	vincent.desmaris@chalmers.se	Chalmers University of Technology
Vladimir Drakinskiy	vladimir.drakinskiy@chalmers.se	Chalmers University of Technology
Eduard Driessen	driessen@iram.fr	IRAM
Jack East	jeast@umich.edu	University of Michigan
Pierre Echternach	pierre.m.echternach@jpl.nasa.gov	JPL, California Institute of Technology
Hajime Ezawa	h.ezawa@nao.ac.jp	NAOJ
Yuner Gan	Y.N.Gan@sron.nl	SRON & TUDelft
Jian Rong Gao	j.r.gao@tudelft.nl	SRON & TUDelft
John Garrett	john.garrett@astro.ox.ac.uk	University of Oxford
Gregory Gay	gregory.gay@obspm.fr	CNRS-LERMA-Paris Observatory
Hugh Gibson	hnosbig@gmail.com	Gibson Microwave Design
Paul Goldsmith	paul.f.goldsmith@jpl.nasa.gov	JPL, California Institute of Technology
Gregory Goltsman	goltsman@mspu-phys.ru	Moscow State Pedagogical University
Sunil Golwala	golwala@caltech.edu	California Institute of Technology

Paul Grimes	pgrimes@cfa.harvard.edu	Harvard-Smithsonian CfA
Christopher Groppi	cgroppi@asu.edu	Arizona State University
Paul Hartogh	hartogh@mps.mpg.de	Max Planck Inst. Sol. Sys. Res.
Darren Hayton	darren.j.hayton@jpl.nasa.gov	JPL, California Institute of Technology
Jeffrey Hesler	hesler@vadiodes.com	Virginia Diodes, Inc.
Ronald Hesper	r.hesper@astro.rug.nl	University of Groningen
Jonathan Hoh	jrhoh@asu.edu	Arizona State University
Heinz-Wilhelm Hübers	heinz-wilhelm.huebers@dlr.de	German Aerospace Center (DLR)
Hiroaki Imada	hiroaki.imada@gmail.com	ISAS & JAXA
Yoshihisa Irimajiri	irimaji@nict.go.jp	NICT
Willem Jellema	W.Jellema@sron.nl	SRON
Cecile Jung-Kubiak	cecile.d.jung@jpl.nasa.gov	JPL, California Institute of Technology
Tsungyu Kao	wilt_kao@longwavephotonics.com	Long Wave Photonics
Boris Karasik	boris.s.karasik@jpl.nasa.gov	JPL, California Institute of Technology
Kirit Karkare	kkarkare@kicp.uchicago.edu	University of Chicago
Irmantas Kasalynas	irmantak@ktl.mii.lt	Center for Physical Sci. and Technol.
Akira Kawakami	kawakami@nict.go.jp	NICT
Jonathan Kawamura	jonathan.h.kawamura@jpl.nasa.gov	JPL, California Institute of Technology
Ali Khalatpour	akhalat@mit.edu	MIT
Andrey Khudchenko	A.Khudchenko@sron.nl	University of Groningen / NOVA
Junhan Kim	junhankim@email.arizona.edu	University of Arizona
Jenna Kloosterman	jkloosterm@usi.edu	University of Southern Indiana
Takafumi Kojima	t.kojima@nao.ac.jp	NAOJ
Daniel Koller	koller@vadiodes.com	Virginia Diodes, Inc.
Jacob Kooi	Jacob.W.Kooi@jpl.nasa.gov	JPL, California Institute of Technology
Mikko Kotiranta	mikko.kotiranta@iap.unibe.ch	University of Bern
Sascha Krause	saschak@chalmers.se	Chalmers University of Technology
Matthias Kroug	matthias.kroug@nao.ac.jp	NAOJ
Artem Kuzmin	artem.kuzmin@kit.edu	Karlsruhe Institute of Technology
W.M. Lauuwen	w.m.laauwen@sron.nl	SRON
Jing Li	lijing@pmo.ac.cn	Purple Mountain Observatory, CAS
Dariusz Lis	darek.lis@obspm.fr	Paris Observatory
Alvydas Lisauskas	alvydas.lisauskas@ff.vu.lt	Vilnius University
Ted Macioce	tmacioce@caltech.edu	California Institute of Technology
Alain Maestrini	alain.maestrini@obspm.fr	Paris Observatory
Tomotake Matsumura	tomotake.matsumura@ipmu.jp	University of Tokyo, Kavli IPMU
Hiroshi Matsuo	h.matsuo@nao.ac.jp	NAOJ
Imran Mehdi	imran.mehdi@jpl.nasa.gov	JPL, California Institute of Technology
Michael Merker	Michael.merker@kit.edu	Karlsruhe Institute of Technology
Wei Miao	wmiao@mwlab.pmo.ac.cn	Purple Mountain Observatory, CAS
Raquel Monje	rmonje@jpl.nasa.gov	JPL, California Institute of Technology
Diego Moro-Melgar	diego.moro-melgar@acst.de	ACST GmbH

Anthony Mroczkowski	moon_pi@hotmail.com	European Southern Observatory
Axel Murk	murk@iap.unibe.ch	University of Bern
Deacon Nemchick	Deacon.J.Nemchick@jpl.nasa.gov	JPL, California Institute of Technology
Robert Newman	TNewman@millitech.com	Millitech, Inc.
Omid Noroozian	o.noroozian@gmail.com	NRAO & NASA GSFC & UVA
Satoshi Ochiai	ochiai@nict.go.jp	NICT
Alexey Pavolotsky	alexey.pavolotsky@chalmers.se	Chalmers University of Technology
Carlos Pérez-Moreno	carlos@gmr.ssr.upm.es	Technical University of Madrid
Sofia Rahiminejad	sofia.rahiminejad@jpl.nasa.gov	JPL, California Institute of Technology
Hawal Rashid	hawal.rashid@gmail.com	Chalmers University of Technology
Theodore Reck	theodore.reck@jpl.nasa.gov	JPL, California Institute of Technology
Jun Ren	jren1@nd.edu	University of Notre Dame
Nicholas Rommelfanger	nrommelfanger@umail.ucsb.edu	JPL, California Institute of Technology
Richard Roy	richard.j.roy@jpl.nasa.gov	JPL, California Institute of Technology
Yuki Sakurai	yuki.sakurai@ipmu.jp	University of Tokyo, Kavli IPMU
Lorene Samoska	Lorene.Samoska@jpl.nasa.gov	JPL, California Institute of Technology
Jack Sayers	jack@caltech.edu	California Institute of Technology
Wenlei Shan	wenlei.shan@nao.ac.jp	NAOJ
Mark Sherwin	sherwin@ucsb.edu	UCSB
Sheng-Cai Shi	scshi@pmo.ac.cn	Purple Mountain Observatory, CAS
Jose Siles	Jose.V.Siles@jpl.nasa.gov	JPL, California Institute of Technology
Jose Silva	j.r.g.d.silva@sron.nl	SRON & KAI, U. of Groningen
Peter Sobis	ps@omnisys.se	Omnisys Instruments AB
Adrian Tang	Adrian.J.Tang@jpl.nasa.gov	JPL, California Institute of Technology
Mark Taylor	mark.taylor@jpl.nasa.gov	JPL, California Institute of Technology
Bertrand Thomas	Bertrand.Thomas@radiometer-physics.de	Radiometer Physics GmbH
Edward Tong	etong@cfa.harvard.edu	Harvard-Smithsonian CfA
Amanda Voegtlin	Voegtlin@pdx.edu	MCECS Research, Portland State U.
Christopher Walker	cwalker@as.arizona.edu	University of Arizona
Andrew Whimster	andrew.whimster@teratechcomponents.com	Teratech Components
Jeremy Whitton	jeremy.d.whitton@gmail.com	Arizona State University
Martina Wiedner	martina.wiedner@obspm.fr	Paris Observatory
Thierry Wiertz	thierry.wiertz@airliquide.com	Air Liquide Advanced Technologies
Benjamin Williams	bswilliams@ucla.edu	UCLA
Ghassan Yassin	ghassan.yassin@physics.ox.ac.uk	University of Oxford
Stephen Yates	s.yates@sron.nl	SRON
Changyun Yoo	cyoo@ucsb.edu	UCSB
Lingzhen Zeng	lingzhen@cfa.harvard.edu	Harvard-Smithsonian CfA
Haotian Zhu	zhuhaotian_2007@126.com	Polytechnic School of Montreal